Perspectives in Neural Computing

W0042381

Springer

London
Berlin
Heidelberg
New York
Barcelona
Hong Kong
Milan
Paris
Santa Clara
Singapore
Tokyo

Also in this series

Maria Marinaro and Roberto Tagliaferri (Eds)
Neural Nets - WIRN VIETRI-96
3-540-76099-7

Adrian Shepherd
Second -Order Methods for Neural Networks
3-540-76100-4

Dimitris C. Dracopoulos
Evolutionary Learning Algorithms for Neural Adaptive Control
3-540-76161-6

John A. Bullinaria, David W. Glasspool and George Houghton (Eds)
4th Neural Computation and Psychology Workshop, London,
9-11 April 1997: Connectionist Representations
3-540-76208-6

Maria Marinaro and Roberto Tagliaferri (Eds)
Neural Nets - WIRN VIETRI-97
3-540-76157-8

Gustavo Deco and Dragan Obradovic
An Information-Theoretic Approach to Neural Computing
0-387-94666-7

Thomas Lindblad and Jason M. Kinser
Image Processing using Pulse-Coupled Neural Networks
3-540-76264-7

L. Niklasson, M. Bodén and T. Ziemke (Eds)
ICANN 98
3-540-76263-9

Maria Marinaro and Roberto Tagliaferri (Eds)
Neural Nets - WIRN VIETRI-98
1-85233-051-1

Amanda J.C. Sharkey (Ed.)
Combining Artificial Neural Nets
1-85233-004-X

Dietmar Heinke, Glyn W. Humphreys and Andrew Olson (Eds)
Connectionist Models in Cognitive Neuroscience
The 5th Neural Computation and Psychology Workshop, Birmingham, 8-10 September 1998
1-85233-052-X

Dirk Husmeier

Neural Networks for Conditional Probability Estimation

Forecasting Beyond Point Predictions

 Springer

Dirk Husmeier, PhD
Neural Systems Group, Department of Electrical & Electronic Engineering, Imperial College,
Exhibition Road, London. SW7.

Series Editor
J.G. Taylor, BA, BSc, MA, PhD, FInstP
Centre for Neural Networks, Department of Mathematics, King's College,
Strand, London WC2R 2LS, UK

ISBN-13:978-1-85233-095-8 Springer-Verlag London Berlin Heidelberg

British Library Cataloguing in Publication Data
Husmeier, Dirk
   Neural networks for conditional estimation :
   forecasting beyond point predictions. - (Perspectives in
   neural computing)
   1.Neural networks (Computer science) 2.Prediction theory
   I.Title
   006.3'2
   ISBN-13:978-1-85233-095-8
Library of Congress Cataloging-in-Publication Data
Husmeier, Dirk, 1964-
   Neural networks for conditional probability estimation :
   forecasting beyond point predictions / Dirk Husmeier.
       p.   cm. -- (Perspectives in neural computing)
   ISBN-13:978-1-85233-095-8   e-ISBN-13:978-1-4471-0847-4
   DOI: 10.1007/978-1-4471-0847-4

   1. Neural networks (Computer science) 2. Distribution
   (probability theory)--Data processing. I. Title. II. Series
   QA76.87.H87 1999   98-48438
   006.3'2--dc21   CIP

Typesetting: Camera-ready by the editor

34/3830-543210 Printed on acid-free paper

To   Ulli

Acknowledgements

Major parts of this book are based on my PhD thesis, which I completed in the Department of Mathematics at King's College London and which was funded by a Postgraduate Knight Studentship from the University of London. I would like to thank my supervisor, John Taylor, for his support, advice and insightful criticism. I am fortunate to have been part of the research group he has led and am grateful to Valeriu Beiu and fellow students Paulo Adeodato, Bart Krekelberg, Oury Monchi, and Rasmus Petersen for their methodological support and inspiring discussions. My thanks also go to Mike Reiss, David Lavis, Mike Freeman, Richard Everson, Iead Rezek, Stephen Hunt, Jon Rogers, and Sreejith Das for various assistance in solving software and hardware problems, as well as to Robert Dale, Rasmus Petersen, William Penny and Stephen Roberts for meticulous proofreading of parts of the manuscript.

Although I have tried to develop a personal point of view on the topics covered by this book, I recognise that I owe much to other authors. In particular my work profited greatly from the study of David MacKay's publications on the Bayesian evidence method, which were a pleasure to read and substantially inspired the work leading to Chapters 9-12.

During the revision of the manuscript and the work on those chapters that were not part of my thesis, I was employed as a research associate in the Department of Electrical Engineering at Imperial College London. I would like to express my gratitude to Stephen Roberts for allowing me to dedicate large parts of my working hours to the completion of this project.

I would also like to acknowledge the support of the editors at Springer Verlag, Rosie Kemp and Vicki Swallow. The interest they took in my work helped me considerably to keep on course, and I am grateful for all their assistance during the laborious process from the first manuscript to the camera-ready copy.

Last but not least, my thanks go to my wife, Ulrike Husmeier. It is needless to point out that the process of writing this book demanded huge sacrifices of our common time, and yet she managed to help me find the necessary distraction and avoid the perils of overly grim determination.

Preface

Conventional applications of neural networks usually predict a single value as a function of given inputs. In forecasting, for example, a standard objective is to predict the future value of some entity of interest on the basis of a time series of past measurements or observations. Typical training schemes aim to minimise the sum of squared deviations between predicted and actual values (the 'targets'), by which, ideally, the network learns the conditional mean of the target given the input. If the underlying conditional distribution is Gaussian or at least *unimodal*, this may be a satisfactory approach. However, for a *multimodal* distribution, the conditional mean does not capture the relevant features of the system, and the prediction performance will, in general, be very poor. This calls for a more powerful and sophisticated model, which can learn the whole conditional probability distribution.

Chapter 1 demonstrates that even for a deterministic system and 'benign' Gaussian observational noise, the conditional distribution of a future observation, conditional on a set of past observations, can become strongly skewed and *multimodal*. In **Chapter 2**, a general neural network structure for modelling conditional probability densities is derived, and it is shown that a *universal approximator* for this extended task requires at least *two* hidden layers. A training scheme is developed from a *maximum likelihood* approach in **Chapter 3**, and the performance of this method is demonstrated on three stochastic time series in **chapters 4** and **5**. Several extensions of this basic paradigm are studied in the following chapters, aiming at both an increased training speed and a better generalisation performance. **Chapter 7** shows that a straightforward application of the *Expectation Maximisation (EM)* algorithm does not lead to any improvement in the training scheme, but that in combination with the *random vector functional link (RVFL)* net approach, reviewed in **Chapter 6**, the training process can be accelerated by about two orders of magnitude. An empirical corroboration for this 'speed-up' can be found in **Chapter 8**. **Chapter 9** discusses a simple *Bayesian* approach to network training, where a conjugate prior distribution on the network parameters naturally results in a penalty term for *regularisation*. However, the hyperparameters still need to be set by intuition or cross-validation, so a consequent extension is presented in **chapters 10** and **11**, where the Bayesian *evidence* scheme, introduced to the neural network community by MacKay for regularisation and model selection in the simple case of Gaussian homoscedastic noise, is generalised to arbitrary conditional probability densities. The Hessian matrix of the error function is calculated with an extended version of the EM algorithm. The resulting update equations for the hyperparameters and the expression for the model evidence are found to reduce to MacKay's results in the above limit of Gaussian noise and thus provide a consequent generalisation of these earlier results. An empirical test of the evidence-based regularisation scheme, presented in **Chapter 12**, confirms that the problem of *overfitting* can be considerably reduced, and that the training process is stabilised with respect to changes in the length of training time. A further improvement of the *generalisation performance* can be achieved by employing network *committees*, for which two weighting schemes – based on either the *evidence* or the *cross-validation* performance – are derived in **Chapter 13**. **Chapters 14** and **16** report the results of extensive simulations on a synthetic and a real-world problem, where the intriguing observation is made that in network committees, *overfitting* of the individual models can be useful and may lead to better prediction results than obtained with an ensemble of properly regularised networks. An explanation for this curiosity can be given in terms of a modified *bias-variance dilemma*, as expounded in **Chapter 13**. The subject of **Chapter 15** is the problem of feature selection and the identification of irrelevant inputs. To this end, the *automatic relevance determination (ARD)* scheme of MacKay and Neal is adapted to learning in comittees of probability-predicting RVFL networks. This method is applied in **Chapter 16** to a real-world benchmark problem, where the objective is the prediction of housing prices in the Boston metropolitan area on the basis of various socio-economic explanatory variables. The book concludes in **Chapter 17** with a brief summary.

Nomenclature

In this book, the following notation has been adopted. Lower-case boldface letters, for example $\mathbf{x}$, are used to denote vectors, while upper-case boldface letters, such as $\mathbf{H}$, denote matrices. The transpose of a vector or a matrix is denoted by a superscript $\dagger$, for example, $\mathbf{x}^{\dagger}$. In most cases, $\mathbf{x}$ denotes a column vector, whereas $\mathbf{x}^{\dagger}$ denotes the corresponding row vector. The upper-case letter P is used to represent both a discrete probability and a probability density. When used with different arguments, P represents different functions. For example, $P(x)$ denotes the distribution of a random variable x and $P(y)$ the distribution of a random variable y, so that these distributions are denoted by the same symbol P even though they represent different functions. Moreover, the notation does not distinguish between a random variable, X, and its realisation, x. I hope these conventions simplify the notation and reduce unnecessary opacity rather than cause confusion.

The symbols used for some frequently occuring quantities in this book are listed below:

$\mathcal{S}$-layer	first hidden layer of sigmoidal nodes
$\mathcal{G}$-layer	second hidden layer of nodes with a bell-shaped transfer function
m	number of nodes in the input layer, also: dimension of the hyperparameter space
H	number of nodes in the $\mathcal{S}$-layer
K	number of nodes in the $\mathcal{G}$-layer
a_k	output weights or prior probabilities
σ_k	kernel width
β_k	parameters determining the kernel width
$\mathbf{q}$	set of all network parameters
$\mathbf{w}_k$	vector of weights in the kth weight group, that is, weights feeding into the kth $\mathcal{G}$-node
α_k	inverse variance of a Gaussian prior distribution on the weights in the kth weight group
γ_k	number of well-determined parameters in the kth weight group [defined in (10.72)]
$\mathcal{S}(.)$	sigmoidal transfer function, i.e. a function that is continuous, monotonically increasing, and bounded from above and below
$\varsigma(.)$	sigmoid function, $\varsigma(x) = (1 + e^{-x})^{-1}$
U	EM error function, defined in (7.3)
E	maximum likelihood error function, defined in (3.8).
E^*	total error function, defined in (10.30)

E_{train} same as E, the training 'error'
E_{cross} cross-validation 'error'
E_{gen} empirical generalisation 'error'
          [estimated on the test set according to (8.4)]
$\mathcal{E}$ theoretical generalisation 'error' [defined in (3.7)]
N, N_{train} number of exemplars in the training set
N_{cross} number of exemplars in the cross-validation set
N_{gen} number of exemplars in the test set
N_{Com} number of networks in a committee of networks
E_{gen}^{Com} empirical generalisation 'error' of a network committee

Table of Contents

List of Figures

1. Introduction

Forecasting is the art of saying what will happen, and then explaining why it didn't! (Chatfield)

1.1 Conventional forecasting and Takens' embedding theorem

One of the central problems in science is forecasting. If we have some knowledge about the past behaviour of a system, how can we make meaningful predictions about its future? A standard approach adopted in physics is to construct a mathematical model based on theoretical considerations and, using measured data for specifying the initial conditions, try to integrate the equations of motion forward in time to predict the future state. This procedure, however, is not always feasible. In fields such as economics we still lack the 'first principles' necessary to make good models. In other cases, such as fluid flow, initial data are difficult to obtain. Finally, in complex nonlinear systems with many degrees of freedom, like a turbulent fluid, the weather, or the economy, it is not possible to solve the equations of dynamics explicitly and to keep track of motion in the high dimensional state space. In these cases model-based forecasting becomes impossible and calls for a different prediction paradigm.

A principled alternative approach was introduced by Yule, who in 1927 attempted to predict the sunspot cycle by building an *ad hoc* linear model directly from the available data. From this the discipline of *time series prediction* has evolved as an attempt to build an *empirical* model directly from the given data and to make predictions on the basis of a sequence of past observations or measurements. More precisely, the past information contained in a time series is encapsulated in a so-called *delay* or *lag vector*, and a functional dependency between this lag vector and a future measurement or observation is inferred from the set of past data. An intuitive reason why this approach should be successful in making predictions is the following. Recent advances in dynamical systems theory have revealed that dissipation can reduce the number of effectively relevant degrees of freedom to a small number, and that the motion of the system, which in principle occurs in a high-dimensional space, becomes confined after some time to a subspace Γ of low dimensionality called an attractor[1]. The problem then is to somehow identify the coordinates which characterise this attractor. This should in principle be possible on the basis of a sequence of past measurements or observations, since most measurements will probe some combination of the relevant variables, and measurements at different times will in general probe different combinations. Thus a sequence of m measurements should contain enough information to predict the motion on the attractor, provided m is sufficiently large compared to the attractor dimensionality d. This intuitive conjecture was formulated in a more rigorous way by Takens [60].

Let s_t denote the state vector of the system at time t. This is a high-dimensional vector which includes the whole information about the system of interest, like the quantum states of all the nuclei in the sun plasma, the positions, momenta and bonds of all atoms in a protein (see Figure 1.4), or the detailed knowledge of the global economic situation. The system evolves in time, symbolised by the scalar index t, according to the dynamics

$$s_{t+\tau} \quad = \quad f^\tau(s_t) \tag{1.1}$$

where a typical sequence of iterates under f lies on the attractor Γ of dimension d. Both the state vector and the dynamics are unknown, and the only information about the system can be obtained by observation or measurement of some low-dimensional entity x_t,

$$x_t \quad = \quad h(s_t) \tag{1.2}$$

where h is called the *measurement function*, and for convenience in the notation we will assume that x_t is a scalar. Since f and h are both unknown, the states s_t cannot be reconstructed in their original form from the observations x_t. However, it may be possible to construct a state space that is in some sense equivalent to the original one. Define the m-dimensional delay vector of the last m observations

$$\mathbf{x}_t \quad = \quad [x_t, x_{t-\tau}, \ldots, x_{t-(m-1)\tau}]. \tag{1.3}$$

With (1.1) and (1.2) this can be written as

$$\begin{aligned}
\mathbf{x}_t \quad &= \quad [x_t, x_{t-\tau}, \ldots, x_{t-(m-1)\tau}] \\
&= \quad \left[h(s_t), h(s_{t-\tau}), \ldots, h(s_{t-(m-1)\tau})\right] \\
&= \quad \left[h(s_t), h \circ f^{-\tau}(s_t), \ldots, h \circ f^{-(m-1)\tau}(s_t)\right]
\end{aligned}$$

---

[1] Note that this attractor can be a fractal object with a non-integer dimension. See, e.g., [20].

$$\begin{aligned}
&= && [h, h \circ f^{-\tau}, \ldots, h \circ f^{-(m-1)\tau}](s_t) \\
&:= && \Phi(s_t),
\end{aligned} \qquad (1.4)$$

where the function Φ maps the states of the d-dimensional dynamical system s_t into m-dimensional delay vectors x_t. The important result found by Takens [60] is that for $m \geq 2d + 1$ and any positive value[2] of τ this function is a diffeomorphism[3], which allows rewriting (1.4) in the form

$$x_t = \Phi(s_t) = \Phi \circ f^t(s_0) = \Phi \circ f^t \circ \Phi^{-1}(x_0). \qquad (1.5)$$

Consequently, for $m \geq 2d + 1$, a smooth dynamics F is induced on the so-called *embedding space* of reconstructed vectors,

$$\begin{aligned}
x_t &= F^t(x_0) && (1.6) \\
F^t &:= \Phi \circ f^t \circ \Phi^{-1} && (1.7)
\end{aligned}$$

which is equivalent to the original dynamics f and can therefore be used for any purpose that the original dynamics could be used for, for instance prediction of future states. The important consequence is that we can predict the future value $x_{t+\tau}$ of a time series solely on the basis of the observed past data $\{x_t, x_{t-\tau}, \ldots\}$, that is, without explicit knowledge of the actual dynamics in the true system. An illustration of this theorem can be found in Figure 1.1.

Although Takens' theorem does not give any hints how to find the reconstructed dynamics F and how to decide on the required dimension m, it provides a theoretical foundation that has encouraged the application of neural networks to the task of time series prediction since the end of the last decade. A lag vector x_t of dimension m is presented at the input layer of a multilayer perceptron (MLP), and the network is trained by an appropriate training scheme, like backpropagation, to predict the value at the next time step, $x_{t+\tau}$, as a function F of x_t. If m has been chosen sufficiently large, Takens' theorem ensures that such a function F exists, and it can in principle be implemented in the network by the universal approximation theorem [24], [23], [43]. A comprehensive overview of various neural network time series applications can, for instance, be found in the report on the 1992 Santa Fe time series prediction competition [65]. However, the prediction performance was found to vary greatly with the particular problem under consideration. While neural networks achieved very satisfactory results on some time series, they completely failed on others. The organisers of the Santa Fe competition wrote:

---

[2] To be precise, it should be added that Takens' theorem only holds *generically*. For a limit cycle, for instance, τ must not be rationally related to the period of the process. Otherwise, the theorem is valid irrespective of the size of τ.

[3] A *diffeomorphism* is a smooth one-to-one map with a smooth inverse.

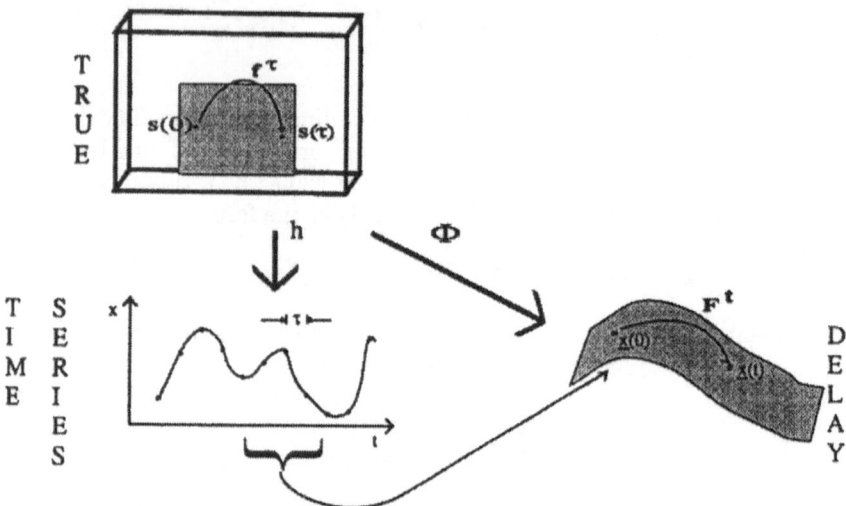

Fig. 1.1. Illustration of Takens' embedding theorem. The true dynamical system f and its states s_t, shown at the top left, are unknown. The system is explored by measurements h, which map the states s_t onto low dimensional entities x_t and produce a time series $\{x_t, x_{t-\tau}, \ldots\}$ (shown at the bottom left). Values of this time series, separated by the lag time τ, form a delay vector $\mathbf{x}_t = [x_t, x_{t-\tau}, \ldots, x_{t-(m-1)\tau}]$ of dimension m. The delay reconstruction map Φ maps the original d-dimensional state s_t into the m-dimensional delay vector $\mathbf{x}_t$. For $m \geq 2d+1$, Φ is a diffeomorphism, and the reconstructed dynamics of the delay vectors, shown at the bottom right, thus becomes equivalent to the true dynamical system f. (Adapted from [12], with permission from Elsevier Science.)

> *These results [...] make clear that a naive application of the techniques that worked so well for Data Set A [...] and to some degree for Data Set D fails for [...] this financial data set.*

Obviously, a major problem in practical neural network applications is the impact on the generalisation performance caused by overfitting, especially when the input dimension m is large and the available training data are sparse. That is, although a network might be able, in principle, to learn the underlying dynamics of a given time series, a naive application of a backpropagation-like training scheme can easily lead to a network configuration that only memorises the particular set of training data presented, but is incapable of generalising to new exemplars not seen before. In fact, many recent contributions in neural network research have focused on this problem, as reviewed, for instance, in [7]. However, there is another, more fundamental problem inherent in the whole approach.

1.2 Implications of observational noise

Takens' theorem assumes that the measurements are arbitrarily precise, resulting in arbitrarily precise states. This requirement, of course, can never be satisfied in practical applications, since noise and imprecisions in the measurements are inevitable and set a limit to the accuracy at which x_t can be determined. As Simon Singh writes in his bestseller 'Fermat's Last Theorem' [56]:

> *Mathematics gives science a rigorous beginning and upon this infallible foundation scientists add inaccurate measurements and imperfect observations.*

The consequence of this observational noise is that the reconstructed states are blurred, that is, many states are consistent with a given series of measurements. One might naively assume that this does not cause too serious a problem. After all, measurement and observational errors tend to be Gaussian distributed, and the measured time series values are consequently symmetrically scattered about their true values. By minimising the commonly employed sum-of-squares error function, a neural network is known to learn the conditional mean of the target conditioned on the input vector[4]. Consequently, the actual prediction should not be affected, and the only additional requirement seems to be the development of sound error estimates, although even this task was only poorly understood by the time the Santa Fe competition was held. As the organisers of this competition wrote [65]:

> *To force competition entrants to address the issue of the reliability of their estimates, we required them to submit [...] both the predicted values and their estimated accuracies. [...] Analyzing the competition entries, we were quite surprised by how little effort was put into estimating the error bars. In all cases, the techniques used to generate the submitted error estimates were much less sophisticated than the models used for the point-predictions.*

Weigend and Nix [64] addressed this deficiency and developed a network architecture for estimating the error bars as a function of the network input. However, this improvement can, in general, not be expected to be sufficient, since the naive argumentation presented above is erroneous.

Depending on the flow of information between the individual degrees of freedom, small fluctuations of an observed variable may correspond to large fluctuations of an unobserved variable. In this case the uncertainty of the

---

[4] See, for instance, [7], Section 6.1.

reconstructed state may be much larger than that of the individual measurements, and the noise in the determination of the true state s from noisy measurements of x may be acutely amplified. An illustrative example for this phenomenon is given by the Lorenz equations[5]

$$
\begin{aligned}
\dot{x} &= \alpha(y - x) \\
\dot{y} &= \beta x - xz - y \\
\dot{z} &= xy - \gamma z
\end{aligned}
\tag{1.8}
$$

in which α, β and γ are constants, and the dot denotes a derivative with respect to time. Let this three-dimensional system be projected onto a one-dimensional time series, and let x be the observed variable (hence the measurement function $h(.)$ is the projection onto the x-axis). As x does not depend on z directly, information about z depends on the flow of information through y: when z changes it causes $\dot{y}$ to change, which causes y and thus $\dot{x}$ to change. Now assume that $x \approx 0$. Since the only coupling to z is through the xz term in the expression for $\dot{y}$, a large change in z causes only a small change in x or, equivalently, a small change in x corresponds to a large change in z. Thus the noise in the determination of the true state $s = [x, y, z]$ from noisy measurements of x is acutely amplified when x is small.

For a more general discussion, let us assume that the noise in the measurements is additive, that is

$$
\mathbf{x}_t = \mathbf{x}_t^0 + \mathbf{n}_t,
\tag{1.9}
$$

where $\mathbf{x}_t$ is the measured delay vector (defined in (1.3)) and $\mathbf{x}_t^0$ the true delay vector due to the state $\mathbf{s}_t$,

$$
\mathbf{x}_t^0 = \Phi(\mathbf{s}_t).
\tag{1.10}
$$

Further assume that the noise $\mathbf{n}_t$ is identically and independently Gaussian distributed with constant standard deviation ε,

$$
P(\mathbf{n}) \propto \exp\left(-\frac{\|\mathbf{n}_t\|^2}{2\varepsilon^2}\right).
\tag{1.11}
$$

Then $P(\mathbf{x}_t | \mathbf{s}_t)$, the conditional probability for observing the delay vector $\mathbf{x}_t$ when the system is in state $\mathbf{s}_t$, is given by

$$
P(\mathbf{x}_t | \mathbf{s}_t) \propto \exp\left(-\frac{1}{2\varepsilon^2}\|\mathbf{x}_t - \Phi(\mathbf{s}_t)\|^2\right),
\tag{1.12}
$$

which is an isotropic Gaussian centred on the true delay vector $\mathbf{x}_t^0 = \Phi(\mathbf{s}_t)$. However, in the prediction task the true state of the system is unknown. The

---

[5] The following example is taken from [12].

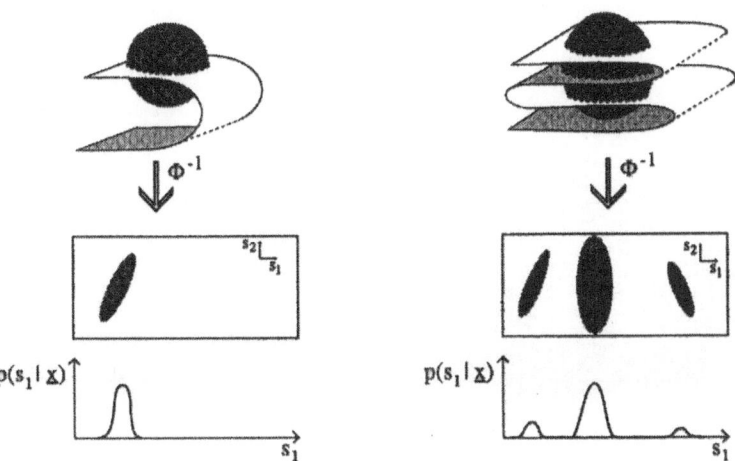

Fig. 1.2. Consequences of observational noise for time series prediction.
The folded sheets in the upper part of the figures represent the mapping Φ of the
true state space Γ into the embedding space of delay vectors $\mathbf{x}_t$. The grey spheres
symbolise isotropic observational noise. Arrows indicate the inverse mapping back
into the true state space Γ, which graphically corresponds to unfolding the sheets
$\Phi(\Gamma)$. The figure on the left shows that a distribution that is symmetric in the
embedding space typically becomes distorted and unisotropic in the original state
space. If $\Phi(\Gamma)$ is tightly folded, as shown on the right, the distribution in state
space $P(\mathbf{s}|\mathbf{x})$ even becomes multimodal. This implies that the true state s can
no longer be located, and conventional approaches to time series prediction break
down. (Adapted from [12], with permission from Elsevier Science.)

probability for it to be $\mathbf{s}_t$ when the time series $\mathbf{x}_t$ has been observed, $P(\mathbf{s}_t|\mathbf{x}_t)$,
can be expressed in terms of $P(\mathbf{x}_t|\mathbf{s}_t)$ by Bayes' theorem,

$$P(\mathbf{s}_t|\mathbf{x}_t) \quad \propto \quad P(\mathbf{x}_t|\mathbf{s}_t)P(\mathbf{s}_t), \qquad (1.13)$$

but this is *not* a Gaussian as $P(\mathbf{x}_t|\mathbf{s}_t)$ has now to be interpreted as a function
of $\mathbf{s}_t$, and $\Phi(.)$ is generally nonlinear. The consequences are illustrated in
Figure 1.2. The folded sheets in the upper part of the figure represent $\Phi(\Gamma)$,
the mapping of the true state space Γ into the embedding space of lag vectors
$\mathbf{x}_t$. The grey spheres symbolise observational noise, which is assumed to be
isotropic in the embedding space. The corresponding distribution in the original state space Γ is obtained by 'unfolding' the sheet, as illustrated in the
bottom part of the figure. On the left-hand side this leads to a distorted and
unisotropic, but still unimodal distribution. A location of the original state
$\mathbf{s}_t$ is still, to a certain extent, possible. However, when $\Gamma(\Phi)$ is tightly folded
relative to the noise variance, as shown on the right-hand side of Figure 1.2,
'unfolding' the sheet leads to a multimodal distribution that spreads out over

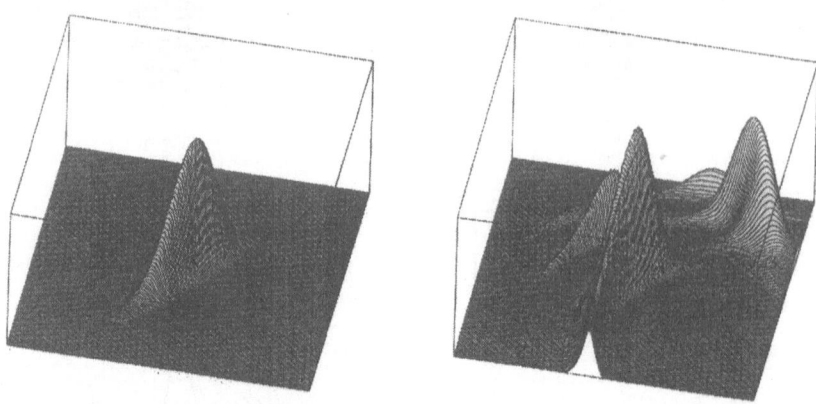

Fig. 1.3. Probability distribution in the state space of the Ikeda map.
The graphs show the probability distribution in state space, $P(\mathbf{s}_t|\mathbf{x}_t)$, for the Ikeda map, with the measurement function $h(\mathbf{s}_t) = h(x_t, y_t) = x_t$ and a 5-dimensional delay vector $\mathbf{x}_t = (x_t, x_{t-1}, \ldots, x_{t-m+1})$, $m = 5$. The value for $P(\mathbf{s}_t|\mathbf{x}_t)$ is plotted vertically (for a fixed delay vector $\mathbf{x}_t$) and the state $\mathbf{s}_t = (x_t, y_t)$ horizontally. On the left, the Gaussian measurement error has a small standard deviation of $\sigma = 0.02$. The distribution in state space is distorted but still unimodal. Increasing the standard deviation of the measurement noise to $\sigma = 0.2$, as shown on the right, leads to a multimodal distribution that defies the location of the correct state. (Reprinted from [12], with permission from Elsevier Science.)

the whole state space. In this case the true state $\mathbf{s}_t$ can no longer be localised, and deterministic predictability breaks down.

A further illustration of this problem is given in Figure 1.3, taken from [12], which shows the distribution $P(\mathbf{s}_t|\mathbf{x}_t)$ in the state space of the Ikeda map[6]. At a low noise level, this distribution is distorted but still unimodal. As the noise level increases, however, the distribution assumes a more complex, multimodal form, which renders the location of the true state $\mathbf{s}_t$ impossible. Figure 1.2, right, shows that, in general, the latter scenario can also happen at low noise levels at the limit of achievable precision if $\Phi(\Gamma)$ is tightly folded (which usually happens if the leading Lyapunov exponent[7] of the dynamical system is large). It is obvious that in this case the conventional approach to time series prediction, which inherently assumes a unimodal distribution of the future value x_{t+1} about some correct value $x_{t+1}^0 = F(\mathbf{x}_t)$, breaks down. A new method, which is capable of modelling the whole conditional probability distribution, is therefore needed.

---

[6] The Ikeda map is $\mathbf{s}_t := (x_{t+1}, y_{t+1}) = (1 + A[x_t \cos b_t - y_t \sin b_t], A[x_t \sin b_t + y_t \cos b_t])$, where $b_t = 0.4 - 6.0/(1 + x_t^2 + y_t^2)$ and $A = 0.7$.

[7] The Lyapunov exponent is a measure of the speed of divergence of neighbouring trajectories in a dynamical system; see, for instance, [33].

1.3 Implications of dynamic noise

The previous section has shown that even for a deterministic system, subject only to 'benign' Gaussian isotropic noise in the measurements, a complicated stochastic process can arise when we want to predict a future time series value x_{t+1} on the basis of a set of past values $\{x_t, x_{t-1}, \ldots\}$ (for notational convenience, the value of the arbitrary lag time τ is henceforth set to unity, $\tau := 1$)[8]. The problem is further complicated by the presence of dynamic noise. Note that Equation (1.1), which is deterministic, describes an autonomous, isolated system. In general, however, systems are in contact with their environment, whose influence modifies the function f in (1.1). Within the frame of reference of the system of interest, this modification appears to be a stochastic perturbation of the dynamics f and is thus equivalent to noise (*dynamic* noise as opposed to the *observational* noise discussed before).

As an example, consider Figure 1.4, which shows the structure of one of the four monomers of hemoglobin, the protein responsible for the transport of oxygen in the blood. Here nodes represent atoms, and edges covalent bonds between atoms. The state vector s_t contains the positions and momenta of all the atoms in the molecule[9]. The evolution of the system, that is, the function f in Equation (1.1), is mainly governed by the forces acting between the atoms (covalent bonds, electric fields between charged side chains, hydrogen bonds etc.). In fact, if the protein was isolated, these internal forces would completely and deterministically define the evolution of the system, and several molecular dynamics studies aiming at a basic understanding of dynamic processes in proteins simulate the hemoglobin monomer in this isolated form (e.g. [53]). In reality, however, our system is *not* isolated. A protein does not exist *in vacuo*, but is contained in a solvent. Consequently, there are long-range interactions between charged groups in protein side chains and the electric fields of the solvent molecules (water, for instance, has a high electric dipole moment). The solvent itself is not isolated either, but in contact with its environment, which has an influence on state variables like temperature and pressure. If we want to retain an autonomous system in the sense of Equation (1.1), we will have to include the environment – and possibly the whole universe – in the state vector s_t, giving rise to a super-system whose dynamics will no longer be confined to a low-dimensional attractor. A more realistic form of description is to leave the original system unchanged. This, however, implies that our system is no longer autonomous or deterministic, since the coupling with the environment will cause a stochastic perturbation

---

[8] Recall that Takens' embedding theorem holds for (nearly) any value of τ. However, in the presence of noise the prediction performance may depend critically on this parameter. In fact, a possible way to reduce the effects of observational noise is to find an optimal value for τ, as discussed, for instance, in [12].

[9] It is assumed that a description of the system within the framework of classical mechanics is sufficient.

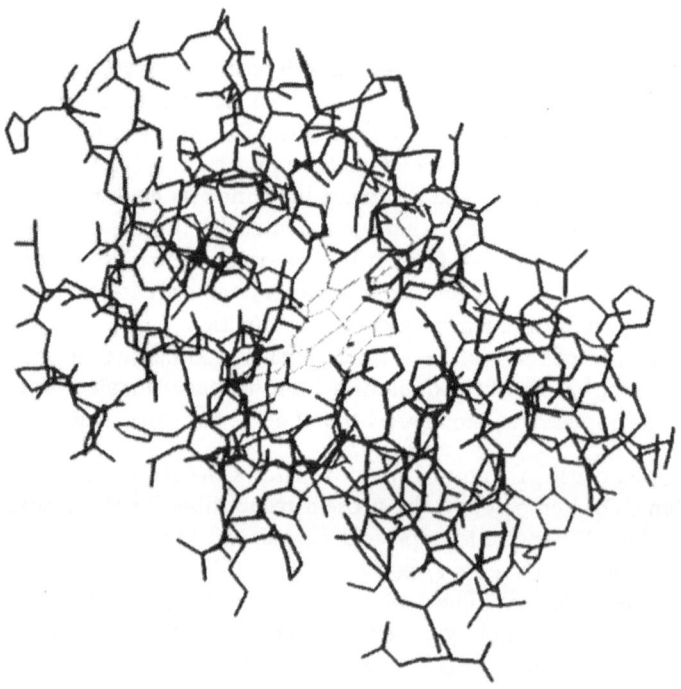

Fig. 1.4. Hemoglobin. The figure shows the structure of the alpha monomer of hemoglobin, the protein responsible for the transport of oxygen in the blood. Nodes represent atoms, and edges covalent bonds between atoms. The heme group, located in the centre of this macro-molecule, is plotted in a lighter colour.

of the function f in Equation (1.1). Our system therefore becomes inherently of a stochastic nature, and the condition of Takens' embedding theorem no longer applies.

1.4 Example

Let us turn again to the example of the hemoglobin monomer, shown in Figure 1.4. The physiological function of this protein is to carry small molecules by binding them as ligands at the Fe atom in the heme plane, which is shown together with its two adjacent helices in Figure 1.5. The main ligand transported in our body is oxygen, O_2, although for the purpose of this demonstration we will consider the case that H_2O is bound as a ligand at the

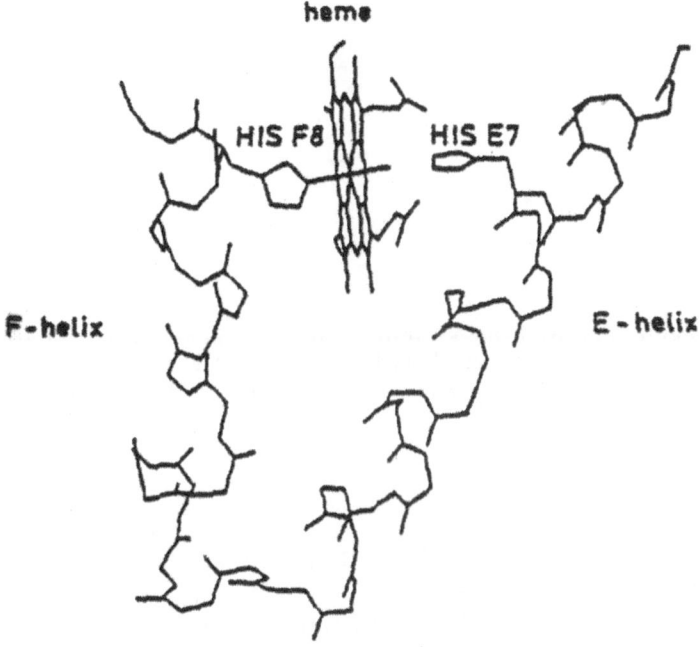

Fig. 1.5. Heme and adjacent helices. The figure shows the heme, the active group of hemoglobin, at which ligands are bound. Also shown are the adjacent helices, E-helix and F-helix. Note the close distance between the heme and the ring *HIS E7* in the E-helix.

heme iron[10], as shown in Figure 1.6. In this case there exists the possibility of a hydrogen bond between the oxygen atom O of the ligand H_2O and the nitrogen atom N in the histidine ring of the E-helix (His E7; see Figure 1.6). Figure 1.7 shows typical trajectories of the distance between one of the two hydrogen atoms H of the ligand and the nitrogen N in His E7. The trajectories were obtained with the method of *Molecular Dynamics* (MD), where based on an empirical potential the equations of motion for the whole set of atoms in the protein are solved numerically; see [25] for details. In the first part of the recorded time segment, both time series show a two-state nature, with high-frequency oscillations around one of two meta-stable states and occasional transitions between these two states. This indicates that the hydrogen bond is closed, being formed alternatingly by one or the other of the two H-atoms (note the strongly anticorrelated nature of the two time series in Figure 1.7). At a later stage, in the second half of the recorded time seg-

---

[10] H_2O is likely to be bound when the iron has been oxidised into its Fe^{3+} form.

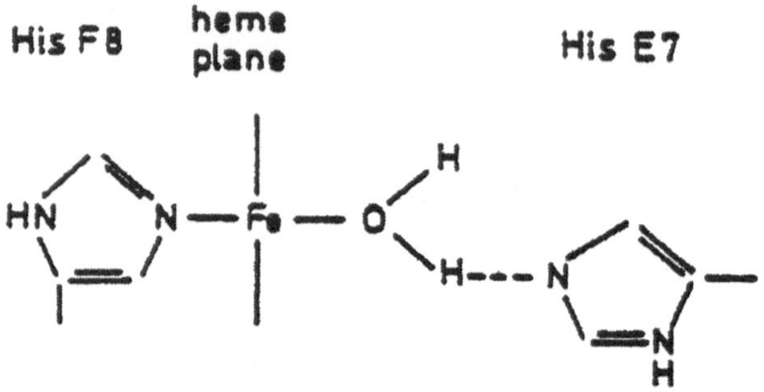

Fig. 1.6. Ligand and hydrogen bond. If H_2O is bound as a ligand at the heme, a hydrogen bond, indicated by the dashed line, can be formed with the N atom of the His E7 ring (which belongs to the E-helix; see Figure 1.5).

ment, this structure disappears, which indicates a break-up of the hydrogen bond.

Now assume that we want to predict the future value x_{t+1} of one of these time series on the basis of past time series values $\{x_t, x_{t-1}, \ldots\}$. In particular we are interested in the exact form of the transitions between the two meta-stable states when the hydrogen bond is closed. If the system were deterministic and the measurements x_t arbitrarily precise, this would, in principle, be possible as a consequence of Takens' embedding theorem. In fact, however, neither of these conditions is satisfied. The measurements $\{x_t\}$ are inevitably subject to observational noise. In a real experiment, measurement errors result from the limited precision and possibly systematic biases of the measurement instruments. In a computer simulation, they are the consequence of the finite length of float numbers and the ensuing roundoff errors. According to the discussion in Section 1.2, this can already lead, under certain conditions, to a time series that is inherently stochastic. More serious, in the present example, is the fact that the deterministic Equation (1.1) does not apply. Since the protein is not *in vacuo* but rather surrounded by a solvent, which itself is in contact with its environment, the dynamics are subject to stochastic perturbations[11]. Consequently, the time series is *per se* of a stochastic nature, and the exact form of the transitions is unpredictable. The best we can do, therefore, is to predict the conditional probability $P(x_{t+1}|\mathbf{x}_t)$, which is the probability that we observe x_{t+1} at the next time step $t+1$ given that we have observed the recent time series values $\mathbf{x}_t = (x_t, x_{t-1}, \ldots, x_{t-m+1})$. For illustration assume that our hemoglobin is in a state where the two time series of Figure 1.7 show a state transition, that is, where the first hydrogen

---

[11] In the MD simulations this was effected by a stochastic force similar to that of equations (4.24) and (4.25), by which the system was coupled to an external heat bath.

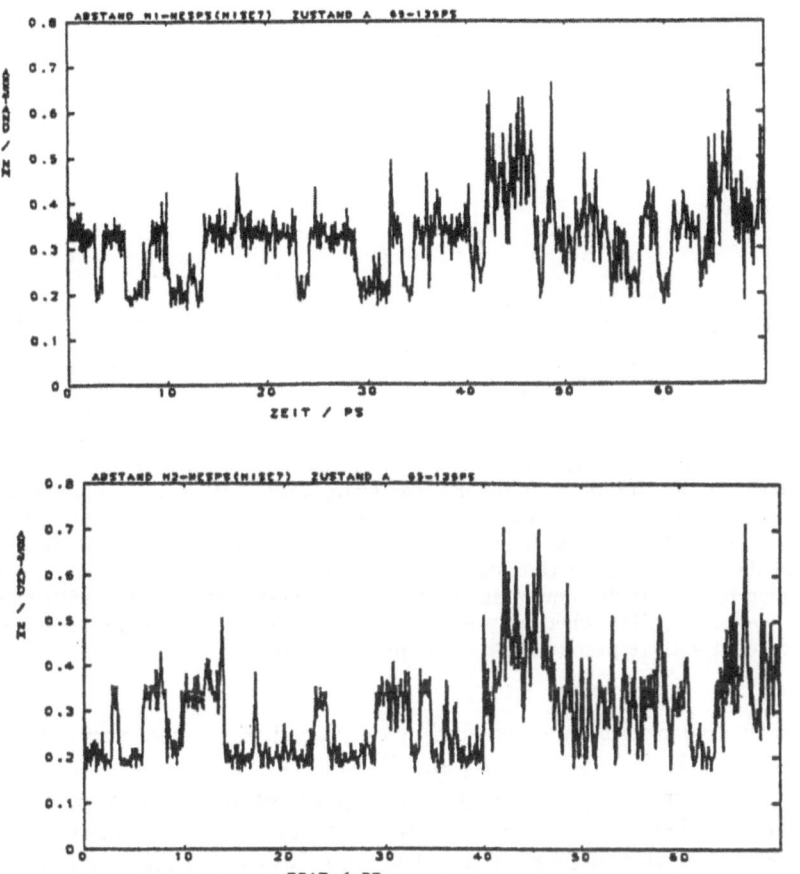

Fig. 1.7. Evolution of the hydrogen-nitrogen distance. The trajectories show the evolution of the distance between one of the two hydrogen atoms (upper graph: $H1$, bottom graph: $H2$) of the ligand (H_2O) and the nitrogen atom in the His E7 ring (see Figure 1.6). The abscissa represents time in picoseconds (ps), the ordinate distance in nanometre (nm). For small times, at the left part of the figures, both graphs show high-frequency oscillations around two meta-stable states and occasional transitions between these two states. This indicates that the hydrogen bond is closed, alternatingly formed by the first ($H1$) or the second ($H2$) hydrogen atom (note the strongly anticorrelated nature of the two trajectories). At a later time, in the right part of the figures, a loss of this distinguished structure indicates a break-up of the hydrogen bond. The trajectories result from *Molecular Dynamics* (MD) simulations, where, based on an empirical potential, the equations of motion for the whole set of atoms in the protein are solved numerically. Details can be found in [4], [25], [63].

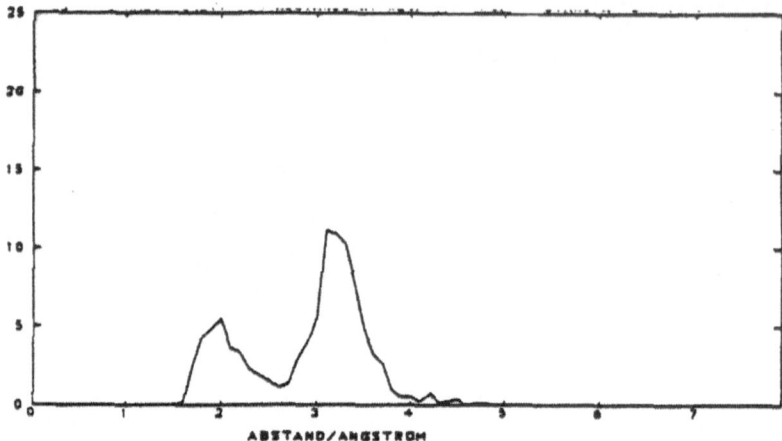

Fig. 1.8. Bimodal distribution of the hydrogen-nitrogen distance. The figure shows the empirical approximation (histogram) of the conditional probability density $P(x_{t+1}|\mathbf{x}_t)$ (ordinate), where x_{t+1} (abscissa) symbolises the $N - H1$ distance 1 ps ahead (that is, at time $t + 1$), and $\mathbf{x}_t$ represents the lag vector of recent distances at time t (measured in the state s_t at time t). The histogram was obtained by repeatedly simulating the dynamics of hemoglobin for 1 ps with *Molecular Dynamics* (MD), where all simulations started from the same initial state s_t, but differed with respect to the random perturbation.

H_1 in the hydrogen bond is replaced by the second, H_2 (or vice versa). Consequently, the distance between H_1 and N, which was oscillating about a value of 0.2 nm, suddenly 'jumps' to the larger value of about 0.32 nm, whereas the distance between H_2 and N is decreased from a value of about 0.32 nm to the new shorter distance of about 0.2 nm. Now recall that the system is *not* deterministic, but that the observed state transition is triggered by external stochastic perturbations. If the external influences had been slightly different, the transition might have occured in a different way, or might not have occured at all. This can be demonstrated in the MD simulations by turning the clock back and repeating the simulation with different stochastic perturbations. That is, given that we are in a state where a transition is likely to occur, we simulate the evolution of the hemoglobin structure for a time interval of, say, $\tau = 1ps$, after which we measure the distance between the hydrogens of the ligand and the nitrogen in the HIS-E7 ring. We then return to the initial state and repeat the procedure several times with different stochastic influences. This allows us to empirically estimate the conditional probability $P(x_{t+1}|\mathbf{x}_t)$ by a histogram over a discretised distance measure, as shown in Figure 1.8. It is seen that the resulting form of the density is bimodal, with two peaks at the most probable distances of 0.2 nm and 0.32 nm (compare with the trajectories in Figure 1.7). Now assume that we are applying a conventional network for the prediction of the time series in Figure 1.7. In this case we can only predict a *single* value $\hat{x}_{t+1}$ as a function of the lag vector $\mathbf{x}_t$, $\hat{x}_{t+1} = g(\mathbf{x}_t)$, where $\hat{x}_{t+1}$ is the predicted value for x_{t+1}. The exact form of $g(\mathbf{x}_t)$, that is the functional dependency of the prediction x_{t+1} on the lag vector $\mathbf{x}_t$, depends on the chosen error function. It is known[12] that with a quadratic error function $E = \sum_t [x_{t+1} - g(\mathbf{x}_t)]^2$, a network learns to predict the conditional mean,

$$g(\mathbf{x}_t) = \langle x_{t+1} | \mathbf{x}_t \rangle := \int x_{t+1} P(x_{t+1} | \mathbf{x}_t) dx_{t+1} \qquad (1.14)$$

and with a linear error function $E = \sum_t |x_{t+1} - g(\mathbf{x}_t)|$, the network predicts the conditional median $g(\mathbf{x}_t) = [x_{t+1} | \mathbf{x}_t]^{med}$, which is defined by the equation

$$\int_{-\infty}^{[x_{t+1}|\mathbf{x}_t]^{med}} x_{t+1} P(x_{t+1}|\mathbf{x}_t) dx_{t+1} = \int_{[x_{t+1}|\mathbf{x}_t]^{med}}^{\infty} x_{t+1} P(x_{t+1}|\mathbf{x}_t) dx_{t+1}$$

$$(1.15)$$

For a bimodal distribution, as in the present example, both predictors, $\langle x_{t+1} | \mathbf{x}_t \rangle$ and $[x_{t+1} | \mathbf{x}_t]^{med}$, lie between the two modes in a region of low probability density. This implies that the actual predicted value is extremely unlikely to occur, and the prediction results are therefore bound to be very poor. Moreover, the structure of the process, with its transitions between the two meta-stable states and its inherent bimodality, cannot be captured by the prediction of only a single value. This underlines again that we need a more powerful model, which is able to predict the entire probability distribution.

Note that the problem described here is very similar to that of predicting financial time series. The hydrogen-nitrogen distance in the above example corresponds to a particular feature of interest, like the exchange rate between two currencies. This is the entity of interest, for which a prediction is to be made on the basis of a time series of past observations. The state of the protein is akin to the economic states of the immediately involved countries; in our case these are the two countries between which the exchange rate is to be predicted. Finally, the surrounding solvent can be likened to the global political and economic situation, which has a perturbing influence on the economic states of the two countries of interest. As is the case with the example presented here, the financial time series will be of a multimodal nature if the system of interest contains various distinct states, and phase transitions between these states are triggered by external stochastic perturbations. A more comprehensive discussion of this subject can be found in [17] and [46].

---

[12] Strictly speaking, the following results only hold if (i) the training set is sufficiently large, (ii) the network is sufficiently complex, and (iii) the optimisation algorithm is optimal. See [7], Chapter 6 for a derivation.

1.5 Conclusion

Conventional approaches to time series prediction model a single value x_{t+1} as a function of an m-dimensional lag vector $\mathbf{x}_t$. The theoretical basis is given by Takens' embedding theorem, according to which, for sufficiently large values of m, the dynamics in the so-called embedding space of lag vectors $\mathbf{x}_t$ are equivalent to the true dynamics of the system. However, Takens' theorem assumes that all observations or measurements are arbitrarily precise, and that the dynamics of the underlying system are autonomous and deterministic. These conditions are usually not satisfied in practice. The consequence is that real-world time series intrinsically tend to be of a stochastic nature. The conventional method of single value predictions typically leads to a model that captures the mean or the median of the conditional probability distribution $P(x_{t+1}|\mathbf{x}_t)$. If the latter is symmetric and unimodal, this may give rather satisfactory results. However, the previous sections have demonstrated that the conditional probability can easily become of a more complex, multimodal form, and that this can even happen for determinsitic systems subject merely to Gaussian (that is, unimodal and symmetric) observational noise. In this case, single value predictions, like the conditional mean or the conditional median, are not representative for the conditional distribution as a whole. In fact, for a symmetric bimodal distribution, both the conditional mean and the conditional median lie between the two modes in a region of low probability density and will therefore predict a value that is extremely unlikely to occur. Consequently, conventional prediction models, including the conventional way neural networks are applied to the problem of time series prediction, may lead to very poor prediction results. This explains the observation made in [65] that the prediction performance varied greatly with the particular problem under consideration, and that all prediction models tended to fail on particular data sets.

1.6 Objective of this book

The objective of this book, therefore, is to study a neural network approach to the prediction of the entire conditional probability distribution (rather than just a single value). This requires the design of a new, more complex network architecture and the derivation of an appropriate training scheme. Further important questions to be addressed concern the time complexity of learning and the problem of generalisation. The following pages give an outline of the topics covered in the various chapters of this book.

Structure

How can we design a neural network that is not restricted to point predictions, as conventional architectures, but which is capable of modelling arbitrary conditional probability densities?

Chapter 2 | Extension of the universal approximation theorem to conditional probability densities, derivation of an appropriate network architecture, comparison with related models.

Learning

Since the true probability density is not known, we do not have a target in the training process. How can we derive an appropriate supervised learning scheme for this generalised task?

Chapter 3 | Derivation of a learning rule for predicting conditional probability densities.

Chapter 4 | Overview of various benchmark problems.

Chapter 5 | Demonstration of the network performance on the benchmark problems.

Model Complexity and Training Times

In Chapter 2, a universal approximator network for predicting arbitrary conditonal probability densities is found to need at least *two* hidden layers. This architectural complexity renders the training process rather slow. How can the standard training algorithm be significantly accelerated?

Chapter 6 | Review of the random vector functional link net (RVFL) approach.

Chapter 7 | Derivation of an improved training scheme, which combines the RVFL approach with the Expectation Maximisation (EM) algorithm.

Chapter 8 | Demonstration and empirical test of the improved training scheme on the benchmark problems.

Generalisation

How can we improve the generalisation performance and prevent overfitting?
What is the appropriate regularisation scheme? Are there any alternative
approaches?

Chapter 9 A simple Bayesian regularisation scheme. This can be interpreted
as a generalised version of weight decay. The regularisation parameters (so-
called 'hyperparameters') have to be set by intuition or cross-validation.

Chapter 10 The Bayesian evidence scheme for regularisation. This approach
allows the hyperparameters to be inferred from the data. The derived hyperparameter update rules are a generalisation of those found by MacKay for
conventional point-predicting networks.

Chapter 11 The Bayesian evidence scheme for model selection: how to assess
the generalisation performance of a network on the basis of the training data.

Chapter 12 Empirical test and demonstration of the Bayesian evidence
scheme for regularisation. Comparison with conventional regularisation methods.

Chapter 13 Network committees: Is regularisation always useful? Derivation of an evidence-based and cross-validation-based weighting scheme.

Chapter 14 Empirical study and demonstration: Committees of networks
trained with different regularisation schemes.

Feature Selection

Is it possible to develop a training method such that the network automatically discards irrelevant inputs and focuses on relevant ones?

Chapter 15 Derivation and comparison of two schemes for automatically
determining the relevance of an input variable.

Real-World Application

Chapter 16 Application of the methods developed in the previous chapters to the prediction of housing prices in the Boston metropolitan area on the basis of various socio-economic explanatory variables.

2. A Universal Approximator Network for Predicting Conditional Probability Densities

The structure of a universal approximator network for predicting conditional probability densities is derived, and it is shown that the resulting architecture can deal with both stochastic and determinstic processes. Two variants, the derivative-of-sigmoid mixture (DSM) and the Gaussian mixture (GM) networks are presented, and their relation to a stochastic kernel expansion is noted. The chapter concludes with a comparison between these models and several relevant alternative approaches which have recently been introduced to the neural network community.

2.1 Introduction

The present chapter introduces a novel approach to the prediction of conditional probability densities with neural networks, following [2], [27], and [30]. Its relation to alternative, independently derived models will be noted in Section 2.9.

Let $\mathbf{x}$ be an $m-$dimensional vector of input variables, and y a scalar target variable. For predicting univariate time series, for example, $\mathbf{x}$ corresponds to a lag vector of m recent time series values, $\mathbf{x}_t = (x_t, x_{t-1}, \ldots, x_{t-m+1})$, and y to the value at the next time step, $y = x_{t+1}$. The following derivations can easily be generalised to include multivariate time series, i.e. those where the target y is a vector rather than a scalar. However, for the sake of simplicity in the notation, this will not be considered here. When the underlying system is deterministic and the observations or measurements are noise-free, then the prediction of a single value as a function of the input vector $\mathbf{x}$ is an appropriate approach. As a consequence of the universal approximation theorem [24], [23], this can be accomplished with a one-hidden-layer feedforward network

$$f(\mathbf{x}) \;=\; \sum_i a_i \mathcal{S}\left(\sum_j w_{ij} x_j\right). \qquad (2.1)$$

Here the a_i denote weights between the hidden layer and the (linear) output node, w_{ij} are the weights feeding into the hidden layer, $S(.)$ is a sigmoidal (i.e. bounded, smooth, and strictly monotonically increasing) transfer function, and the input vector $\mathbf{x}$ is taken to be augmented, i.e., to contain an additional element of constant activity 1 for including biases. However, as discussed earlier in Chapter 1, this approach is not satisfactory when either the system is non-deterministic, or the system is deterministic but the measurements are subject to observational noise. In this case, our objective should be to develop a model that can predict the conditional probability density $P(y|\mathbf{x})$. This is clearly a generalisation of the previous task, which is included in this more general approach as the special case of a δ-peaked conditional density

$$P(y|\mathbf{x}) \quad \rightarrow \quad \delta\Big(y - f(\mathbf{x})\Big). \tag{2.2}$$

In what follows, a network model will be derived that can deal with both the noisy and the deterministic case, i.e., which is a universal approximator for the conditional probability density $P(y|\mathbf{x})$ as well as for a deterministic function $f(\mathbf{x})$. For this derivation, it is convenient to formulate the problem in terms of the *cumulative* conditional probability distribution

$$F(y|\mathbf{x}) \quad = \quad \int\limits_{-\infty}^{y} P(y'|\mathbf{x})dy' \tag{2.3}$$

as this function does not become singular in the deterministic limit, but rather reduces to the Heaviside or step-function

$$F(y|\mathbf{x}) \quad \rightarrow \quad \Theta(y - f(\mathbf{x})) \tag{2.4}$$

where

$$\Theta(x) \quad = \quad \begin{cases} 1 & \text{if} \quad x \geq 0 \\ 0 & \text{if} \quad x < 0 \end{cases} \tag{2.5}$$

Note that $F(y|\mathbf{x})$ is positive and monotonically increasing in y from $F(-\infty|\mathbf{x}) = 0$ to $F(\infty|\mathbf{x}) = 1$. The conditional probability *density* is regained by taking the derivative:

$$P(y|\mathbf{x}) \quad = \quad \frac{\partial}{\partial y}F(y|\mathbf{x}). \tag{2.6}$$

2.2 A single-hidden-layer network

Let us start with a one-hidden-layer network for predicting $F(y|\mathbf{x})$:

$$F(y|\mathbf{x}) \quad = \quad \sum_i a_i S\left(\sum_j \tilde{w}_{ij}x_j + \beta_i y\right). \tag{2.7}$$

Both the target y and the lag vector $\mathbf{x}$ are presented to the input of the network, where β_i denotes the weight between y and the ith hidden node, and $\tilde{w}_{ij}$ the weight between x_j and the ith hidden node (the remaining parameters a_i are the same as in (2.1)). One can arrive at a slight re-interpretation of this by factoring β_i out and defining $w_{ij} := -\tilde{w}_{ij}/\beta_i$. This yields

$$F(y|\mathbf{x}) \quad = \quad \sum_i a_i \mathcal{S}\left(\beta_i\left[y - \sum_j w_{ij}x_j\right]\right), \qquad (2.8)$$

where the β_i's are now to be interpreted as steepness parameters for the transfer functions of the hidden units. In order to ensure that $F(y|\mathbf{x})$ is a cumulative distribution function, the conditions stated above under equation (2.5) have to be satisfied. This can be accomplished by choosing a sigmoidal function $\mathcal{S}(.)$ with range $[0,1]$, e.g. the sigmoid function

$$\mathcal{S}(x) \quad \rightarrow \quad \varsigma(x) := \frac{1}{1 + e^{-x}}, \qquad (2.9)$$

and by imposing the following constraints on the parameters:

$$\beta_i > 0, \quad a_i \geq 0, \quad \sum_i a_i = 1 \qquad (2.10)$$

The resulting network structure is depicted in Fig.2.1 top. Let us now analyse how the network behaves in the deterministic limit, at which one arrives when the output weights become of the Kronecker delta form, $a_i \rightarrow \delta_{i,i^\bullet}$, and when the steepness parameter in the non-vanishing branch becomes very large, $\beta_{i^\bullet} \rightarrow \infty$. From (2.8) one then obtains, using the fact that $\lim_{\beta \rightarrow \infty} \varsigma(\beta x) = \Theta(x)$,

$$F(y|\mathbf{x}) \quad \rightarrow \quad \Theta\left(y - \sum_j w_{i^\bullet j}x_j\right), \qquad (2.11)$$

which corresponds to the point prediction

$$y \quad = \quad \sum_j w_{i^\bullet j}x_j. \qquad (2.12)$$

This shows that, in the deterministic limit, the network can only predict a *linear process*. A one-hidden-layer network is therefore too restricted for the problem considered here.

2.3 An additional hidden layer

This deficiency can be remedied by connecting the x_i to an additional hidden layer, as shown in Fig.2.1 middle. Denoting the additional weights by u_{jk}, we get the following expression for the network output:

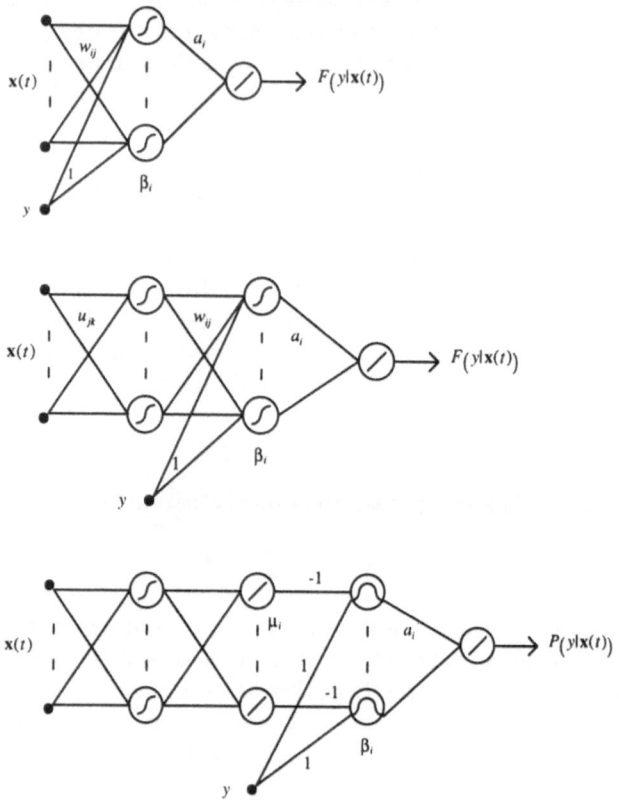

Fig. 2.1. Network architectures for predicting conditional probability distributions and densities. The figure shows different network architectures for predicting cumulative conditional probability distributions and conditional probability densities. **Top:** A single-hidden-layer network for predicting cumulative conditional probability distributions $F(y|x)$. In the deterministic limit this model is restricted to the prediction of *linear* processes. **Middle:** Universal approximator network for predicting cumulative conditional distributions $F(y|x)$. As described in the text, the additional hidden layer overcomes the restriction to linear processes. **Bottom:** Architecture similar to that of the second network, but for predicting conditional probability *densities* $P(y|x)$. The additional hidden layer of linear nodes does not perform any active processing and has only been introduced for didactic purposes. An interpretation of the parameters a_i, β_i and the outputs of the linear nodes, μ_i, is given in the text. Reprinted from [30], with permission from Elsevier Science.

$$F(y|\mathbf{x}) \;=\; \sum_i a_i \varsigma \left(\beta_i \left[y - \sum_j w_{ij} \mathcal{S} \left(\sum_k u_{jk} x_k \right) \right] \right), \qquad (2.13)$$

where the symbols have the same meaning as before. Defining

$$\mu_i(\mathbf{x}) \;:=\; \sum_j w_{ij} \mathcal{S} \left(\sum_k u_{jk} x_k \right), \qquad (2.14)$$

Equation (2.13) can be re-written as

$$F(y|\mathbf{x}) \;=\; \sum_i a_i \varsigma \Big(\beta_i \left[y - \mu_i(\mathbf{x}) \right] \Big). \qquad (2.15)$$

In the deterministic limit ($a_i \to \delta_{i,i^*}$, $\beta_{i^*} \to \infty$), this reduces to

$$F(y|\mathbf{x}) \;\to\; \Theta \big[y - \mu_{i^*}(\mathbf{x}) \big] \qquad (2.16)$$

that is, to the point prediction $y = \mu_{i^*}(\mathbf{x})$. However, since from (2.14) $\mu_{i^*}(\mathbf{x})$ is the output of a one-hidden-layer network, it can now, by the universal approximation theorem, be any continuous *nonlinear* function. Thus the restriction to the linear case in the deterministic limit has been overcome.

2.4 Regaining the conditional probability density

In order to regain the conditional probability *density*, we have to take the derivative of $F(y|\mathbf{x})$ with respect to y. From (2.15), we obtain

$$
\begin{aligned}
P(y|\mathbf{x}) \;&=\; \frac{\partial}{\partial y} F(y|\mathbf{x}) \\
&=\; \frac{\partial}{\partial y} \sum_i a_i \varsigma \Big(\beta_i \left[y - \mu_i(\mathbf{x}) \right] \Big) \\
&=\; \sum_i a_i \beta_i \varsigma' \Big(\beta_i \left[y - \mu_i(\mathbf{x}) \right] \Big), \qquad (2.17)
\end{aligned}
$$

where the prime denotes a derivative with respect to the whole argument. Let us define

$$\mathcal{G}_\beta(x) \;:=\; \beta \varsigma'(\beta x). \qquad (2.18)$$

It is interesting to note that this transfer function is very similar to a Gaussian, as shown in Fig.2.2. Inserting (2.18) into (2.17) gives

$$P(y|\mathbf{x}) \;=\; \sum_i a_i \mathcal{G}_{\beta_i} \big[y - \mu_i(\mathbf{x}) \big]. \qquad (2.19)$$

The corresponding network structure is shown in Fig.2.1c, and is seen to contain *two* active hidden layers. The first hidden layer of sigmoidal units will be referred to as the $\mathcal{S}$-layer, the final hidden layer of units with a 'bell-shaped' transfer function as the $\mathcal{G}$-layer. The additional hidden layer of linear nodes, referred to as μ-nodes, has no computational efficiency and has been introduced solely for didactic reasons. (In this way, the 'total input to the ith $\mathcal{G}$-node' can be referred to simply as $\mu_i(\mathbf{x})$. It is recalled, from (2.14), that this is the output of a one-hidden-layer network.)

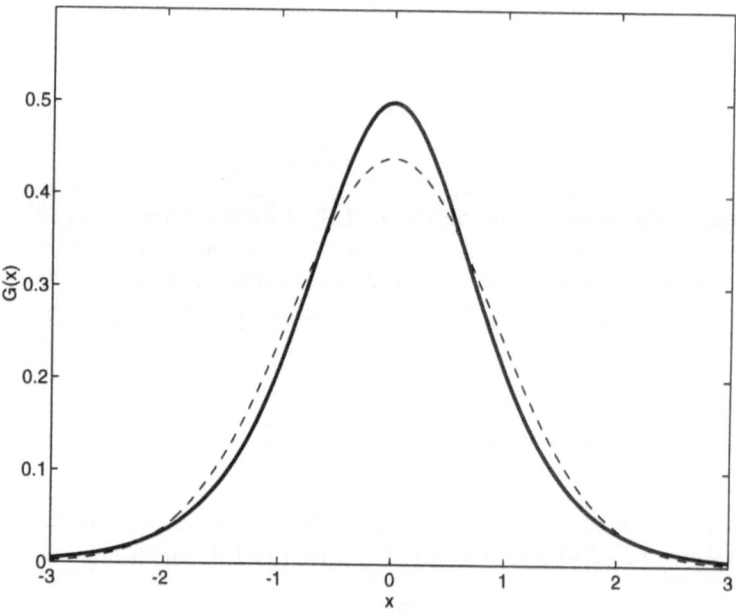

Fig. 2.2. Bell-shaped transfer functions. The solid line shows the bell-shaped transfer function of the DSM network, (2.18), for $\beta = 2$. This is compared with a Gaussian of the same standard deviation, $\sigma = \frac{\pi}{\sqrt{3}\beta} = \frac{\pi}{2\sqrt{3}} = 0.91$, shown by the dashed line.

2.5 Moments of the conditional probability density

An attractive property of (2.19) is the fact that one can obtain a closed form solution for the moment generating function,

$$M(s) = \langle\exp(sy)|\mathbf{x}\rangle := \int_{-\infty}^{\infty} \exp(sy)P(y|\mathbf{x})dy, \qquad (2.20)$$

from which all the (conditional) moments can be derived by

$$\langle y^m|\mathbf{x}\rangle := \int_{-\infty}^{\infty} y^m P(y|\mathbf{x})dy = \lim_{s\to 0} \frac{d^m M(s)}{ds^m}, \qquad (2.21)$$

as seen from a Taylor series expansion of the exponential function in (2.20),

$$M(s) = \sum_{i=0}^{\infty} \frac{s^i}{i!}\langle y^i|\mathbf{x}\rangle \qquad (2.22)$$

(e.g. [22], p.58-59). Inserting the expression for $P(y|\mathbf{x})$ from (2.19) into (2.21) yields, after some algebra (see the appendix to this chapter, Section 2.11),

$$M(s) = \sum_i a_i \exp\left(s\mu_i(\mathbf{x})\right)\frac{\pi s/\beta_i}{\sin(\pi s/\beta_i)}. \qquad (2.23)$$

Using (2.21) and (2.23), the following expressions for the first two moments are obtained (see appendix, Section 2.11):

$$\langle y|\mathbf{x}\rangle = \sum_i a_i\mu_i(\mathbf{x}), \qquad (2.24)$$

$$\langle y^2|\mathbf{x}\rangle = \sum_i a_i\left(\left(\mu_i(\mathbf{x})\right)^2 + \frac{1}{3}\left(\frac{\pi}{\beta_i}\right)^2\right). \qquad (2.25)$$

The first moment is the conditional mean, and is given by a convex sum over the μ-unit outputs. From the second moment, one can derive an expression for the conditional standard deviation (the 'error bar'):

$$\sigma_{y|\mathbf{x}}^2 = \langle y^2|\mathbf{x}\rangle - \langle y|\mathbf{x}\rangle^2 = \sum_i a_i\left(\left[\mu_i(\mathbf{x})\right]^2 + \frac{1}{3}\left(\frac{\pi}{\beta_i}\right)^2\right) - \left(\sum_i a_i\mu_i(\mathbf{x})\right)^2$$

$$= \sum_i a_i\left(\left[\mu_i(\mathbf{x})\right]^2 + \frac{1}{3}\left(\frac{\pi}{\beta_i}\right)^2\right) - 2\left(\sum_i a_i\mu_i(\mathbf{x})\right)^2 + \left(\sum_i a_i\mu_i(\mathbf{x})\right)^2$$

$$= \sum_i a_i\left(\left[\mu_i(\mathbf{x})\right]^2 + \frac{1}{3}\left(\frac{\pi}{\beta_i}\right)^2\right) - 2\left(\sum_i a_i\mu_i(\mathbf{x})\right)^2 + \sum_k a_k\left(\sum_i a_i\mu_i(\mathbf{x})\right)^2$$

$$= \sum_i a_i\left[\left(\left[\mu_i(\mathbf{x})\right]^2 + \frac{1}{3}\left(\frac{\pi}{\beta_i}\right)^2\right) - 2\mu_i(\mathbf{x})\left(\sum_k a_k\mu_k(\mathbf{x})\right) + \left(\sum_k a_k\mu_k(\mathbf{x})\right)^2\right]$$

$$= \sum_i a_i\left[\frac{1}{3}\left(\frac{\pi}{\beta_i}\right)^2 + \left(\mu_i(\mathbf{x}) - \sum_k a_k\mu_k(\mathbf{x})\right)^2\right],$$

whence

$$\sigma_{y|x} \;=\; \sqrt{\sum_i a_i \left[\frac{1}{3}\left(\frac{\pi}{\beta_i}\right)^2 + \left(\mu_i(\mathbf{x}) - \sum_k a_k \mu_k(\mathbf{x})\right)^2 \right]}. \qquad (2.26)$$

2.6 Interpretation of the network parameters

Assume that only one output weight is different from zero, i.e. $a_i = \delta_{ik}$. From (2.26), one obtains

$$\sigma_k \;=\; \sqrt{\frac{1}{3}\left(\frac{\pi}{\beta_k}\right)}, \qquad (2.27)$$

which is the standard deviation or 'width' of a single bell-shaped transfer function $\mathcal{G}_\beta(.)$. Since the output of the latter is always non-negative, $\mathcal{G}_\beta(y) \geq 0 \; \forall y$, and its integral over $[-\infty, \infty]$ is unity due to (2.9) and (2.18),

$$\int_{-\infty}^{\infty} \mathcal{G}_\beta(y)dy \;=\; \int_{-\infty}^{\infty} \beta\varsigma'(\beta y)dy \;=\; \int_{-\infty}^{\infty} \varsigma'(z)dz \;=\; \varsigma(\infty)-\varsigma(-\infty) \;=\; 1-0 \;=\; 1,$$

the function $\mathcal{G}_\beta\big(y-\mu_k(\mathbf{x})\big)$ may be interpreted as the conditional probability density of y conditioned on $\mathbf{x}$ for the kth component in the mixture (2.19):

$$P(y|\mathbf{x}, k) \;:=\; \mathcal{G}_\beta\big(y - \mu_k(\mathbf{x})\big). \qquad (2.28)$$

Moreover, (2.10) allows us to interpret the output weights a_k as the prior probabilities for the different components:

$$P(k) \;:=\; a_k. \qquad (2.29)$$

Inserting (2.28) and (2.29) into (2.19) gives

$$P(y|\mathbf{x}) \;=\; \sum_k P(k)P(y|\mathbf{x}, k), \qquad (2.30)$$

which is recognised as a special case of a general kernel expansion:

$$P(y|\mathbf{x}) \;=\; \sum_k P(y, k|\mathbf{x}) \;=\; \sum_k P(k|\mathbf{x})P(y|\mathbf{x}, k). \qquad (2.31)$$

The difference between (2.31) and the neural network model (2.30) is that, for the latter, the prior probabilities $P(k)$ are chosen $\mathbf{x}$-independent. According to the previous discussion, however, this is sufficient to ensure universal approximation capability.

2.7 Gaussian mixture model

The use of the sigmoid function $\varsigma(.)$, (2.9), for the second hidden layer nodes in the network model defined by (2.13) is not compelling, and it can in fact be replaced by any other sigmoidal function that is confined to the interval $[0, 1]$. One can therefore replace ς by the function *erf*, whose derivative is a Gaussian. Hence, (2.18) is replaced by

$$\mathcal{G}_\beta(x) \quad := \quad \sqrt{\frac{\beta}{2\pi}} \exp\left(-\frac{\beta x^2}{2}\right), \tag{2.32}$$

and the network model given by inserting (2.32) into (2.19) will be referred to as the *Gaussian mixture model* (GM). Note that the standard deviation or width of a single kernel $\mathcal{G}_{\beta_k}(.)$ is related to β_k by

$$\sigma_k \quad = \quad \sqrt{\frac{1}{\beta_k}}, \tag{2.33}$$

whereas for the sigmoid the dependence was found to be $\sigma_k \propto 1/\beta_k$ (see 2.27). The parameter β will henceforth be referred to as the *inverse kernel width* or *precision*. Let us now repeat the derivations of Section 2.5 for the GM model. The moment-generating function for a single Gaussian distribution with mean μ_k and standard deviation $\sigma_k = \sqrt{1/\beta_k}$ is known (e.g. [22], pp.77-80) to be

$$P_k(y) = \sqrt{\frac{\beta_k}{2\pi}} \exp\left(-\frac{\beta_k(y-\mu_k)^2}{2}\right) \quad \Rightarrow \quad M_k(s) = \exp\left(\mu_k s + \frac{s^2}{2\beta_k}\right). \tag{2.34}$$

The total moment generating function is then obtained as

$$M(s) = \int_{-\infty}^{\infty} P(y|\mathbf{x})\exp(sy)dy = \int_{-\infty}^{\infty} \sum_k a_k \mathcal{G}_{\beta_k}\left[y - \mu_k(\mathbf{x})\right]\exp(sy)dy$$

$$= \sum_k a_k \int_{-\infty}^{\infty} \mathcal{G}_{\beta_k}\left[y - \mu_k(\mathbf{x})\right]\exp(sy)dy = \sum_k a_k \int_{-\infty}^{\infty} P(y|\mathbf{x}, k)\exp(sy)dy$$

$$= \sum_k a_k M_k(s) = \sum_k a_k \exp\left(\mu_k(\mathbf{x})s + \frac{s^2}{2\beta_k}\right). \tag{2.35}$$

A Taylor series expansion gives

$$M(s) = \sum_k a_k \left(1 + \left[\mu_k(\mathbf{x})s + \frac{s^2}{2\beta_k}\right] + \frac{1}{2}\left[\mu_k(\mathbf{x})s + \frac{s^2}{2\beta_k}\right]^2 + \dots\right)$$

$$= 1 + s\sum_k a_k\mu_k(\mathbf{x}) + \frac{s^2}{2}\sum_k a_k\left(\left[\mu_k(\mathbf{x})\right]^2 + \frac{1}{\beta_k}\right) + \mathcal{O}(s^3), \tag{2.36}$$

leading to the following expressions for the first two moments:

$$\langle y|\mathbf{x}\rangle \;=\; \lim_{s\to 0}\frac{dM(s)}{ds} \;=\; \sum_k a_k \mu_k(\mathbf{x}), \qquad (2.37)$$

$$\langle y^2|\mathbf{x}\rangle \;=\; \lim_{s\to 0}\frac{d^2 M(s)}{ds^2} \;=\; \sum_k a_k \left(\left[\mu_k(\mathbf{x})\right]^2 + \frac{1}{\beta_k}\right). \qquad (2.38)$$

The derivation of the conditional standard deviation, $\sigma_{y|\mathbf{x}}$, follows the same line as that of (2.26), with $\frac{1}{3}\left(\frac{\pi}{\beta_k}\right)^2$ replaced by $\frac{1}{\beta_k}$, yielding

$$\sigma_{y|\mathbf{x}} \;=\; \sqrt{\sum_k a_k \left[\frac{1}{\beta_k} + \left(\mu_k(\mathbf{x}) - \sum_i a_i \mu_i(\mathbf{x})\right)^2\right]}. \qquad (2.39)$$

2.8 Derivative-of-sigmoid versus Gaussian mixture model

The previous derivations have led to two mixture models, which differ with respect to the kernel functions employed in the $\mathcal{G}$-layer. When the derivative of the sigmoid is chosen (2.18), the corresponding model will henceforth be referred to as the *derivative-of-sigmoid mixture* (DSM), whereas the model employing the Gaussian, (2.32), has already been introduced as the *Gaussian mixture* (GM) network. However, since these two transfer functions are very similar, as seen from Fig.2.2, the modelling capabilities of the two approaches do not differ significantly. What is more important, is the choice of an appropriate training scheme, and the remainder of this book will focus on the development and comparative evaluation of different such schemes. In earlier work, [27], [30], presented in Chapter 5, the DSM model was applied, as it arises naturally along the line of derivation adopted here and presented above. Later, it was superseded by the GM model, mainly for didactic purposes. Since *Gaussian* mixture models are well-known in the statistics community and are increasingly being employed in the neural network community, the introduction of new, but similar transfer functions only adds unnecessary intricacy that is not helpful in communicating the advantages and disadvantages of different training schemes. Moreover, the parameter update rules to be derived in Chapter 3, though not more efficient, are found to be easier to interpret for the GM model. Last but not least, the GM model, combined with a network modification to be discussed in Chapter 7, allows straightforward application of the EM algorithm.

2.9 Comparison with other approaches

2.9.1 Predicting local error bars

Weigend and Nix [64] introduced a combination of two one-hidden-layer networks for predicting input-dependent standard deviations (local error bars). The first network with weights $\mathbf{w}$ takes the input vector $\mathbf{x}$ and generates an output $\mu(\mathbf{x}; \mathbf{w})$ which represents the regression function. The second network with weights $\mathbf{v}$ also takes the input vector $\mathbf{x}$, and generates an output representing the variance of the noise distribution, $\sigma^2(\mathbf{x}, \mathbf{v})$. The parameters of the combined network, $(\mathbf{w}, \mathbf{v})$, are trained by a maximum likelihood scheme (described in Chapter 3). Bishop and Qazaz [8] improved this scheme by first integrating out the weights of the first network, $\mathbf{w}$, before optimising the weights of the second network, $\mathbf{v}$. This is similar to the Bayesian evidence scheme discussed in Chapter 10, and helps to avoid overfitting of the second network[1]. However, the whole approach is still based on the assumption of a Gaussian distribution of the noise (although with input-dependent variance), and cannot deal with skewed and multimodal conditional distributions $P(y|\mathbf{x})$.

2.9.2 Indirect method

An indirect way of modelling the conditional probability density $P(y|\mathbf{x})$ is by applying the definition

$$P(y|\mathbf{x}) \quad = \quad \frac{P(y, \mathbf{x})}{P(\mathbf{x})}. \tag{2.40}$$

Two networks are employed, one for modelling the distribution of the input vector, $P(\mathbf{x})$, the other for modelling the joint distribution of the input vector and the target, $P(y,\mathbf{x})$. The advantage is that this requires only the learning of *unconditional* probabilities, which can be accomplished by a *one-hidden-layer* network, thereby considerably reducing the complexity of the optimisation problem. Several models have been studied by Ormoneit and Neuneier et al., [45], [46], the simplest one being the Specht network [58], which centres a Gaussian kernel on each data point and treats their variances as (the only) adjustable parameters[2]. However, the authors found that the predictions achieved were rather poor, and that this indirect approach was significantly outperformed by direct approaches to predicting conditional densities.

---

[1] A joint optimisation of $\mathbf{w}$ and $\mathbf{v}$ leads to a biased result since the regression function inevitably fits part of the noise on the data, leading to an under-estimation of $\sigma(\mathbf{x})$.

[2] This model is thus equivalent to a Parzen estimator, briefly described in Section 4.3.

2.9.3 Complete kernel expansion: Conditional Density Estimation Network (CDEN) and Mixture Density Network (MDN)

As discussed in Section 2.6, the network model introduced in this chapter follows the kernel expansion (2.30). Bishop [6], [7], Neuneier et al. [45] and Ormoneit [46] extended this approach to the more general expansion (2.31) by choosing $\mathbf{x}$-dependent prior probabilities $a_k = a_k(\mathbf{x}) = P(k|\mathbf{x})$. Moreover, the kernel widths were also modelled as $\mathbf{x}$-dependent functions, $\beta_k = \beta_k(\mathbf{x})$. The network, referred to as the *Mixture Density Network* (MDN) by Bishop and the *Conditional Density Estimation Network* (CDEN) by Neuneier and Ormoneit, thus remains essentially the same, except that the $\{a_k, \beta_k\}$, which are treated as adaptable, weight-like parameters in the models introduced earlier in this chapter (DSM, GM), are now themselves given as outputs of a (seperate) neural network. Since this leads to a considerable increase in the overall network complexity, the conditional probability density is typically modelled by a small number of complex kernels $\mathcal{G}_\beta(.)$, whereas the GM and DSM networks employ a large number of simple kernels. An empirical comparison between these two approaches will be discussed later, in Chapter 5.

A major disadvantage of these direct methods is the large network complexity, which is equivalent to that of a two-hidden-layer model. Consequently, backpropagation-like training schemes, to be discussed in the next chapter, are rather slow. Moreover, the Expectation Maximisation (EM) algorithm, reviewed in Section 7.1, which usually offers a considerable alleviation in this respect, cannot be applied directly. Hence a main part of this book will be devoted to discussing a modification of the network structure such that the EM scheme can be applied (Section 7.3). Two alternative modifications which allow the application of the EM algorithm are briefly reviewed in the following two sub-sections.

2.9.4 Distorted Probability Mixture Network (DPMN)

The Distorted Probability Mixture Network (DPMN) of Neuneier et al. [45] and Ormoneit [46] applies the kernel expansion (2.31) with the simplification

$$P(y|\mathbf{x}, k) \quad := \quad P(y|k). \tag{2.41}$$

This leads to

$$P(y|\mathbf{x}) = \sum_k P(y|k)P(k|\mathbf{x}), \tag{2.42}$$

where the second term on the right follows from Bayes' rule,

$$P(k|\mathbf{x}) = \frac{P(\mathbf{x}|k)P(k)}{\sum_i P(\mathbf{x}|i)P(i)}. \tag{2.43}$$

The conditional densities for the targets and input vectors, $P(y|k)$ and $P(\mathbf{x}|k)$, are modelled by Gaussians (in the latter case by a multivariate Gaussian with a diagonal covariance matrix), and the priors $P(k)$ are given by adaptable output weights $P(k) := a_k$. While the training process can be accelerated considerably in this way, a major disadvantage of this approach is its incapacity to deal with the deterministic limit (2.2). The model is therefore not a proper universal approximator, which probably accounts for the poor performance in one of the applications reported in [45].

2.9.5 Mixture of Experts (ME) and Hierarchical Mixture of Experts (HME)

The *Mixture of Experts* (ME) network of Jordan and Jacobs [34] is similar to the CDEN or MDN, except that, in order to apply the EM algorithm, the $\mu_k(\mathbf{x})$ and $a_k(\mathbf{x})$ are given by linear models rather than by the output of a one-hidden-layer network:

$$\mu_k(\mathbf{x}) := \mathbf{w}_k^\dagger \mathbf{x} = \sum_i w_{ki} x_i, \tag{2.44}$$

$$a_k(\mathbf{x}) := \frac{\xi_k(\mathbf{x})}{\sum_i \xi_i(\mathbf{x})}, \quad \xi_k(\mathbf{x}) := \mathbf{v}_k^\dagger \mathbf{x} = \sum_i v_{ki} x_i \tag{2.45}$$

(the kernel width parameters β_k are chosen $\mathbf{x}$-independent). The overall conditional mean is given by

$$\mu(\mathbf{x}) = \langle y|\mathbf{x} \rangle = \sum_k a_k(\mathbf{x})\mu_k(\mathbf{x}), \tag{2.46}$$

(according to (2.24) and (2.37)). Since the restriction (2.44) implies severe limitations for the approximation capability of the model, a hierarchical extension is introduced, leading to the so-called *Hierarchical Mixture of Experts* (HME) network:

$$\mu_k(\mathbf{x}) := \sum_i a_{ki}(\mathbf{x})\mu_{ki}(\mathbf{x}), \quad \mu_{ki}(\mathbf{x}) := \mathbf{w}_{ki}^\dagger \mathbf{x}, \tag{2.47}$$

$$\xi_{ki}(\mathbf{x}) := \mathbf{v}_{ki}^\dagger \mathbf{x}, \quad a_{ki}(\mathbf{x}) := \frac{\xi_{ki}(\mathbf{x})}{\sum_j \xi_{kj}(\mathbf{x})} \tag{2.48}$$

2.9.6 Soft histogram

Finally, the *soft histogram* approach of Weigend and Srivastava [59], [66] can be shown to be a special case of the general kernel expansion of Section 2.9.3.

In a preprocessing step, the range of the target data is partitioned into k bins, each containing approximately the same number of exemplars. The bin centres, $\mu_1 \leq \ldots \leq \mu_k \leq \ldots \leq \mu_K$, are obtained by computing the mean of the targets within each bin. The network contains k normalised exponential units in its output layer, which represent the $\mathbf{x}$-dependent output weights $a_k(\mathbf{x})$. The conditional probability density at a target value y, $P(y|\mathbf{x})$, is obtained by linearly interpolating the output between adjacent bin centres, the latter being divided by their corresponding bin width to ensure normalisation. This can be interpreted as a mixture model of the form (2.31) with triangular-shaped, $\mathbf{x}$-independent kernels

$$P(y|\mathbf{x}, k) = P(y|k) := \begin{cases} \frac{y - \mu_{k-1}}{\mu_k - \mu_{k-1}} h_k & \text{if} \quad \mu_{k-1} \leq y \leq \mu_k \\ \frac{\mu_{k+1} - y}{\mu_{k+1} - \mu_k} h_k & \text{if} \quad \mu_k \leq y \leq \mu_{k+1} \\ 0 & \text{otherwise} \end{cases} \tag{2.49}$$

with the condition

$$h_k = \frac{2}{\mu_{k+1} - \mu_{k-1}} \tag{2.50}$$

to be satisfied by the bin heights, h_k, in order to ensure normalisation of $P(y|\mathbf{x})$,

$$\int_{-\infty}^{\infty} P(y|\mathbf{x})dy = \int_{-\infty}^{\infty} \sum_k P(k|\mathbf{x})P(y|k)dy = \sum_k P(k|\mathbf{x}) \int_{-\infty}^{\infty} P(y|k)dy$$

$$= \sum_k P(k|\mathbf{x}) \frac{h_k}{2} \left(\mu_{k+1} - \mu_{k-1} \right) dy = \sum_k P(k|\mathbf{x}) = 1. \tag{2.51}$$

2.10 Summary

In the present chapter, a neural network architecture for modelling conditional probability densities has been introduced. The objective was to derive the structure of a network that can deal with both stochastic and deterministic data. It was shown that a universal approximator network which retains its universal approximation capability in the singular low-noise limit requires at least two hidden layers. The resulting architecture can be interpreted in terms of a kernel expansion of the conditional probability density. This gives rise to a natural interpretation of the network parameters in terms of prior probabilities for different components, kernel centres and kernel widths. A similar kernel expansion is the basis of several alternative models independently developed by others at about the same time. The approaches differ mainly with respect to the specific choice of the kernel function and the input dependence of the model parameters. In the network studied in this book, only the kernel centres are modelled as x-dependent functions, whereas both the prior probabilities and the kernel widths are given by adaptable, but x-independent parameters. An approach which treats all model parameters as x-dependent functions, as adopted in the CDEN and the MDN, is in principle more powerful. However, given that the total network complexity is required to be limited - either for practical computational reasons or due to a limitation in the amount of available training data - the latter approach typically employs only a few complex kernels, whereas the approach developed in this book employs a larger number of simple kernels. An empirical comparison between the two methods on a time-series prediction problem will be studied later in Section 5.4.3.

Given the network architecture for modelling conditional probability densities, we now need to derive a training scheme for adapting the network parameters. This seems, at first, to be more difficult than for the conventional approach of predicting only a single value, since we do not have a target with which the network output could be compared. The solution to this problem is to adopt the method of *maximum likelihood*. This will be discussed in the following chapter.

2.11 Appendix: The moment generating function for the DSM network

The moment generating function was defined in (2.20). Inserting the expression for the conditional probability density from (2.17) into (2.20) gives

$$M(s) \;=\; \int\limits_{-\infty}^{\infty} P(y|\mathbf{x}) \exp(sy) dy \;=\; \int\limits_{-\infty}^{\infty} \sum_k a_k \frac{\partial}{\partial y} \varsigma\Big(\beta_k\big[y - \mu_k(\mathbf{x})\big]\Big) \exp(sy) dy$$

$$=\; \sum_k a_k \int\limits_{-\infty}^{\infty} \exp(sy) \frac{\partial}{\partial y} \varsigma\Big(\beta_k\big[y - \mu_k(\mathbf{x})\big]\Big) dy \;=\; \sum_k a_k \int\limits_{0}^{1} \exp(sy) d\varsigma$$

Using the substitution $y = y(\varsigma)$ with

$$\varsigma(y) \;=\; \frac{1}{1 + \exp\Big(-\beta_k[y - \mu_k]\Big)} \quad \Rightarrow \quad 1 + \exp\Big(-\beta_k[y - \mu_k]\Big) \;=\; \frac{1}{\varsigma} \quad \Rightarrow$$

$$\exp\Big(-\beta_k[y - \mu_k]\Big) \;=\; \frac{1 - \varsigma}{\varsigma} \quad \Rightarrow \quad \exp(-\beta_k y) \;=\; \frac{1 - \varsigma}{\varsigma} \exp(-\beta_k \mu_k) \quad \Rightarrow$$

$$\exp(sy) \;=\; \Big(\frac{1 - \varsigma}{\varsigma}\Big)^{-s/\beta_k} \exp(s\mu_k),$$

the moment generating function becomes

$$M(s) \;=\; \sum_k a_k \int_0^1 \exp(sy)d\varsigma \;=\; \sum_k a_k \exp(\mu_k s) \int_0^1 \left(\frac{1-\varsigma}{\varsigma}\right)^{-s/\beta_k} d\varsigma.$$

Now

$$\int_0^1 \left(\frac{1-\varsigma}{\varsigma}\right)^{-s/\beta_k} d\varsigma \;=\; \frac{\Gamma(1+s/\beta_k)\Gamma(1-s/\beta_k)}{\Gamma(2)}$$

(e.g. [11], p.70), where $\Gamma(.)$ is the gamma function

$$\Gamma(x) \;=\; \int_0^\infty e^{-u} u^{x-1} du,$$

which has the following properties:

$$\Gamma(1) \;=\; 1, \qquad x \neq 0 \;\Rightarrow\; \Gamma(x+1) \;=\; x\Gamma(x), \qquad \Gamma(x)\Gamma(1-x) \;=\; \frac{\pi}{\sin(\pi x)}$$

(e.g. [11] p.103). Hence

$$\int_0^1 \left(\frac{1-\varsigma}{\varsigma}\right)^{-s/\beta_k} d\varsigma \;=\; \frac{\pi s/\beta_k}{\sin(\pi s/\beta_k)},$$

and $M(s)$ thus reduces to

$$M(s) \;=\; \sum_k a_k \exp(\mu_k s) \frac{\pi s/\beta_k}{\sin(\pi s/\beta_k)}.$$

In order to obtain the first two moments from $M(s)$ by use of (2.21), $M(s)$ is expanded in a power series up to second order in s, using

$$\exp(x) \;=\; 1 + x + \frac{x^2}{2} + \dots$$

and

$$\sin^{-1}(x) \;=\; \frac{1}{x} + \frac{x}{6} + \dots$$

(for $0 < |x| < \pi$, see [11], p.33). This leads to

$$M(s) \;=\; \sum_k a_k \left[1 + \mu_k s + \frac{(\mu_k s)^2}{2} + \dots\right]\left[1 + \frac{1}{6}\left(\frac{\pi s}{\beta_k}\right)^2 + \dots\right]$$

$$=\; \sum_k a_k \;+\; s \sum_k a_k \mu_k \;+\; \frac{s^2}{2} \sum_k a_k \left[(\mu_k)^2 + \frac{1}{3}\left(\frac{\pi}{\beta_k}\right)^2\right] + \mathcal{O}(s^3)$$

Hence, the first two moments are obtained as

$$\langle y|\mathbf{x}\rangle \;=\; \lim_{s\to 0}\frac{dM(s)}{ds} \;=\; \sum_{k} a_k \mu_k(\mathbf{x}),$$

$$\langle y^2|\mathbf{x}\rangle \;=\; \lim_{s\to 0}\frac{d^2 M(s)}{ds^2} \;=\; \sum_{k} a_k \left(\big(\mu_k(\mathbf{x})\big)^2 + \frac{1}{3}\left(\frac{\pi}{\beta_k}\right)^2 \right).$$

3. A Maximum Likelihood Training Scheme

An error function E for a mixture model is derived from a maximum likelihood approach. The derivation of a gradient descent scheme is performed for both the DSM and the GM networks, and leads to a modified form of the backpropagation algorithm. However, a straightforward application of this method is shown to suffer from considerable inherent convergence problems due to large curvature variations of the error surface. A simple rectification scheme based on a curvature-based shape modification of E is presented.

3.1 The cost function

In the previous chapter, a neural-network architecture for modelling conditional probability densities has been derived. We now need an appropriate training algorithm for adapting the network parameters

$$\mathbf{q} \quad := \quad [\mathbf{w}, a_1, \ldots, a_{K-1}, \beta_1, \ldots, \beta_K], \tag{3.1}$$

where K denotes the number of kernel nodes in the $\mathcal{G}$-layer, β_k are the (inverse) kernel width parameters, a_k are the output weights or prior probabilities (note that due to the constraint $\sum_k a_k = 1$ only $K - 1$ parameters are free), and $\mathbf{w}$ denotes the set of remaining weights in the network. The derivation of such a training scheme requires some cost or objective function $E(\mathbf{q})$, with respect to which the optimisation of the network parameters can be defined. When a conventional network for point-predictions is given, one usually compares the predicted values, $\mu(\mathbf{x}_t; \mathbf{w})$, with the actual targets, y_t, and adapts the weights $\mathbf{w}$ so as to minimise the sum of squared deviations,

$$E(\mathbf{w}) \quad = \quad \frac{1}{2N} \sum_{t=1}^{N} \left[y_t - \mu(\mathbf{x}_t, \mathbf{w}) \right]^2 \tag{3.2}$$

(mean-of-squares error). This method, however, is not applicable to the problem considered here, since the target to be learned is the whole probability distribution $P(y|\mathbf{x})$, which is not known. Fortunately, there is a theorem which allows us to set up an alternative cost function.

THEOREM (Kullback)
Consider a random variable y and two probability densities $P_0(.)$ and $P(.)$. Then

$$-\Big\langle \ln(P_0) \Big\rangle_{P_0} \;\leq\; -\Big\langle \ln(P) \Big\rangle_{P_0} \tag{3.3}$$

where

$$-\Big\langle \ln(P_0) \Big\rangle_{P_0} \;:=\; -\int_{-\infty}^{\infty} P_0(y) \ln\Big(P_0(y)\Big)\,dy \tag{3.4}$$

$$-\Big\langle \ln(P) \Big\rangle_{P_0} \;:=\; -\int_{-\infty}^{\infty} P_0(y) \ln\Big(P(y)\Big)\,dy \tag{3.5}$$

PROOF
The theorem is a direct consequence of the following inequality:

$$\ln(x) \;\leq\; x-1 \quad \forall x \tag{3.6}$$

Substituting $x := \frac{P(y)}{P_0(y)}$ yields

$$\ln\left(\frac{P(y)}{P_0(y)}\right) \leq \frac{P(y)}{P_0(y)} - 1 \;\;\Rightarrow\;\; P_0(y)\ln\left(\frac{P(y)}{P_0(y)}\right) \leq P(y) - P_0(y) \;\;\Rightarrow$$

$$P_0(y)\ln\Big(P(y)\Big) - P_0(y)\ln\Big(P_0(y)\Big) \;\leq\; P(y) - P_0(y) \;\;\Rightarrow$$

$$\int_{-\infty}^{\infty} P_0(y)\ln\Big(P(y)\Big)\,dy - \int_{-\infty}^{\infty} P_0(y)\ln\Big(P_0(y)\Big)\,dy \leq \int_{-\infty}^{\infty} P(y)\,dy - \int_{-\infty}^{\infty} P_0(y)\,dy \Rightarrow$$

$$\int_{-\infty}^{\infty} P_0(y)\ln\Big(P(y)\Big)\,dy - \int_{-\infty}^{\infty} P_0(y)\ln\Big(P_0(y)\Big)\,dy \;\leq\; 1-1 = 0 \;\;\Rightarrow$$

$$\Big\langle \ln(P) \Big\rangle_{P_0} - \Big\langle \ln(P_0) \Big\rangle_{P_0} \leq 0 \;\;\Rightarrow\;\; -\Big\langle \ln(P_0) \Big\rangle_{P_0} \leq -\Big\langle \ln(P) \Big\rangle_{P_0}$$

◇

Assume that $P_0(y|\mathbf{x})$ is the correct probability density of the data-generating process, and let $P(y|\mathbf{x},\mathbf{q})$ denote the probability density predicted by the network. Then, by Kullback's theorem, the function

$$\mathcal{E}(\mathbf{q}) := -\Big\langle \ln(P) \Big\rangle_{P_0} \;=\; -\int_{-\infty}^{\infty}\int_{-\infty}^{\infty} P_0(y,\mathbf{x})\ln\Big(P(y|\mathbf{x},\mathbf{q})\Big)\,dy\,d\mathbf{x}$$

$$= - \int_{-\infty}^{\infty} \left[\int_{-\infty}^{\infty} P_0(y|\mathbf{x}) \ln \left(P(y|\mathbf{x}, \mathbf{q}) \right) dy \right] P_0(\mathbf{x}) d\mathbf{x}$$

$$\geq - \int_{-\infty}^{\infty} \left[\int_{-\infty}^{\infty} P_0(y|\mathbf{x}) \ln \left(P_0(y|\mathbf{x}) \right) dy \right] P_0(\mathbf{x}) d\mathbf{x} = - \left\langle \ln(P_0) \right\rangle_{P_0} (3.7)$$

has its global minimum at $P(y|\mathbf{x}, \mathbf{q}) = P_0(y|\mathbf{x})$ and is therefore, in principle, an appropriate cost function for network training. In practice, however, the integral in (3.7) cannot be solved analytically, since the true distribution, $P_0(y|\mathbf{x})$, is not known (if it was, there would be no need for a network in the first place.) One way to proceed in these circumstances, therefore, is to use a simple Monte Carlo approximation to the integral, based on the given training data. That is, given that the input-target pairs $(\mathbf{x}_t, y_t)$ in the training set $\mathbf{D} = \{(\mathbf{x}_t, y_t)\}_{t=1}^{N}$ are *independent*, and drawing on the fact that they are distributed according to the true probability density $P_0(y, \mathbf{x}) = P_0(y|\mathbf{x})P_0(\mathbf{x})$, we can make the approximation

$$\mathcal{E}(\mathbf{q}) \quad \approx \quad E(\mathbf{q}) \quad := \quad -\frac{1}{N} \sum_{t=1}^{N} \ln \left(P(y_t|\mathbf{x}_t, \mathbf{q}) \right). \tag{3.8}$$

When N is large, the deviation of E from $\mathcal{E}$ in (3.8) is small, and the minimisation of $E(\mathbf{q})$ leads to a close approximation of the true conditional probability density $P_0(y|\mathbf{x})$ (given that the network is sufficiently complex to model P_0). However, for sparse data sets $\mathbf{D}$ with a small number of training pairs N, $\mathcal{E}$ and E in (3.8) can differ significantly. In this case, a training process that minimises the cost function E does *not* lead to a good estimate of P_0. This effect is known as *overfitting*.

The set of parameters $\mathbf{q}$ that minimise (3.8) is equivalent to what is called a *maximum likelihood estimator* in statistics. Consider again the set of training data $\mathbf{D} = \{(\mathbf{x}_t, y_t)\}_{t=1}^{N}$, whose distribution is approximated by the model distribution $P(\mathbf{D}|\mathbf{q})$. Then the maximum likelihood estimate $\hat{\mathbf{q}}$ is given by the parameters $\mathbf{q}$ that maximise $P(\mathbf{D}|\mathbf{q})$ or, equivalently, $L(\mathbf{q}) := \ln P(\mathbf{D}|\mathbf{q})$. Given that the model $P(\mathbf{D}|\mathbf{q})$ is sufficiently complex to be capable of approximating the true distribution (realisable case), this estimator can be shown to be asymptotically (i.e. for $N \to \infty$) unbiased and to have minimum variance (e.g. [49], p.262). If we assume that the samples in $\mathbf{D}$ are independent, then

$$L(\mathbf{q}) = \ln \left(P(\mathbf{D}|\mathbf{q}) \right) = \ln \left(\prod_{t=1}^{N} P(\mathbf{x}_t, y_t|\mathbf{q}) \right) = \ln \left(\prod_{t=1}^{N} P(y_t|\mathbf{x}_t, \mathbf{q}) \prod_{t=1}^{N} P(\mathbf{x}_t) \right). \tag{3.9}$$

The last step follows from the fact that the distribution of the input vectors $\mathbf{x}_t$ is *not* modelled by the neural network and is therefore independent of $\mathbf{q}$, $P(\mathbf{x}_t|\mathbf{q}) = P(\mathbf{x}_t)$. Re-writing (3.9) and comparing the result with (3.8) thus gives

$$L(\mathbf{q}) = \sum_{t=1}^{N} \ln\left(P(y_t|\mathbf{x}_t, \mathbf{q})\right) + C = -NE(\mathbf{q}) + C \qquad (3.10)$$

Hence minimising $E(\mathbf{q})$ is equivalent to maximising $L(\mathbf{q})$, i.e. the argument of the minimum of $E(\mathbf{q})$ is the maximum likelihood estimate $\hat{\mathbf{q}}$.

It is noted in brief that the maximum-likelihood error function $E(\mathbf{q})$ in (3.8) is a generalisation of the mean-square-error function (3.2), and reduces to the latter if the conditional probability density $P(y|\mathbf{x}, \mathbf{q})$ is a Gaussian with constant variance $\sigma^2 := 1/\beta$:

$$P(y_t|x_t) = \sqrt{\frac{\beta}{2\pi}} \exp\left(-\frac{\beta}{2}[y_t - \mu(\mathbf{x}_t; \mathbf{w})]^2\right) \qquad (3.11)$$

Inserting (3.11) into (3.8) yields[1]

$$E(\mathbf{w}) = -\frac{1}{N}\sum_{t=1}^{N} \ln\left(P(y_t|\mathbf{x}_t, \mathbf{q})\right) = -\frac{1}{N}\sum_{t=1}^{N}\left[-\frac{\beta}{2}[y_t - \mu(\mathbf{x}_t; \mathbf{w})]^2 + \ln\left(\sqrt{\frac{\beta}{2\pi}}\right)\right]$$

$$= \frac{\beta}{2N}\sum_{t=1}^{N}[y_t - \mu(\mathbf{x}_t; \mathbf{w})]^2 - \ln\left(\sqrt{\frac{\beta}{2\pi}}\right) \propto \sum_{t=1}^{N}[y_t - \mu(\mathbf{x}_t; \mathbf{w})]^2 + \text{const}$$

which, when minimised with respect to $\mathbf{w}$, gives the same parameters as a minimisation of (3.2).

Moreover, it is noted that for a *time series* the above condition of *independence* is satisfied if the system can be modelled by an mth order Markov process. Let $\mathbf{D} = \{x_{-m+1}, \ldots, x_{N-1}, x_N\}$, where for convenience in the following notation initial observations are taken at negative times. Also, recall that $\mathbf{x}_t = (x_t, x_{t-1}, \ldots, x_{t-m+1})$. Then

$$P(\mathbf{D}|\mathbf{q}) = P(x_{-m+1}, \ldots, x_{N-1}, x_N|\mathbf{q}) = P(\mathbf{x}_0)P(x_1|\mathbf{x}_0, \mathbf{q})\ldots P(x_N|\mathbf{x}_{N-1}, \mathbf{q})$$
$$(3.12)$$

holds due to the Markov property. Taking the logarithm on both sides thus gives

$$\ln\left(P(\mathbf{D}|\mathbf{q})\right) = \sum_{t=1}^{N} \ln\left(P(x_t|\mathbf{x}_{t-1}, \mathbf{q})\right) + \ln\left(P(\mathbf{x}_0)\right) = \sum_{t=1}^{N} \ln\left(P(x_t|\mathbf{x}_{t-1}, \mathbf{q})\right) + C.$$
$$(3.13)$$

The independence of $P(\mathbf{x}_0)$ from $\mathbf{q}$ follows again from the fact that the input distribution is not modelled by the network. Defining $y_t := x_{t+1}$ (the target at time t is the value at the next time step x_{t+1}) and shifting the summation index t by one leads to (3.10).

As a final remark, a convention about the notation will be introduced. Since the constant in (3.10) has no influence on the optimisation problem, it will be convenient to neglect it altogether. This corresponds to setting the

---

[1] The same derivation can be found in [7], Chapter 6, and [64].

second term in the argument of the logarithm on the right-hand side of (3.9) to unity. The ensuing notation is, strictly speaking, imprecise. However, this flaw is irrelevant for all practical purposes, and the notation will become more transparent. In the remainder of this book, the expression $P(\mathbf{D}|\mathbf{q})$ will therefore be understood as

$$P(\mathbf{D}|\mathbf{q}) \quad = \quad \prod_{t=1}^{N} P(y_t|\mathbf{x}_t, \mathbf{q}), \qquad (3.14)$$

and the factor containing the unconditional probability density of the input vectors will be omitted.

3.2 A gradient-descent training scheme

A standard training method for adapting the network parameters $\mathbf{q}$ is a simple gradient-descent scheme (with momentum):

$$q_i^{(n+1)} \quad = \quad q_i^{(n)} + \Delta q_i^{(n)} \qquad (3.15)$$

$$\Delta q_i^{(n)} \quad = \quad -\gamma \frac{\partial E^{(n)}}{\partial q_i} + \eta \Delta q_i^{(n-1)}, \qquad (3.16)$$

where the superscript (n) indicates the iteration step, $\eta \in [0, 1[$ is the so-called momentum parameter, and $\gamma > 0$ the learning rate. In general, it will be advantageous to introduce different learning rates for the three different parameter classes, namely γ_a for the priors or output weights a_k, γ_β for the kernel widths σ_k (or inverse kernel widths β_k), and γ_w for the remaining weights. In this section, a derivation of the gradient

$$\frac{\partial E}{\partial q_i} \quad = \quad \frac{1}{N} \sum_{t=1}^{N} \frac{\partial \varepsilon_t}{\partial q_i} \qquad (3.17)$$

$$\varepsilon_t \quad := \quad -\ln \left(P(y_t|\mathbf{x}_t, \mathbf{q}) \right) \qquad (3.18)$$

for $\{a_k\}$, $\{\beta_k\}$, and $\mathbf{w}$, will be given. Recall from Chapter 2 that two different types of bell-shaped transfer functions for the nodes in the $\mathcal{G}$-layer have been introduced:

$$\text{GM} \quad : \quad \mathcal{G}_\beta(y - \mu) := \sqrt{\frac{\beta}{2\pi}} \exp \left(-\frac{\beta[y - \mu]^2}{2} \right) \qquad (3.19)$$

$$\text{DSM} \quad : \quad \breve{\mathcal{G}}_\beta(y - \mu) := \beta\varsigma' \left(\beta[y - \mu] \right) \qquad (3.20)$$

For means of distinguishing between the two models, a breve will henceforth be used for the transfer function of the DSM model. Moreover, recall that

$\varsigma'(.)$ symbolises a derivative with respect to the whole argument. The sigmoid function $\varsigma(.)$ was defined in (2.9). From this definition the following relations can easily be verified:

$$\varsigma'(x) = \varsigma(x)\Big[1 - \varsigma(x)\Big], \qquad \varsigma''(x) = \varsigma(x)\Big[1 - \varsigma(x)\Big]\Big[1 - 2\varsigma(x)\Big] \quad (3.21)$$

$$\Rightarrow \frac{\varsigma''(x)}{\varsigma'(x)} = 1 - 2\varsigma(x) = 1 - \frac{2}{1 + e^{-x}} = \frac{e^{-x/2} - e^{x/2}}{e^{-x/2} + e^{x/2}} = -\tanh\left(\frac{x}{2}\right) \quad (3.22)$$

Making use of (3.22), the derivatives of $\check{\mathcal{G}}_\beta$ with respect to β and μ are given by

$$\frac{\partial \check{\mathcal{G}}_\beta}{\partial \beta} = \frac{\partial}{\partial \beta}\Big[\beta\varsigma'\Big(\beta[y - \mu]\Big)\Big] = \varsigma'\Big(\beta[y - \mu]\Big) + \beta[y - \mu]\varsigma''\Big(\beta[y - \mu]\Big)$$

$$= \beta\varsigma'\Big(\beta[y - \mu]\Big)\left(\frac{1}{\beta} + [y - \mu]\frac{\varsigma''\Big(\beta[y - \mu]\Big)}{\varsigma'\Big(\beta[y - \mu]\Big)}\right)$$

$$= \check{\mathcal{G}}_\beta(y - \mu)\left(\frac{1}{\beta} - [y - \mu]\tanh\left(\frac{\beta(y - \mu)}{2}\right)\right) \quad (3.23)$$

$$\frac{\partial \check{\mathcal{G}}_\beta}{\partial \mu} = \frac{\partial}{\partial \mu}\Big[\beta\varsigma'\Big(\beta[y - \mu]\Big)\Big] = -\beta^2\varsigma''\Big(\beta[y - \mu]\Big)$$

$$= -\beta^2\varsigma'\Big(\beta[y - \mu]\Big)\frac{\varsigma''\Big(\beta[y - \mu]\Big)}{\varsigma'\Big(\beta[y - \mu]\Big)} = \beta^2\varsigma'\Big(\beta[y - \mu]\Big)\tanh\left(\frac{\beta[y - \mu]}{2}\right)$$

$$= \beta\check{\mathcal{G}}_\beta(y - \mu)\tanh\left(\frac{\beta[y - \mu]}{2}\right) \quad (3.24)$$

From (3.19), we obtain for the derivatives of $\mathcal{G}_\beta$ with respect to β and μ:

$$\frac{\partial \mathcal{G}_\beta}{\partial \beta} = \left[\frac{1}{2\beta} - \frac{[y - \mu]^2}{2}\right]\mathcal{G}_\beta(y - \mu) \quad (3.25)$$

$$\frac{\partial \mathcal{G}_\beta}{\partial \mu} = \beta[y - \mu]\mathcal{G}_\beta(y - \mu) \quad (3.26)$$

The similarities between these derivatives can be seen when taking the limit of small deviations $|y - \mu| \ll 1$, which allows the approximation $\tanh(x) \approx x$ to be made. From (3.23) and (3.24) we obtain

$$\frac{\partial \check{\mathcal{G}}_\beta}{\partial \beta} \approx \beta\left[\frac{1}{\beta^2} - \frac{[y - \mu]^2}{2}\right]\check{\mathcal{G}}_\beta(y - \mu) \quad (3.27)$$

$$\frac{\partial \check{\mathcal{G}}_\beta}{\partial \mu} \approx \frac{\beta^2}{2}[y - \mu]\check{\mathcal{G}}_\beta(y - \mu). \quad (3.28)$$

which is of a similar form as (3.25) and (3.26).

The following derivations of the gradient-descent update equations will lead to terms which include a factor of the form $a_k \mathcal{G}_{\beta_k}/P$. Now recall from (2.28) that $\mathcal{G}_{\beta_k}(y - \mu_k(\mathbf{x}; \mathbf{w}))$ denotes the (class-)conditional probability density of y given $\mathbf{x}$ for the kth component in the mixture, and from (2.29) that a_k is the *prior* probability of a data point having been generated from component k of the mixture. We can therefore apply Bayes' rule to obtain

$$\frac{a_k \mathcal{G}_{\beta_k}(y_t - \mu_k(\mathbf{x}_t; \mathbf{w}))}{P(y_t|\mathbf{x}_t)} = \frac{P(k)P(y_t|\mathbf{x}_t, k)}{P(y_t|\mathbf{x}_t)} = P(k|y_t, \mathbf{x}_t). \qquad (3.29)$$

The expression on the right is called the *posterior* probability for the kth component after observing data point $(y_t, \mathbf{x}_t)$. This is the probability of a data point having been generated from component k, conditional on the additional information given by the observation of $(y_t, \mathbf{x}_t)$. Since all the probabilities in (3.29) are modelled by the network and thus depend on the network parameters $\mathbf{q}$, this should be written more precisely

$$\frac{a_k \mathcal{G}_{\beta_k}(y_t - \mu_k(\mathbf{x}_t; \mathbf{w}))}{P(y_t|\mathbf{x}_t, \mathbf{q})} = \frac{P(k|\mathbf{q})P(y_t|\mathbf{x}_t, k, \mathbf{q})}{P(y_t|\mathbf{x}_t, \mathbf{q})} = P(k|y_t, \mathbf{x}_t, \mathbf{q}). \qquad (3.30)$$

However, here and in the remainder of this book the posterior distribution will be abbreviated as

$$\pi_k(t) \quad := \quad P(k|y_t, \mathbf{x}_t, \mathbf{q}). \qquad (3.31)$$

Moreover, in order to simplify the notation for the forthcoming derivations, the abbreviations $\mu_k := \mu_k(\mathbf{x}_t; \mathbf{w})$, $P := P(y_t|\mathbf{x}_t, \mathbf{q})$, and $\mathcal{G}_k := \mathcal{G}_{\beta_k}(y - \mu_k(\mathbf{x}_t; \mathbf{w}))$ will be used.

3.2.1 Output weights

With (2.19) and (3.18), the derivative of ε_t with respect to a_k is obtained as

$$-\frac{\partial \varepsilon_t}{\partial a_k} = -\frac{\partial \varepsilon_t}{\partial P}\frac{\partial P}{\partial a_k} = -\frac{\partial}{\partial P}[-\ln(P)]\frac{\partial}{\partial a_k}\sum_i a_i \mathcal{G}_i = \frac{\mathcal{G}_k}{P}. \qquad (3.32)$$

On multiplying both numerator and denominator by a_k and making use of (3.30) and definition (3.31), this leads to

$$-\frac{\partial \varepsilon_t}{\partial a_k} = \frac{\pi_k(t)}{a_k}. \qquad (3.33)$$

The constraint $\sum_k a_k = 1$ can be satisfied by adding its derivative, multiplied by a Lagrange parameter λ, to the gradient (3.17):

$$-\frac{\partial E}{\partial a_k} = -\frac{1}{N}\sum_{t=1}^{N}\frac{\partial \varepsilon_t}{\partial a_k} + \lambda\frac{\partial}{\partial a_k}\sum_{i=1}^{K} a_i = \frac{1}{N}\sum_{t=1}^{N}\frac{\pi_k(t)}{a_k} + \lambda. \qquad (3.34)$$

Setting the derivative to zero, multiplying both sides by $a_k N$, and then summing over k gives for the Lagrange parameter

$$0 = \sum_{k=1}^{K}\sum_{t=1}^{N} \pi_k(t) + \lambda N \sum_{k=1}^{K} a_k = \sum_{t=1}^{N}\sum_{k=1}^{K} \pi_k(t) + \lambda N = N + \lambda N$$

$$\Rightarrow \quad \lambda = -1 \tag{3.35}$$

and thus

$$-\frac{\partial E}{\partial a_k} = \frac{1}{N}\left(\sum_{t=1}^{N} \frac{\pi_k(t)}{a_k} - N\right) = \frac{1}{a_k N}\sum_{t=1}^{N}[\pi_k(t) - a_k]. \tag{3.36}$$

Note that this derivative cannot simply be set to zero to solve for $a_k = \frac{1}{N}\sum_t \pi_k(t)$, since $\pi_k(t) = P(k|\mathbf{q}, t) = P(k|a_1, \dots a_{K-1}, \beta_1, \dots, \beta_K, \mathbf{w}, t)$ itself is a nonlinear function of a_k. The only feasible approach is therefore an iterative scheme, like the gradient descent method described above. A problem with the direct application of (3.36), however, is that when adapting the a_k's along the gradient with a fixed learning rate γ_a, it is not necessarily ensured that the new output weights will obey the second constraint, $a_k \in [0, 1]$. For this reason it is advantageous to write the output weight a_k as the function of a new variable φ_i, $i = 1, \dots, K$, such that both constraints $a_k \in [0, 1]$ and $\sum_k a_k = 1$ are always satisfied, e.g.

$$a_k \quad := \quad \frac{\exp(\varphi_k)}{\sum_i \exp(\varphi_i)} \tag{3.37}$$

or

$$a_k \quad := \quad \frac{(\varphi_k)^2}{\sum_i (\varphi_i)^2}. \tag{3.38}$$

The gradient descent search is then carried out in the space of the new variables φ_k. For the so-called softmax function (3.37) we obtain[2]

$$\frac{\partial a_k}{\partial \varphi_i} = \frac{\delta_{ki}\exp(\varphi_i)}{\sum_j \exp(\varphi_j)} - \frac{\exp(\varphi_i)\exp(\varphi_k)}{\left(\sum_j \exp(\varphi_j)\right)^2} = \delta_{ik}a_i - a_i a_k$$

$$= \delta_{ik}a_k - a_i a_k = a_k(\delta_{ik} - a_i). \tag{3.39}$$

Inserting (3.33) and (3.39) into

$$-\frac{\partial \varepsilon_t}{\partial \varphi_i} = -\sum_k \frac{\partial \varepsilon_t}{\partial a_k}\frac{\partial a_k}{\partial \varphi_i} \tag{3.40}$$

yields

---

[2] The derivation given in [7], pp.215-217, is followed.

$$-\frac{\partial \varepsilon_t}{\partial \varphi_i} \;=\; \sum_k \pi_k(t)(\delta_{ik} - a_i) \;=\; \pi_i(t) - a_i \sum_k \pi_k(t) \;=\; \pi_i(t) - a_i \quad (3.41)$$

and thus, from (3.17),

$$-\frac{\partial E}{\partial \varphi_k} \;=\; -\frac{1}{N}\sum_t \frac{\partial \varepsilon_t}{\partial \varphi_k} \;=\; \frac{1}{N}\sum_t [\pi_k(t) - a_k]. \quad (3.42)$$

The simulations reported in this book (Chapter 5), however, were carried out with the alternative function (3.38). The derivation of the corresponding update rule, which is slightly more complex, will be given in the appendix to this chapter, Section 3.4.

3.2.2 Kernel widths

By application of the chain rule, we obtain

$$-\frac{\partial \varepsilon_t}{\partial \beta_k} \;=\; -\frac{\partial \varepsilon_t}{\partial \mathcal{G}_k}\frac{\partial \mathcal{G}_k}{\partial \beta_k}. \quad (3.43)$$

The second term, $\frac{\partial \mathcal{G}_k}{\partial \beta_k}$, has already been calculated in (3.23) and (3.25). The first term gives

$$-\frac{\partial \varepsilon_t}{\partial \mathcal{G}_k} \;=\; -\frac{\partial \varepsilon_t}{\partial P}\frac{\partial P}{\partial \mathcal{G}_k} \;=\; -\frac{\partial}{\partial P}[-\ln(P)]\frac{\partial}{\partial \mathcal{G}_k}\sum_i a_i \mathcal{G}_i \;=\; \frac{a_k}{P}. \quad (3.44)$$

Inserting (3.23), (3.25) and (3.44) into (3.43) yields

$$\text{DSM}\;:\;\; -\frac{\partial \varepsilon_t}{\partial \beta_k} \;=\; \frac{a_k \breve{\mathcal{G}}_k}{P}\left[\frac{1}{\beta_k} - [y_t - \mu_k]\tanh\left(\frac{\beta_k(y_t - \mu_k)}{2}\right)\right] \quad (3.45)$$

$$\text{GM}\;:\;\; -\frac{\partial \varepsilon_t}{\partial \beta_k} \;=\; \frac{1}{2}\frac{a_k \mathcal{G}_k}{P}\left[\frac{1}{\beta_k} - (y_t - \mu_k)^2\right] \quad (3.46)$$

With (3.30) and (3.31) these expressions simplify to

$$\text{DSM}\;:\;\; -\frac{\partial \varepsilon_t}{\partial \beta_k} \;=\; \pi_k(t)\left[\frac{1}{\beta_k} - [y_t - \mu_k]\tanh\left(\frac{\beta_k(y_t - \mu_k)}{2}\right)\right] \quad (3.47)$$

$$\text{GM}\;:\;\; -\frac{\partial \varepsilon_t}{\partial \beta_k} \;=\; \frac{1}{2}\pi_k(t)\left[\frac{1}{\beta_k} - (y_t - \mu_k)^2\right] \quad (3.48)$$

Thus the gradient (recall (3.17)) is obtained as

$$\text{DSM}\;:\; -\frac{\partial E}{\partial \beta_k} \;=\; \frac{1}{N}\sum_{t=1}^{N}\pi_k(t)\left[\frac{1}{\beta_k} - [y_t - \mu_k]\tanh\left(\frac{\beta_k[y_t - \mu_k]}{2}\right)\right] \quad (3.49)$$

$$\text{GM}\;:\; -\frac{\partial E}{\partial \beta_k} \;=\; \frac{1}{2N}\sum_{t=1}^{N}\pi_k(t)\left[\frac{1}{\beta_k} - [y_t - \mu_k]^2\right] \quad (3.50)$$

Strictly speaking, β_k should be written as an appropriate function of another variable, e.g. $\beta_k = \exp(\varphi_k)$, in order to ensure that it satisfies the positivity constraint (2.10). For the simulations reported later in Chapter 5, however, this was found *not* to be necessary. The reason is that one normally initialises the network with rather large kernel widths σ_k, which then shrink during training so as to allow the different kernels to focus on different, localised regions in the state space (discussed in more detail later). From the relations $\beta_k \propto 1/\sigma_k$ for the DSM model (2.27) and $\beta_k \propto 1/\sigma_k^2$ for the GM model (2.33) we obtain

$$\frac{\partial E}{\partial \beta_k} = \left(\frac{\partial \beta_k}{\partial \sigma_k}\right)^{-1} \frac{\partial E}{\partial \sigma_k} \propto \begin{cases} \sigma^3 \frac{\partial E}{\partial \sigma_k} & \text{(GM)} \\ \sigma^2 \frac{\partial E}{\partial \sigma_k} & \text{(DSM)} \end{cases}. \tag{3.51}$$

It can be seen that expressing the training scheme in terms of β_k rather than σ_k gives rise to a sigma-dependent *effective* learning rate that scales like $\gamma_{eff} \propto \sigma^3$ for the GM model and $\gamma_{eff} \propto \sigma^2$ for the DSM model. Consequently, the adaptation process slows down as the kernel widths become very narrow, $\gamma_{eff} \to 0$ as $\sigma_k \to 0$, which considerably reduces the risk of 'overshooting' to negative values.

3.2.3 Remaining weights

For the adaptation of the remaining weights $\mathbf{w}$, it is sufficient to calculate the derivative of E with respect to the μ-unit outputs, $\frac{\partial E}{\partial \mu_k}$. This is the usual definition of the error of a network (in this case, the sub-network defined by the S-layer and the respective μ-node), from which the weights $\mathbf{w}$ feeding into and out of the S-layer can be adapted by application of the chain rule (leading to the standard backpropagation algorithm)

$$-\frac{\partial E}{\partial w_i} = -\frac{1}{N} \sum_{t=1}^{N} \sum_{k=1}^{K} \frac{\partial \varepsilon_t}{\partial \mu_k} \frac{\partial \mu_k(\mathbf{x}_t; \mathbf{w})}{\partial w_i} \tag{3.52}$$

From (3.44), we get

$$-\frac{\partial \varepsilon_t}{\partial \mu_k} = -\frac{\partial \varepsilon_t}{\partial \mathcal{G}_k} \frac{\partial \mathcal{G}_k}{\partial \mu_k} = \frac{a_k}{P} \frac{\partial \mathcal{G}_k}{\partial \mu_k}. \tag{3.53}$$

Inserting (3.24) and (3.26) into (3.53) and making use of (3.30) and (3.31) results in

$$\text{DSM} \quad : \quad -\frac{\partial \varepsilon_t}{\partial \mu_k} = \pi_k(t) \beta_k \tanh\left(\frac{\beta_k [y_t - \mu_k]}{2}\right) \tag{3.54}$$

$$\text{GM} \quad : \quad -\frac{\partial \varepsilon_t}{\partial \mu_k} = \pi_k(t) \beta_k [y_t - \mu_k] \tag{3.55}$$

This leads to the gradients

$$\text{DSM:} -\frac{\partial E}{\partial w_i} = \frac{1}{N} \sum_{k=1}^{K} \beta_k \sum_{t=1}^{N} \pi_k(t) \tanh\left(\frac{\beta_k[y_t - \mu_k(\mathbf{x}_t; \mathbf{w})]}{2}\right) \frac{\partial \mu_k(\mathbf{x}_t; \mathbf{w})}{\partial w_i} \quad (3.56)$$

$$\text{GM:} -\frac{\partial E}{\partial w_i} = \frac{1}{N} \sum_{k=1}^{K} \beta_k \sum_{t=1}^{N} \pi_k(t)[y_t - \mu_k(\mathbf{x}_t; \mathbf{w})] \frac{\partial \mu_k(\mathbf{x}_t; \mathbf{w})}{\partial w_i} \quad (3.57)$$

The derivatives $\frac{\partial \mu_k}{\partial w_i}$ are obtained by application of the chain rule, leading to expressions known from standard backpropagation.

3.2.4 Interpretation of the parameter adaptation rules

Equation (3.36) expresses the intuitively plausible result that each data point $(\mathbf{x}_t, y_t)$ contributes a term to the overall update rule for the priors a_k which corrects these parameters towards the *posterior* $\pi_k(t)$. Consequently, if a network branch $\mu_k(\mathbf{x}; \mathbf{w})$ fits some relevant region in the state space of the data-generating process, the posteriors will always be large and a_k will increase. If, conversely, $\mu_k(\mathbf{x}; \mathbf{w})$ fits some marginal region with small posteriors, the corresponding prior a_k will decrease. The update rules for the kernel widths, (3.49) and (3.50), and the remaining weights $\mathbf{w}$ (which determine the kernel centres, $\mu_k(\mathbf{x}; \mathbf{w})$), (3.56) and (3.57), contain the posterior $\pi_k(t)$ as a factor. Consequently, only data points $(\mathbf{x}_t, y_t)$ for which $\pi_k(t)$ is sufficiently large contribute significantly to updating the parameters related to the kth kernel. Since due to

$$\pi_k(t) = P(k|y_t, \mathbf{x}_t, \mathbf{q}) \propto P(y_t|\mathbf{x}_t, k, \mathbf{q}) = \mathcal{G}_{\beta_k}(y - \mu_k(\mathbf{x}_t; \mathbf{w}))$$

the posterior $\pi_k(t)$ is uni-modal, this requires that y_t is sufficiently close to the kernel centre $\mu_k(\mathbf{x}_t; \mathbf{w})$. In this way the network gradually splits up the state space into different smoothly overlapping regions, with each kernel focusing on one of them (demonstrated in Chapter 5). The update rules for the GM-model are intuitively plausible. The weights $\mathbf{w}$ are adapted such that the kth kernel interpolates those data points $(\mathbf{x}_t, y_t)$ for which the posterior $\pi_k(t)$ is large, (3.55) and (3.57). In the same way, (3.48) and (3.50) show that data points $(\mathbf{x}_t, y_t)$ with a large posterior $\pi_k(t)$ adjust the variance of the kth kernel, $\sigma_k^2 = \frac{1}{\beta_k}$, towards the squared deviation $[y_t - \mu_k(\mathbf{x}_t; \mathbf{w})]^2$). For interpretation of the corresponding update rules for the DSM model, we can make use of the aforementioned fact that only those y_t close to $\mu_k(\mathbf{x}_t; \mathbf{w})$ give rise to significant contributions, which allows approximating tanh(.) by the first term of the Taylor series expansion, $\tanh(x) \approx x$, and thus gives, from (3.49),

$$\text{DSM} \quad : \quad -\frac{\partial E}{\partial \beta_k} \approx \frac{\beta_k}{N} \sum_{t=1}^{N} \pi_k(t) \left[\frac{1}{(\beta_k)^2} - \frac{1}{2}[y_t - \mu_k(\mathbf{x}_t; \mathbf{w})]^2\right] \quad (3.58)$$

and from (3.54),

$$\text{DSM} \quad : \quad -\frac{\partial \varepsilon_t}{\partial \mu_k} \approx \frac{1}{2}\pi_k(t)(\beta_k)^2[y_t - \mu_k(\mathbf{x}_t; \mathbf{w})]. \qquad (3.59)$$

A comparison with (3.50) and (3.55) shows that the main difference from the corresponding update rules for the GM-model is the replacement of β_k by $(\beta_k)^2$. This can be understood from the fact that for the DSM model, β_k is proportional to the inverse standard deviation (2.27), whereas, for the GM-model, β_k is proportional to the inverse variance (2.33),

$$\frac{1}{\beta_k} \quad \propto \quad \begin{cases} \sigma_k & \text{(DSM)} \\ (\sigma_k)^2 & \text{(GM)} \end{cases} \qquad (3.60)$$

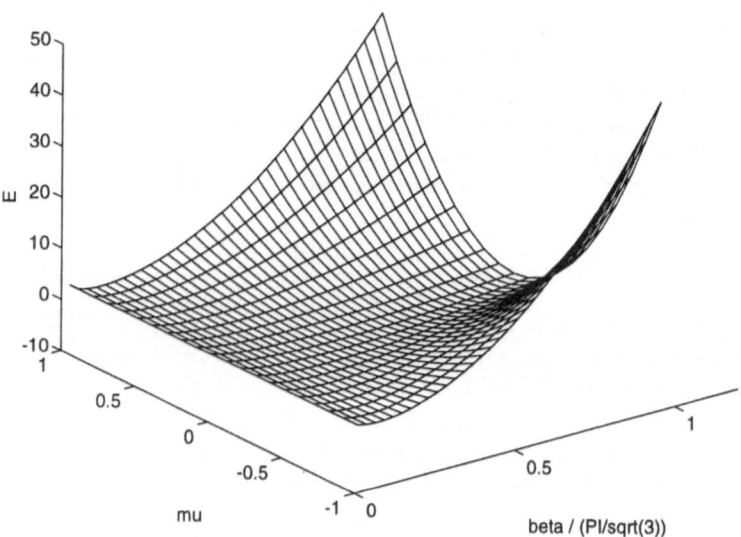

Fig. 3.1. Error surface for a simple Gaussian process. The figure shows the error surface $E = -\frac{1}{N}\sum_t \ln P(y_t|\mathbf{x}_t, \mu, \beta)$ for a network applied to learning the simple Gaussian process (3.61). The network architecture contains only one μ-node and no hidden units. In this way E becomes a function of only two parameters, μ and β. It can be seen that the curvature along μ, $\frac{\partial^2 \varepsilon}{\partial \mu^2}$, depends strongly on β, rendering a gradient descent training scheme with constant learning rates highly inefficient.

3.2.5 Deficiencies of gradient descent and their remedy

A major drawback of a gradient-descent training scheme is its susceptibility to local minima and its inherent slowness in 'flat' regions of the error surface $E(\mathbf{q})$. In fact, in the simulations reported later in this book (Chapter 5), it was found that a naive application of the adaptation rules derived in the previous sections *always* ended up in a sub-optimal solution, i.e., the training scheme *never* succeeded in learning the conditional probability density of the data-generating process. Typical failures of the training scheme are: (1) γ_a too large: all but one of the output weights, a_k, decay to zero before the mappings implemented in the respective network branches, $\mu_k(\mathbf{x}; \mathbf{w})$, can be improved. The network with a single remaining kernel can thus only learn the conditional mean. (2) γ_β too large: several kernel widths σ_k decay immediately to very small values so that the respective kernel centres, $\mu_k(\mathbf{x}; \mathbf{w})$, focus exclusively on a few outliers (overfitting). (3) γ_a, γ_β too small: all the network branches $\mu_k(\mathbf{x}, \mathbf{w})$ become symmetric and fit the same function, namely the conditional mean.

It was found that these problems could not be solved by an appropriate ratio of the learning rates, since the 'optimal' ratio itself seemed to evolve during training (see below for an explanation). In the earliest stages[3] of the work resulting in this book the training scheme was therefore split up into different phases. In the first phase, only the weights $\mathbf{w}$ were adjusted ($\gamma_a = \gamma_\beta = 0$) in order to allow all network branches $\mu_k(\mathbf{x}, \mathbf{w})$ to learn 'something relevant' (preventing the switching-off of kernels at an early stage of the training process). In the second and third phase, only the kernel widths ($\gamma_a = \gamma_w = 0$) and output weights ($\gamma_\beta = \gamma_w = 0$), respectively, were adjusted. Finally, in the last phase, all parameters were adapted simultaneously. Although this scheme proved to be successful, it is completely heuristic and misses out an inherent flaw in the gradient descent adaptation rules. Further details will therefore not be presented here. It is to be noted, though, that similar problems and 'solutions' are reported in [64].

In order to understand the inherent flaws of the above adaptation rules, consider the simple case of learning a one-dimensional constant-mean, constant-variance Gaussian process,

$$P(y_t|x_t) \quad = \quad P(y_t) \quad = \quad \sqrt{\frac{\beta}{2\pi}} \exp\left(-\frac{\beta}{2}[y_t - \mu]^2\right). \qquad (3.61)$$

This distribution can be modelled by a network with $K = 1$ nodes in the $\mathcal{G}$-layer and no $\mathcal{S}$-layer. The example is considered here because the error

---

[3] Presented at the *International Symposium on Forecasting* (ISF), Toronto, June 1995.

function, depending only on two parameters, $E(\mu, \beta)$, can easily be visualised. From (3.8) and (3.61) we obtain

$$
\begin{aligned}
E(\mu, \beta) &= -\frac{1}{N} \sum_{t=1}^{N} \ln\left(P(y_t|\mathbf{x}_t, \mu, \beta)\right) = \frac{\beta}{2N} \sum_{t=1}^{N} (y_t - \mu)^2 - \frac{1}{2} \ln\left(\frac{\beta}{2\pi}\right) \\
&= \frac{\beta}{2}(\bar{y} - \mu)^2 + \frac{\beta}{2N} \sum_{t=1}^{N} (y_t - \bar{y})^2 - \frac{1}{2} \ln\left(\frac{\beta}{2\pi}\right),
\end{aligned}
\tag{3.62}
$$

where $\bar{y}$ is the empirical mean, $\bar{y} = \frac{1}{N} \sum_t y_t$, which for sufficiently large N can be approximated by the true mean μ_0, $\bar{y} \approx \mu_0$. Moreover, the true variance $\sigma_0^2 = \frac{1}{\beta_0}$ can be approximated by $\frac{1}{\beta_0} \approx \frac{1}{N} \sum_t (y_t - \bar{y})^2$, giving rise to the overall approximation

$$
E(\mu, \beta) \approx \frac{\beta}{2}(\mu - \mu_0)^2 + \frac{\beta}{2\beta_0} - \frac{1}{2} \ln\left(\frac{\beta}{2\pi}\right).
\tag{3.63}
$$

A function plot can be found in Fig.3.1. It can be seen that the curvature along μ,

$$
\frac{\partial^2 E}{\partial \mu^2} = \beta
\tag{3.64}
$$

depends on β and becomes larger as β increases. If the system starts off from some state with a small value of β (front left region in the figure) and has its learning rate γ_μ for the adaptation of μ matched properly to the curvature of E, then the agorithm is bound to become unstable as the system moves towards regions with larger values of β (back right). Conversely, if the learning rate is chosen appropriately for a large-β region (back right), then the convergence of the algorithm in the small-β region (front left) will be extremely slow. For this reason a gradient-descent scheme with a fixed ratio of the learning rates, $\gamma_\beta/\gamma_\mu =$ constant, or, even worse, a fixed set of learning rates, $\gamma_\beta =$ constant, $\gamma_\mu =$ constant, is highly inefficient.

This example is very simple, but helpful in intuitively indicating a straightforward method for rectifying the algorithm. Consider the following modified error function $\tilde{E}$, which results from the original error function E on multiplication by $1/\beta$:

$$
\tilde{E}(\mu, \beta) := \frac{1}{\beta} E(\mu, \beta) = \frac{1}{2}(\mu - \mu_0)^2 + \frac{1}{2\beta_0} - \frac{1}{2\beta} \ln\left(\frac{\beta}{2\pi}\right).
\tag{3.65}
$$

The derivatives are given by

$$
\frac{\partial \tilde{E}}{\partial \beta} = \frac{\partial}{\partial \beta}\left(\frac{E}{\beta}\right) = -\frac{E}{\beta^2} + \frac{1}{\beta} \frac{\partial E}{\partial \beta}
\tag{3.66}
$$

$$
\frac{\partial \tilde{E}}{\partial \mu} = \frac{1}{\beta} \frac{\partial E}{\partial \mu} = 0 \quad \Rightarrow \quad \frac{\partial E}{\partial \mu} = 0
\tag{3.67}
$$

$$
\frac{\partial^2 \tilde{E}}{\partial \mu^2} = 1
\tag{3.68}
$$

From (3.66), it is seen that $\frac{\partial E}{\partial \beta} = 0$ does not imply that $\frac{\partial \bar{E}}{\partial \beta} = 0$. Consequently, minimisation of E and $\bar{E}$ lead, in general, to different states $(\hat{\mu}, \hat{\beta})$. However, the minimisation with respect to μ only does lead to the same state for E and $\bar{E}$ (3.67). Moreover, due to (3.68), the curvature of the modified error function $\bar{E}$ does *not* depend on β. This suggests that the problems discussed above can be solved with a combined scheme that adapts β by gradient descent on the original error function E and μ by gradient descent on the 're-shaped' error function $\bar{E}$.

This simple analysis carries over to the full training scheme. In equation (3.55) it is seen that the terms $\frac{\partial \varepsilon_t}{\partial \mu_k}$, which contribute to the gradient $\frac{\partial E}{\partial w_i}$ via (3.57), contain the factor β_k. This inflicts the same problem of strong curvature variation on the training algorithm as discussed above for the simple two-dimensional example. The problem can be re-interpreted as follows. As a consequence of the prior ignorance about the system of interest, a standard training process starts off from some rather large kernel widths. As training progresses, the different kernels focus on different parts of the function space and, simultaneously with an improvement of the mappings $\mu_k(\mathbf{x}; \mathbf{w})$, the kernel widths σ_k shrink to smaller values (this will become clearer in Chapter 5). Now, it is known from the theory of standard backpropagation that, as the mapping implemented in the network becomes increasingly more accurate, the learning rate should decrease so as to allow the algorithm to converge. In (3.55), the factor β_k can be combined with the learning rate γ to give a new effective learning rate $\gamma_{eff} := \beta_k \gamma$. This effective learning rate, however, shows exactly the opposite behaviour to that required: as the mapping implemented in the kth network branch, $\mu_k(\mathbf{x}; \mathbf{w})$, becomes more accurate, the kernel width σ_k decreases, which leads to an increase of β_k, which causes an increase of the effective learning rate γ_{eff}. Consequently, either the algorithm becomes unstable at a later stage of the training process, or it requires extremely long training times (if γ is chosen sufficiently small to prevent these instablilities, in which case it is far too small for the initial phase when the kernel widths are still large).

However, this deficiency can be rectified in the way described above. For the GM model, β_k enters the equation for calculating the gradient, (3.55), as a factor. It can therefore simply be omitted, which corresponds to a division of ε_t by β_k. The proper shape modification of the error function for the DSM model is less obvious when the original update equations, (3.47) and (3.54), are inspected, but is easily found when inspecting the approximate expressions, (3.58) and (3.59). The latter suggest division of ε_t by β_k for adapting the β_k's (thus getting rid of the overall factor β_k in (3.58)), and of ε_t by $(\beta_k)^2$ for adapting the weights $\mathbf{w}$ (thus removing this factor in (3.59)). This leads to the following shape modifications of the error surfaces:

$$\text{GM} \; : \; \frac{\partial \varepsilon_t}{\partial \mu_k} \; \rightarrow \; \frac{\partial \tilde{\varepsilon}_t}{\partial \mu_k} = \frac{1}{\beta_k} \frac{\partial \varepsilon_t}{\partial \mu_k} \tag{3.69}$$

$$\text{DSM} : \quad \frac{\partial \varepsilon_t}{\partial \mu_k} \rightarrow \frac{\partial \tilde{\varepsilon}_t}{\partial \mu_k} = \frac{1}{(\beta_k)^2} \frac{\partial \varepsilon_t}{\partial \mu_k}, \quad \frac{\partial \varepsilon_t}{\partial \beta_k} \rightarrow \frac{\partial \tilde{\varepsilon}_t}{\partial \beta_k} = \frac{1}{\beta_k} \frac{\partial \varepsilon_t}{\partial \beta_k} \quad (3.70)$$

For the DSM model, (3.16), (3.49), (3.56), and (3.70) lead to

$$\Delta^{(n)}\beta_k = \frac{\gamma_\beta}{N\beta_k} \sum_{t=1}^{N} \pi_k(t) \left[\frac{1}{\beta_k} - [y_t - \mu_k(\mathbf{x}_t; \mathbf{w})] \tanh\left(\frac{\beta_k[y_t - \mu_k(\mathbf{x}_t; \mathbf{w})]}{2} \right) \right]$$
$$+ \eta \Delta^{(n-1)}\beta_k \tag{3.71}$$

$$\Delta^{(n)}w_i = \frac{\gamma_w}{N} \sum_{k=1}^{K} \sum_{t=1}^{N} \pi_k(t) \tanh\left(\frac{\beta_k[y_t - \mu_k(\mathbf{x}_t; \mathbf{w})]}{2} \right) \frac{\partial \mu_k(\mathbf{x}_t; \mathbf{w})}{\partial w_i}$$
$$+ \eta \Delta^{(n-1)}w_i \tag{3.72}$$

For the GM model, (3.16), (3.50), (3.57), and (3.69) give

$$\Delta^{(n)}\beta_k = \frac{\gamma_\beta}{2N} \sum_{t=1}^{N} \pi_k(t) \left[\frac{1}{\beta_k} - [y_t - \mu_k(\mathbf{x}_t; \mathbf{w})]^2 \right] + \eta \Delta^{(n-1)}\beta_k \tag{3.73}$$

$$\Delta^{(n)}w_i = \frac{\gamma_w}{N} \sum_{k=1}^{K} \sum_{t=1}^{N} \pi_k(t)[y_t - \mu_k(\mathbf{x}_t; \mathbf{w})] \frac{\partial \mu_k(\mathbf{x}_t; \mathbf{w})}{\partial w_i} + \eta \Delta^{(n-1)}w_i \tag{3.74}$$

3.3 Summary

An error function for network training has been derived from a maximum likelihood approach. The new function can be regarded as a generalisation of the conventional sum-of-squares error function and reduces to the latter if the probability distribution of the target conditional on the input vector is a Gaussian with constant standard deviation. The network parameters can, in principle, be adapted by gradient descent, leading to a backpropagation-like training algorithm. However, due to large variations in the local curvature of the error function, the direct application of such a scheme is likely to result in extremely long convergence times and/or get stuck in sub-optimal configurations. For this reason, a rectification of the standard algorithm, equivalent to a curvature-based shape modification of the error function, was introduced and led to intuitively plausible corrections of the original parameter-update equations.

An empirical evaluation of the proposed model and training scheme can be found in Chapter 5. The benchmark problems applied in this study (and in the remainder of this book) will be discussed in the following chapter.

3.4 Appendix

Writing the output weights or priors in the form (3.38), $a_k = (\varphi_k)^2 / \sum_i (\varphi_i)^2$, gives

$$\frac{\partial a_k}{\partial \varphi_i} = \frac{\partial}{\partial \varphi_i} \left(\frac{(\varphi_k)^2}{\sum_j (\varphi_j)^2} \right) = \frac{2\varphi_k \delta_{ik}}{\sum_j (\varphi_j)^2} - \frac{2\varphi_i (\varphi_k)^2}{\left(\sum_j (\varphi_j)^2 \right)^2}$$

$$= \frac{2\varphi_i}{\sum_j (\varphi_j)^2} \left(\delta_{ik} - \frac{(\varphi_k)^2}{\sum_j (\varphi_j)^2} \right).$$

Inserting this expression into

$$-\frac{\partial E}{\partial \varphi_i} = -\sum_k \frac{\partial E}{\partial a_k} \frac{a_k}{\partial \varphi_i}$$

yields

$$\frac{\partial E}{\partial \varphi_i} = \frac{2\varphi_i}{\sum_j (\varphi_j)^2} \sum_k \left[\left(\delta_{ik} - \frac{(\varphi_k)^2}{\sum_j (\varphi_j)^2} \right) \frac{\partial E}{\partial a_k} \right]$$

$$= \frac{2\varphi_i}{\sum_j (\varphi_j)^2} \left[\frac{\partial E}{\partial a_i} - \frac{\sum_k (\varphi_k)^2 \partial E / \partial a_k}{\sum_j (\varphi_j)^2} \right]$$

$$= \frac{2\varphi_i}{\sum_j (\varphi_j)^2} \left[\frac{\partial E}{\partial a_i} - (\nabla_{\mathbf{a}} E)^\dagger \mathbf{a} \right] \tag{3.75}$$

where

$$\mathbf{a} = (a_1, \ldots, a_K)^\dagger, \qquad \nabla_{\mathbf{a}} E = \left(\frac{\partial E}{\partial a_1}, \ldots, \frac{\partial E}{\partial a_1} \right)^\dagger.$$

4. Benchmark Problems

This chapter gives an overview of the benchmark problems employed for assessing the prediction performance of the neural network models studied in this book. The first problem is a time series generated from the logistic map with intrinsic noise in the structural parameter α. This gives rise to a stochastic dynamical system that continually switches between the different regimes of fixed-point behaviour, stable limit cycle and chaotic attractor. In the second problem, the noisy logistic map is stochastically coupled to a second stochastic dynamical system where, due to the stochastic coupling, bimodality of the conditional probability distribution arises. The third problem is taken from Ormoneit and Neuneier. A particle moves in a double-well potential subject to Brownian dynamics. The resulting time series shows fast oscillation around one of two metastable states and occasional phase transitions between these two states. As a consequence of the latter, long-term predictions require a model that can capture bimodality.

4.1 Logistic map with intrinsic noise

The first problem is based on the logistic map

$$x_{t+1} = \alpha x_t(1 - x_t), \qquad \alpha \in [0, 4], \qquad x_t \in [0, 1]. \qquad (4.1)$$

This is a classical, well-studied example of a chaotic dynamical system where, depending on the value of the parameter α, the system converges to a stable fixed point, a limit cycle, or a chaotic (fractal) attractor [15], [33]. In the last decade, in a time of great interest in chaos theory and the prediction of deterministic chaotic time series, many researchers in the neural network community applied their models to a synthetic time series generated from (4.1), so this map has since become a widely-applied benchmark problem (see e.g. [1]). In the following study, the system is complicated by subjecting the parameter α to *intrinsic* noise. That is, rather than choosing α as

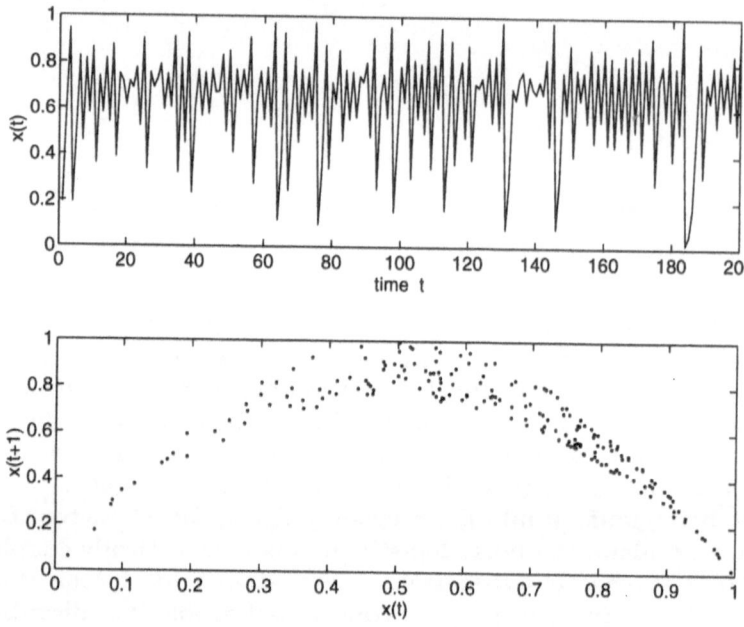

Fig. 4.1. Logistic map with intrinsic noise. *Upper figure:* Time series generated from the stochastic logistic map, (4.2), plotted in the time domain. *Bottom figure:* State-space plot of the same time series.

a fixed value in advance, it will be drawn at any particular time from some distribution $P(\alpha)$. In this way, a series of $\{\alpha_t\}$ becomes the realisation of a stochastic process, causing continual transitions of the system between the different regimes of fixed-point, limit-cycle and chaotic behaviour. The distribution $P(\alpha_t)$ must be chosen with some care, though, since a violation of the condition $\alpha \in [0, 4]$ leads to the divergence of $x_t \to -\infty$. $P(\alpha_t)$ was therefore chosen as the uniform distribution over the interval $[3, 4]$, leading to

$$x_{t+1} = \alpha_t x_t (1 - x_t), \qquad P(\alpha_t) = \begin{cases} 1 & \text{if } \alpha_t \in [3, 4] \\ 0 & \text{otherwise} \end{cases} \qquad (4.2)$$

Figure 4.1 top shows the segment of a time series obtained from (4.2). Since the time series is a first-order Markov process, its state-space plot can easily be visualised, as shown in Fig.4.1 bottom. This will be used extensively in later chapters, since it will allow us to attain a deeper understanding of the learning process in the neural network. Note that a characteristic feature of the time series is a state-space dependence of the noise-level, as seen from Fig.4.1 bottom, which shows that the variance is largest for $x_t = 0.5$, and decreases monotonically to zero as $x_t \to 0, 1$. For the later objective of assessing the prediction performance of the neural network models studied in this book, derivation of an expression for the conditional distribution $P(y|x_t)$ is required, where y has been introduced as an abbreviation for x_{t+1} (henceforth referred to as the *target*). Since for a given x_t the mapping from α to $y = \alpha x_t(1 - x_t)$ is one-to-one, the transformation of the probability densities is given by (see, for example, [49], Chapter 5)

$$P(y|x_t) \;\; = \;\; \left|\frac{dy}{d\alpha}\right|^{-1} P(\alpha). \tag{4.3}$$

From $\frac{dy}{d\alpha} = x_t(1 - x_t)$ and by substituting $\alpha = \frac{y}{x_t(1-x_t)}$, we obtain

$$P(y|x_t) \;\; = \;\; \frac{1}{x_t(1 - x_t)} P_\alpha\left(\frac{y}{x_t(1 - x_t)}\right), \tag{4.4}$$

where the subscript α has been introduced to identify the probability density of (4.2),

$$P_\alpha(x) = \begin{cases} 1 & \text{if} \quad \alpha_t \in [3,4] \\ 0 & \text{otherwise} \end{cases} \tag{4.5}$$

As an example, consider the case $x_t = 0.5$, which leads to

$$P(y|x_t = 0.5) \;\; = \;\; 4P_\alpha(4y). \tag{4.6}$$

Now making use of (4.5), we get

$$P(y|x_t = 0.5) \;\; = \;\; \begin{cases} 4 & \text{if} \quad y \in [3/4, 1] \\ 0 & \text{otherwise} \end{cases} \tag{4.7}$$

The disadvantage of this distribution is the existence of two discontinuities, with which, in general, a neural netwok cannot cope. One might therefore think of smoothing the distribution by adding, say, Gaussian noise to x_t. Let ε_t denote a random variable drawn from a zero-mean Gaussian distribution, let x_t denote the time series obtained from (4.2), and let $\tilde{x}_t := x_t + \varepsilon_t$. Then $P(\tilde{x}_{t+1}|x_t)$ has the desired form of being a smoothed version of the distribution (4.4). However, the distribution of practical interest is $P(\tilde{x}_{t+1}|\tilde{x}_t)$, since *all* time series values are equally subjected to the additive noise, rather than just the target. This distribution is of a much more complicated form, and for Gaussian noise there is in fact no analytic solution. This approach was therefore discarded and the distribution (4.4) left unchanged, especially since the simulations to be discussed in the following chapters showed that the network succeeded in attaining a satisfactory approximation to it in spite of the discontinuities.

Given the distribution (4.4), it is easy to derive expressions for the conditional mean

$$\mu_{y|x_t} \;\; = \;\; \langle y|x_t\rangle \tag{4.8}$$

and the conditional variance

$$\sigma^2_{y|x_t} \quad = \quad \langle y^2|x_t\rangle - \langle y|x_t\rangle^2 \qquad (4.9)$$

where the definition

$$\langle y^m|x_t\rangle \quad := \quad \int y^m P(y|x_t)dy \qquad (4.10)$$

has been used. Inserting (4.4) into (4.8) and making use of (4.5) yields

$$
\begin{aligned}
\mu_{y|x_t} &= \int y P(y|x_t)dy = \frac{1}{x_t(1-x_t)}\int y P_\alpha\left(\frac{y}{x_t(1-x_t)}\right)dy \\
&= \left[\frac{y^2}{2x_t(1-x_t)}\right]_{\frac{y}{x_t(1-x_t)}=3}^{\frac{y}{x_t(1-x_t)}=4} = \frac{4^2-3^2}{2}x_t(1-x_t) \\
&= 3.5x_t(1-x_t) \qquad (4.11)
\end{aligned}
$$

In the same way, the second conditional moment is given by

$$
\begin{aligned}
\langle y^2|x_t\rangle &= \int y^2 P(y|x_t)dy = \frac{1}{x_t(1-x_t)}\int y^2 P_\alpha\left(\frac{y}{x_t(1-x_t)}\right)dy \\
&= \left[\frac{y^3}{3x_t(1-x_t)}\right]_{\frac{y}{x_t(1-x_t)}=3}^{\frac{y}{x_t(1-x_t)}=4} = \frac{4^3-3^3}{3}\left(x_t(1-x_t)\right)^2 \\
&= 12.\bar{3}\left(x_t(1-x_t)\right)^2 \qquad (4.12)
\end{aligned}
$$

Inserting (4.11) and (4.12) into (4.9) gives

$$\sigma^2_{y|x_t} = \langle y^2|x_t\rangle - \langle y|x_t\rangle^2 = \left(12.\bar{3}-3.5^2\right)\left(x_t(1-x_t)\right)^2 = \frac{1}{12}\left(x_t(1-x_t)\right)^2. \qquad (4.13)$$

The conditional standard deviation, too, is thus of a parabola shape,

$$\sigma_{y|x_t} \quad = \quad \frac{1}{\sqrt{12}}x_t(1-x_t), \qquad (4.14)$$

with its peak at $x_t = 0.5$, $\sigma_{y|x_{t=0.5}} = 0.072$, and a decay to zero for $x_t \to 0, 1$ (see Fig.5.5).

4.2 Stochastic combination of two stochastic dynamical systems

As a further complication, the noisy logistic map is coupled to a second stochastic dynamical system. First, consider the stochastic map

$$x_{t+1} = 1 - x_t^{\kappa_t}, \qquad \kappa_t \in [0.5, 1.25] \qquad (4.15)$$

where the κ_t are independently chosen subject to the uniform distribution over the indicated interval. Henceforth this map will be referred to as the *kappa map*. Next, a stochastic switch is introduced, given by the uniformly distributed random variable $\xi_t \in [0, 1]$, a threshold $\theta \in [0, 1]$, and the Heaviside function $\Theta(\xi_t - \theta)$ (where $\Theta(x) = 1$ if $x > 0$, and 0 otherwise). Then a stochastic coupling between (4.15) and the noisy logistic map (4.2) can be constructed via

$$x_{t+1} = \Theta(\xi_t - \theta)\Big[\alpha_t x_t (1 - x_t)\Big] + [1 - \Theta(\xi_t - \theta)]\Big[1 - x_t^{\kappa_t}\Big] \qquad (4.16)$$

where $\alpha_t \in [3, 4]$, $\kappa_t \in [0.5, 1.25]$ and $\xi_t \in [0, 1]$ are all uniformly distributed in the respective interval. Whenever $\xi_t \geq \theta$, the dynamics are according to the logistic map, otherwise they follow the kappa map. This results in a bimodal distribution, as seen from Fig. 4.2. The prior probabilities of the two sub-processes are determined by the threshold constant θ, which was set to $\theta = 1/3$, giving a ratio of 2 : 1 in favour of the logistic map. Since the whole process is first-order Markov again, the state-space plot can easily be visualised, as has been done in Fig.4.2, bottom graph. The plot of the time series in the time domain can be found in the upper graph of Fig.4.2.

In order, in later chapters, to assess the neural network prediction performance, we need to derive an expression for the true probability density $P(y|x_t)$ again. For the noisy logistic map this has already been done in the previous section. The same derivation needs now to be repeated for the kappa map. Since for a given x_t the mapping from κ to $y = 1 - x_t^\kappa$ is one-to-one, the transformation of the probability densities is given by

$$P(y|x_t, kappa) = \left|\frac{dy}{d\kappa}\right|^{-1} P_\kappa(\kappa), \qquad (4.17)$$

where the index 'κ' on the right identifies the distribution of κ,

$$P_\kappa(\kappa) = \begin{cases} 4/3 & \text{if } \kappa \in [0.5, 1.25] \\ 0 & \text{otherwise} \end{cases} \qquad (4.18)$$

and the expression '*kappa*' in the conditioning part on the left indicates partitioning into the regime of the kappa map (i.e. $\xi_t < \theta$). From

$$\frac{dy}{d\kappa} = \frac{d}{d\kappa}(1 - x_t^\kappa) = -\ln(x_t) x_t^\kappa \qquad (4.19)$$

and by substituting

$$y = 1 - x_t^\kappa \quad \Rightarrow \quad \kappa = \frac{\ln(1 - y)}{\ln(x_t)} \qquad (4.20)$$

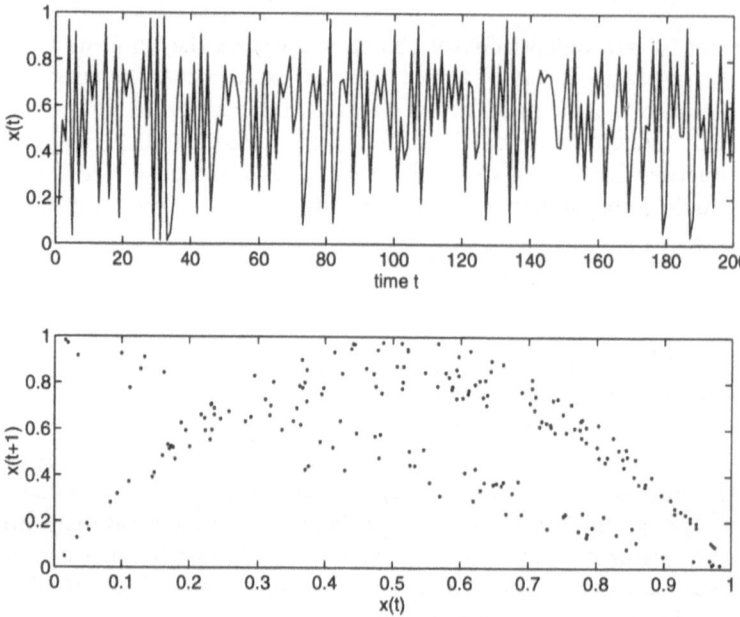

Fig. 4.2. **Stochastic combination of two stochastic dynamical systems.**
Upper figure: Time series generated from the stochastic coupling between the noisy logistic map and the kappa map, (4.16), plotted in the time domain. *Bottom figure:* State-space plot of the same time series.

we obtain

$$P(y|x_t, kappa) = \frac{1}{x_t^\kappa|\ln x_t|}P_\kappa\left(\frac{\ln(1-y)}{\ln(x_t)}\right) = \frac{1}{1-y}P_\kappa\left(\frac{\ln(1-y)}{\ln(x_t)}\right),$$
(4.21)

where $P_\kappa(.)$ is the uniform distribution on the interval $[0.5, 1.25]$, (4.18). The total distribution $P(y|x_t)$ is given by

$$P(y|x_t) = P(y|x_t, kappa)P(kappa) + P(y|x_t, logistic)P(logistic), \quad (4.22)$$

with $P(y|x_t, logistic)$ given by (4.4), $P(y|x_t, kappa)$ given by (4.21), while $P(logistic)$ and $P(kappa)$ depend on θ (the choice $\theta = 1/3$ leads to $P(logistic) = 2/3$ and $P(kappa) = 1/3$). A plot of $P(y|x_t)$ will be shown in Fig.5.6, where this theoretical result is compared with the network prediction obtained from the simulation discussed later in Chapter 5.

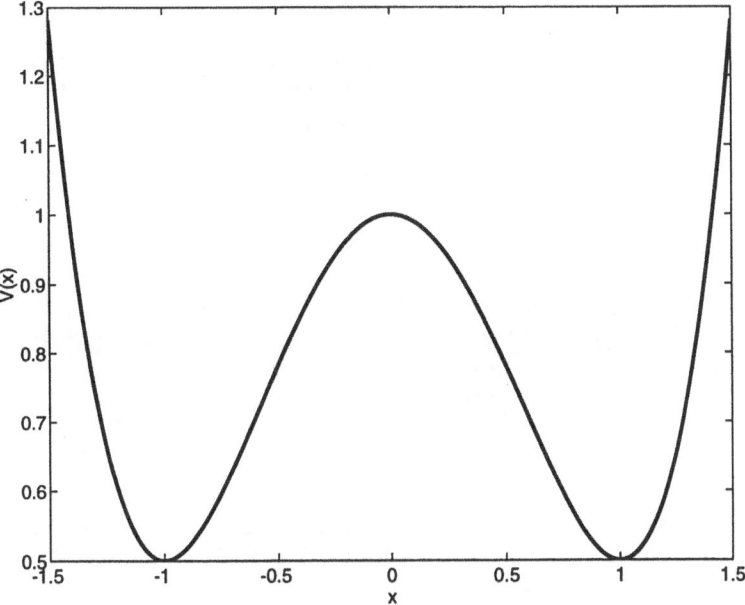

Fig. 4.3. **Double-well potential.** The graph shows a plot of the double-well potential defined in (4.23). Reprinted with permission from [28].

4.3 Brownian motion in a double-well potential

The third time series, first applied by Ormoneit [46] and Neuneier et al. [45], serves as a benchmark problem to compare the prediction performance of the network models studied in this book with different alternative models studied by others. A particle of mass m moves in a double-well potential

$$V(x) \quad = \quad 0.5x^4 - x^2 + 1 \qquad (4.23)$$

subject to the Brownian dynamics

$$\frac{d^2x}{dt^2} = -\frac{1}{m}\frac{dV}{dx} - \alpha\frac{dx}{dt} + R(t), \qquad (4.24)$$

where $R(t)$ is a Gaussian stochastic variable with zero mean and intensity

$$\langle R(t)R(t')\rangle \quad = \quad 2\alpha kT\delta(t - t'). \qquad (4.25)$$

The last term can be interpreted as a coupling between the system and a heat bath of temperature T, where k is the Boltzmann constant and α a friction coefficient. The parameters were chosen as in the aforementioned studies,

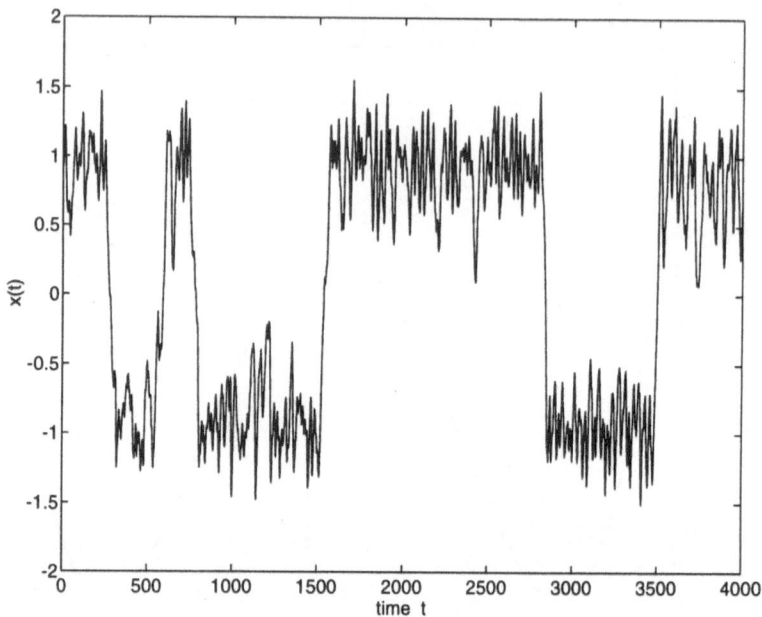

Fig. 4.4. Brownian motion of a particle in a double-well potential. The time series was obtained by numerical integration of (4.24) with the leapfrog algorithm. On a fast time scale, the system shows fluctuations around a semi-stable state, which results in a unimodal conditional distribution. On a slower time scale, however, transitions between the two states occur, rendering the conditional distribution bimodal. Reprinted from [28], with permission of the publisher.

namely $m = 1, \alpha = 1, kT = 1$. Equation (4.24) was integrated numerically with the leapfrog algorithm (see, for instance, [5]), that is, let $\dot{x} := \frac{dx}{dt}$ and $F(t) := -\frac{1}{m}\frac{dV(t)}{dx} - \alpha\dot{x}(t - \delta/2) + R(t)$, then

$$\dot{x}(\frac{\delta}{2}) = \dot{x}(0) + F(0)\frac{\delta}{2}$$

$$\dot{x}(t + \frac{\delta}{2}) = \dot{x}(t - \frac{\delta}{2}) + F(t)\delta$$

$$x(t + \delta) = x(t) + \dot{x}(t + \frac{\delta}{2})\delta \tag{4.26}$$

The discrete stepsize δ needs to be selected sufficiently small so as to prevent instabilities (the results presented here were obtained with $\delta = 0.1$). Figure 4.4 depicts a typical segment of the resulting time series, showing fluctuations around two semi-stable states on a fast time scale and random transitions *between* the two states on a much slower one. Note that this is very similar to the time series of Fig. 1.7. In fact, the double-well potential

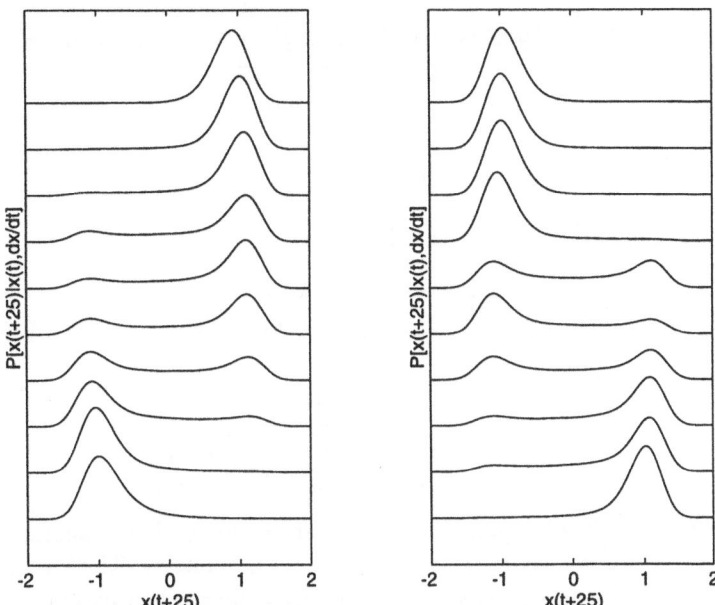

Fig. 4.5. Evolution of the conditional probability density during two phase transitions. The graphs show the evolution of the conditional probability density $P(x_{t+25}|x_t, \dot{x}_t)$ during the phase transitions at (i) $t = 255 - 300$ and (ii) $t = 1510 - 1555$ in the time series segment plotted in Fig.4.4. The time difference between two contiguous graphs is $\Delta t = 5$, and the arrow of time points from the bottom to the top.

presented here can be regarded as a very simple model for the hemoglobin protein, with each of the two minima corresponding to one of the two possible hydrogen bonds between the ligand and the nitrogen of the HisE7 group (compare with Section 1.4). For an economic interpretation of the time series in Fig. 4.4, see [46].

The problem posed to the network was to predict the probability distribution for the particle 25 time steps ahead, conditioned on its current position and velocity,

$$P\left(x_{t+25}|x_t, \dot{x}_t\right). \tag{4.27}$$

In order to assess the network performance, it is desirable to obtain an explicit expression for (4.27). Since this is analytically intractable, an empirical approximation to (4.27) was estimated as follows. Given the state of the system at time t, $\binom{x_t}{\dot{x}_t}$, the equation of motion (4.24) was integrated numerically N times forward in time to arrive at a set of N (in general distinct) states $\binom{x_{t+25}}{\dot{x}_{t+25}}(1), \ldots, \binom{x_{t+25}}{\dot{x}_{t+25}}(N)$. Discarding the velocities $\dot{x}_t$, one can obtain an approximation to (4.27) by applying a Parzen estimator to the set of postitions $\{x_{t+25}(1), \ldots, x_{t+25}(N)\}$. In brief, a Parzen estimator functions in the following way. The probability density for a given value of x can be estimated by centering an interval of length σ on x, counting the number of points that lie within this interval, $N(x)$, and then approximating $P(x)$ by

$$\tilde{P}(x) \quad = \quad \frac{N(x)}{\sigma N}. \tag{4.28}$$

With the definition of the kernel function

$$h(x) \quad := \quad \begin{cases} 1 & \text{if } |x| < 1/2 \\ 0 & \text{otherwise} \end{cases} \tag{4.29}$$

this can be written as

$$\tilde{P}(x) \quad = \quad \frac{1}{\sigma N} \sum_{i=1}^{N} h\left(\frac{x - x_i}{\sigma}\right). \tag{4.30}$$

The density is thus estimated by the superposition of N rectangles of side σ, with each rectangle centred on one of the data points x_i. The discontinuities inherent in the use of rectangles are a disadvantage in this approach. This deficiency can be remedied by using a smoother kernel instead of (4.29), e.g. a Gaussian,

$$h(x) \quad := \quad \frac{1}{\sqrt{2\pi}} \exp\left(-\frac{x^2}{2}\right). \tag{4.31}$$

Inserting (4.31) into (4.30) leads to the Gaussian form of the Parzen estimator, which was employed in this study,

$$\tilde{P}(x) \quad = \quad \frac{1}{N} \sum_{i=1}^{N} \frac{1}{\sqrt{2\pi}\sigma} \exp\left[-\frac{1}{2}\left(\frac{x - x_i}{\sigma}\right)^2\right]. \tag{4.32}$$

The kernel width σ acts as a smoothing parameter. When σ is too small, a great deal of structure is present in the estimated density which represents the properties of the particular data set rather than true structure in the underlying distribution (note that for $\sigma \to 0$, $\tilde{P}$ becomes a set of δ-functions centred on the data points). Conversely, when σ is too large, the estimated density is over-smoothed and the bimodal nature of the distribution, as shown in Fig.4.5, is lost. (See [7], Chapter 2 for a more detailed discussion on this aspect.) For the given problem, $\sigma = 0.1$ was chosen. This choice was based on a visual inspection of the resulting estimate, which proved to be sufficiently smooth, contain the relevant structure of bimodality, and be robust with respect to a variation of σ around this value over quite a large range. The result can be seen in Fig.4.5, which shows the estimated probability densities $P(x_{t+25}|x_t, \dot{x}_t)$ for ten steps of $\Delta t = 5$ during two state-transitions around $t = 300$ and $t = 1500$ in the time series segment of Fig.4.4.

4.4 Summary

In this chapter, three stochastic time series have been discussed, on which the prediction performance of various network models and training schemes will be tested in forthcoming chapters of this book. All problems allow the true conditional probability density to be obtained from the equations of dynamics - either analytically, or by numerical integration. In this way, the network predictions can always be compared with the correct function, which is an important advantage over real-world problems.

It is important to point out that none of the given problems can be tackled successfully by a conventional network for point predictions. In the first example, the standard deviation of the intrinsic noise is state-space dependent, giving rise to a heteroscedastic time series that requires at least the prediction of the *input-dependent* variance. The second and third problems show a *bimodal* conditional probability distribution, where the conventional approach of predicting only the conditional mean turns out to be completely insufficient (compare with the discussion in Chapter 1).

5. Demonstration of the Model Performance on the Benchmark Problems

The DSM network of Chapter 2 and the training algorithm derived in Chapter 3 are applied to the benchmark problems described in Chapter 4. A state-space plot of the network predictions allows the attainment of a deeper understanding of the training process. For the double-well problem, the prediction performance of the DSM network is compared with different alternative approaches, and is found to achieve results comparable to those of the best alternative schemes applied to this problem.

5.1 Introduction

The network model and the training scheme derived in the previous chapters were applied to the benchmark problems described in Chapter 4. The present chapter summarises the results presented in [30], which were obtained with the DSM model. Repeating the simulations with the GM model led to similar results, confirming that the particular choice of model is unimportant. What is important, however, is the selection of an appropriate training scheme. As already mentioned in Section 3.2.5, a naive gradient descent approach is bound to fail due to large variations in the local curvature of the error function. The simulations presented here were therefore performed with the rectified algorithm (3.70), (3.71), (3.72), which corresponds to a curvature-based shape-modification of the error function $E(\mathbf{q})$. Although solving the problems discussed in Chapter 3, this training scheme still depends sensitively on the choice of the ratios between the learning rates γ_w, γ_β, and γ_a. This critical dependence can be understood from Fig.5.1., which shows the state-space plot of the stochastic logistic map, together with the network implementation of a DSM with $K = 2$ kernels. The bold lines show the kernel centres $\mu_k(\mathbf{x}_t; \mathbf{w})$, while the narrow lines show the kernel widths $\mu_k(\mathbf{x}_t; \mathbf{w}) \pm \sigma_k$. Obviously one of the kernels, $\mu_1(\mathbf{x}_t; \mathbf{w})$, fits a relevant region in the state space, whereas the training process for the weights determining the second kernel, $\mu_2(\mathbf{x}_t; \mathbf{w})$, apparently has not yet converged. In fact, however, the plot is taken from a network *after* convergence, where the output

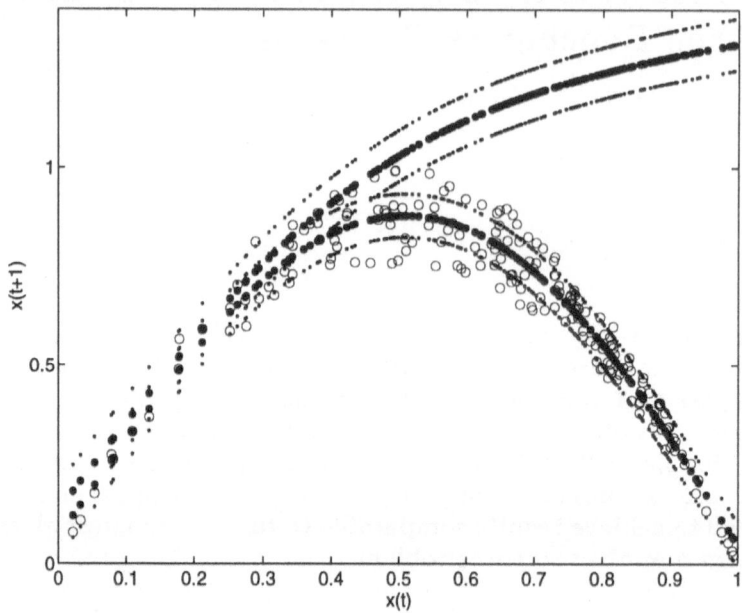

Fig. 5.1. Sub-optimal prediction of the stochastic logistic map. The *circles* show a state-space plot of a time series generated from the stochastic logistic map (4.2), similar as in Fig.4.1. The *bold lines* represent the kernel centres $\mu_k(x_t; \mathbf{w})$ (after training) of a DSM network with $K = 2$ network branches (that is, 2 nodes in the $\mathcal{G}$-layer). The kernel widths, $\mu_k(x_t, \mathbf{w}) \pm \sigma_k$, are indicated by the *narrow lines*. The figure shows a sub-optimal solution where the relevant region in the state space is only fitted by *one* network branch. The output weight of the other branch, a_2, had decayed to zero before the mapping implemented in this branch, $\mu_2(x_t; \mathbf{w})$, could be significantly improved.

weight a_2 for the mis-matching branch $\mu_2(\mathbf{x}_t; \mathbf{w})$ has decayed to zero. This highlights the problem of too large a learning rate γ_a. Recall from (3.36) that the priors a_k are corrected towards their posteriors, $\pi_k(t)$. This implies a competitive process, since

$$\pi_k(t) \quad = \quad \frac{P(k)P(y_t|\mathbf{x}_t, k)}{\sum_i P(i)P(y_t|\mathbf{x}_t, i)} \quad = \quad \frac{a_k \mathcal{G}_{\beta_k}\Big(y_t - \mu_k(\mathbf{x}_t; \mathbf{w})\Big)}{\sum_i a_i \mathcal{G}_{\beta_k}\Big(y_t - \mu_i(\mathbf{x}_t; \mathbf{w})\Big)} \quad (5.1)$$

At the beginning of the training process, immediately after initialisation, the mappings $\mu_k(\mathbf{x}_t; \mathbf{w})$ of all kernels are rather poor, and the posteriors $\pi_k(t)$ are usually of equal size. However, as soon as the mapping in one network branch, $\mu_k(\mathbf{x}_t; \mathbf{w})$, has improved significantly, as shown in Fig.5.1, its average class-conditional probability $P(y_t|\mathbf{x}, k) = \mathcal{G}_{\beta_k}\Big(y_t - \mu_k(\mathbf{x}_t; \mathbf{w})\Big)$ increases by a considerable amount. Consequently, as a result of the competition inherent in (5.1), the posteriors $\pi_k(t)$ for those kernels that have not yet fitted relevant regions in the state space will decrease drastically. This leads to two different scenarios. In the first, the mis-mapping network branches can 'catch up' with the 'pre-runner', i.e. the weights $\mathbf{w}$ are adjusted such that the other network branches $\mu_k(\mathbf{x}_t, \mathbf{w})$ also learn to fit some relevant regions in the state space. In this way the class-conditional density $\mathcal{G}_{\beta_k}\Big(y_t - \mu_k(\mathbf{x}_t, \mathbf{w})\Big)$ and consequently the posterior $\pi_k(t)$ increase again, so that the conditional probability density $P(y|\mathbf{x})$ will be modelled with the whole network structure. However, if the learning rate γ_a has been chosen too large, then the priors or output weights a_k of the mis-mapping kernels decay *before* the implemented mappings $\mu_k(\mathbf{x}_t; \mathbf{w})$ can be significantly improved. Consequently, network branches that are not successful at an early stage of the training process are simply 'switched off'. This might seem akin to a pruning scheme, but there is a major difference. In network pruning, parts of a network structure are disposed of deliberately in order to reduce overcomplexity and thus improve the generalisation performance. In contrast, the scenario described above is a pure artefact of the training scheme and will, in general, only *decrease* the generalisation performance. This is again seen from Fig.5.1. The remaining kernel predicts the conditional mean of the process, $\mu(y_t|\mathbf{x}_t)$, perfectly, and the kernel width seems to be a good estimate for the mean standard deviation. However, the actual standard deviation of the process is state-space dependent, i.e. $\sigma_{y|\mathbf{x}} = \sigma_{y|\mathbf{x}}(\mathbf{x})$. This cannot be predicted with a single kernel since its width parameter β is $\mathbf{x}$-independent. Consequently, the loss of the second kernel has evidently resulted in a sub-optimal state.

5.2 Logistic map with intrinsic noise

5.2.1 Method

In a first study, the network was applied to a time series generated from the stochastic logistic map (4.2). A DSM network with $K = 2$ kernels in the $\mathcal{G}$-layer, $H = 5$ tanh-units in the $\mathcal{S}$-layer, and a single input node (presenting the previous time series value x_t) was employed. All nodes in the $\mathcal{S}$- and $\mathcal{G}$-layer were connected to a bias element of constant activity 1. The network parameters were initialised as follows. All output weights a_k were set to the inverse number of nodes in the $\mathcal{G}$-layer, $a_k = 1/K$. All kernel widths were set to $\sigma_k = 1$, i.e. the inverse kernel widths were set to $\beta_k = \pi/\sqrt{3}$ $\forall k$ (see (2.27)). The remaining weights $\mathbf{w}$, which determine the output of the kernel centres $\mu_k(x_t; \mathbf{w})$, were drawn independently from a zero-mean Gaussian distribution of standard deviation $\sigma = 0.1$. A set of $N_{\text{train}} = 200$ data points

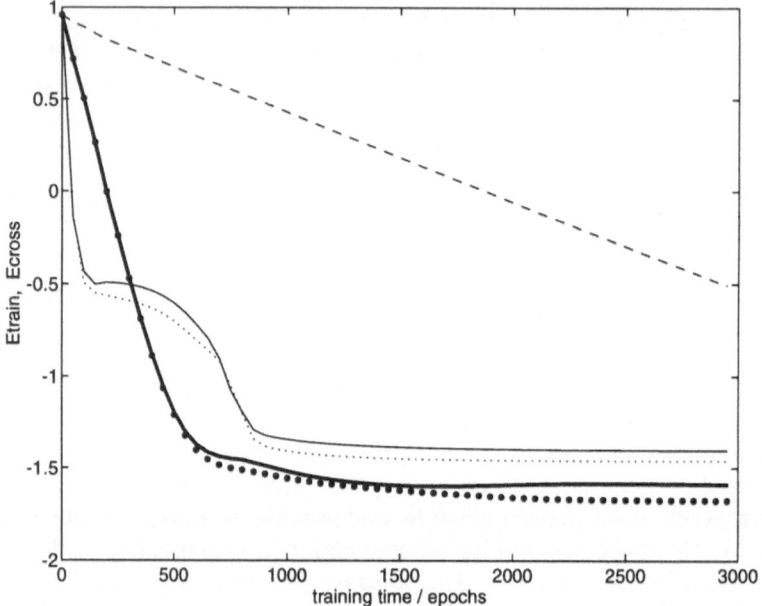

Fig. 5.2. Dependence of the training scheme on the learning rates. The graphs show the evolution of the training 'error' E_{train} and the cross-validation 'error' E_{cross} for a DSM network trained on a time series generated from the stochastic logistic map (4.2). Three different settings of the learning rates were employed. *Dashed line:* $\gamma_a = \gamma_\beta = 0.001$. The learning rates are too small, rendering the convergence of the training process extremely slow. Since no relevant structure is learned by the nework, the graphs for E_{train} and E_{cross} overlap (no overfitting). This is not the case for the other graphs, where the *dotted lines* present E_{train}, and the *solid lines* E_{cross}. *Narrow lines:* $\gamma_a = \gamma_\beta = 0.1$. The learning rate for the output weights is chosen too large. The system gets trapped in a local minimum, as discussed in Section 5.1 and shown in Fig.5.1. *Bold lines:* $\gamma_a = \gamma_\beta = 0.01$. Satisfactory choice of the learning rates, leading to the network performance depicted in Fig.5.4.

was generated from (4.2) for network training, and a second independent set of the same size, $N_{cross} = 200$, was created for monitoring the generalisation performance during training (cross-validation set). A constant of $C = 0.5$ was subtracted from the data, which leads approximately to a shift about the unconditional mean μ_{x_t} (centering)[1]. The objective was to predict the conditional probability density $P(x_{t+1}|x_t)$, where the true distribution is

---

[1] Normalisation, i.e. subtracting the mean from the data (centering) and dividing the latter by the standard deviation (re-scaling), is a standard procedure for alleviating the training process; see e.g. [7], Chapter 8. However, the re-scaling step was omitted here since the x_t are confined to the interval $[0, 1]$, hence the standard deviation is already of the order $\mathcal{O}(1)$.

given by (4.4). All training simulations were carried out with a momentum parameter of $\mu = 0.9$. The dependence of the training scheme on the learning rate is shown in Fig.5.2, which depicts the training 'error'[2]

$$E_{train} = -\frac{1}{N_{train}} \sum_{t=1}^{N_{train}} \ln P(y_t | \mathbf{x}_t) \tag{5.2}$$

(solid line) and the cross-validation 'error'

$$E_{cross} = -\frac{1}{N_{cross}} \sum_{t=1}^{N_{cross}} \ln P(y_t | \mathbf{x}_t) \tag{5.3}$$

(dashed line) for three different settings of the learning rate: (A) $\gamma_a = \gamma_\beta = 0.001$, (B) $\gamma_a = \gamma_\beta = 0.01$, and (C) $\gamma_a = \gamma_\beta = 0.1$. The learning rate for the remaining weights, γ_w, was the same in each case, $\gamma_w = 1$. The last scenario (C) has already been discussed in the previous section. As the learning rate γ_a is too large, the competition imposed on the output weights a_k is too strong, resulting in the decay of one of them, $a_2 \to 0$, and consequent switching off of the respective network branch. The graphs (narrow lines) show a rather fast convergence of the training process within about 1000 epochs[3], which leads, however, only to a local minimum. The network prediction by the corresponding network configuration is plotted in Fig.5.1, and was discussed above, on page 71. The other extreme, scenario (A), shows the problem of too small a learning rate for the adaptation of the kernel widths, γ_β. The convergence of the training simulation, shown by the upper dashed line in Fig.5.2, is seen to slow down considerably, and at the end of the training simulation, after 3000 epochs, the system is still far from convergence. In the corresponding state-space plot, the mappings implemented in the two network branches, $\mu_1(x_t, \mathbf{w})$ and $\mu_2(x_t, \mathbf{w})$, were found to be basically overlapping, $\mu_1(x_t, \mathbf{w}) \approx \mu_2(x_t, \mathbf{w})$. This points to the rather obvious fact that for splitting up the state space among the different kernels, a sufficiently fast kernel width adaptation is required. A successful training process, finally, is given by option (B), with the evolution of E_{train} and E_{cross} shown by the bold lines in Fig.5.2.

5.2.2 Results

For testing the network performance, the parameter configuration was taken at the minimum of the cross-validation 'error', E_{cross}, after about 1500

---

[2] Here and in what follows the term 'error' will be used for the value of the error function $E(\mathbf{q})$ defined in (3.8). Since this is a generalisation of the standard usage of the word 'error', it will be placed in inverted commas throughout.

[3] An 'epoch' is a traverse of the whole training set.

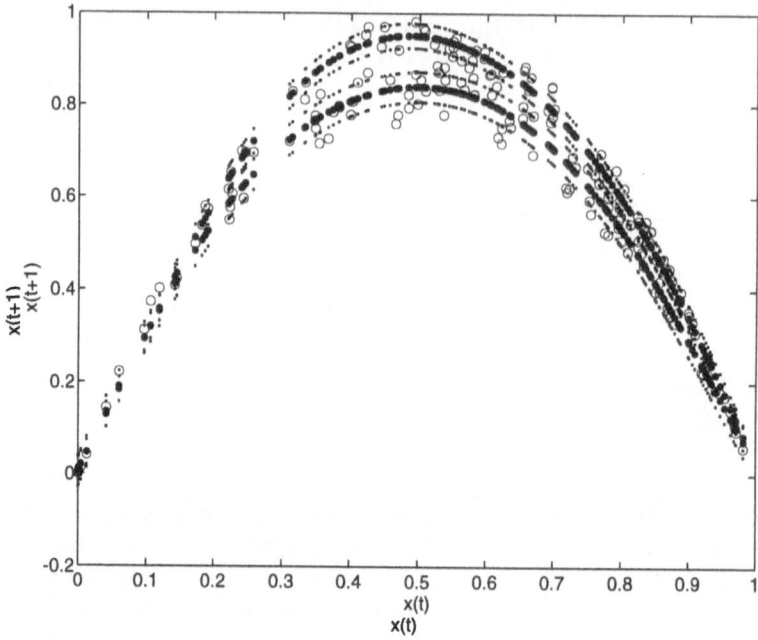

Fig. 5.3. Kernel centres and kernel widths for a DSM network applied to the stochastic logistic map problem. *Circles:* State-space plot of a time series generated from (4.2). *Bold lines:* Kernel centres $\mu_k(x_t, \mathbf{w})$ of a DSM network with $K = 2$ branches. *Narrow lines:* Kernel widths, $\mu_k(x_t, \mathbf{w}) \pm \sigma_k$. The figure illustrates how different network branches focus on different regions of the state space.

epochs. Figure 5.3 shows a function plot of the kernel centres, i.e. the mappings implemented in the two network branches, $\mu_1(x_t, \mathbf{w})$, $\mu_2(x_t, \mathbf{w})$, together with the respective kernel widths, $\mu_k(x_t, \mathbf{w}) \pm \sigma_k$. This demonstrates clearly how the network branches focus on distinct regions of the state space, and how they spread out so as to cover the whole relevant state-space domain. Figure 5.4 shows a state-space plot of the predicted conditional mean with its 1σ error bar, $\langle x_{t+1}|x_t \rangle \pm \sigma_{x_{t+1}|x_t}$, where $\langle x_{t+1}|x_t \rangle$ is given by (2.24) and $\sigma_{x_{t+1}|x_t}$ by (2.26). In Fig.5.5, the predictions for $\langle x_{t+1}|x_t \rangle$ (left) and $\sigma_{x_{t+1}|x_t}$ (right) are compared seperately with the theoretical values, which are given by (4.11) and (4.14). The graphs suggest that the network has successfully captured the underlying dynamics of the stochastic process, including the state-space dependence of the conditional standard deviation $\sigma_{x_{t+1}|x_t}$. Differentiation between the predicted and true conditional mean is, in fact, hardly possible. For the conditional standard deviation, the prediction is rather accurate for most x_t. It is only at the margins of the domain, $x_t \rightarrow 0$, $x_t \rightarrow 1$, where the correct standard deviation tends to zero and the true conditional probability density $P(x_{t+1}|x_t)$ becomes singular, that deviations occur.

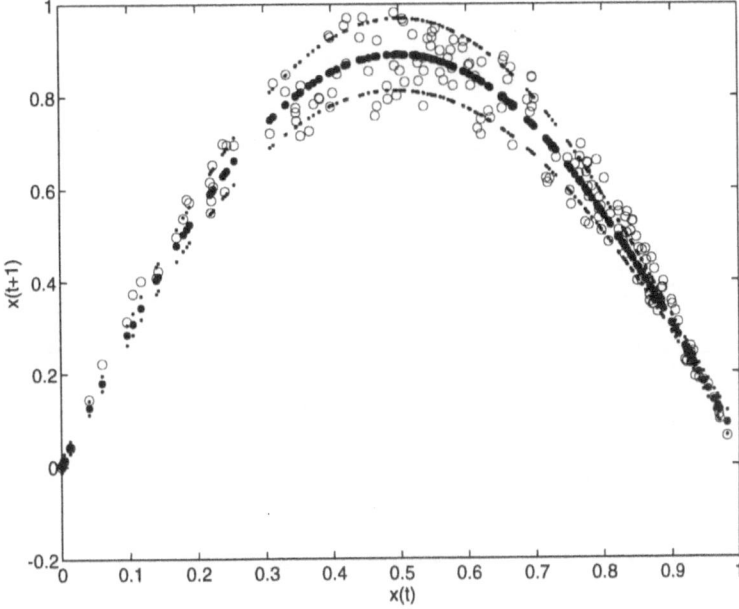

Fig. 5.4. Network prediction of the conditional mean and the conditional standard deviation for the stochastic logistic map problem. The figure shows the conditional mean $\mu(x_t; \mathbf{w}) = \langle x_{t+1}|x_t \rangle$ (*bold line*) with its 1σ-width, $\mu(x_t; \mathbf{w}) \pm \sigma_{x_{t+1}|x_t}$ (*narrow line*), predicted by a DSM network ($K = 2$) that was trained on a time series generated from (4.2). *Circles:* State-space plot of the time series. Note that the network has captured the state-space dependence of the conditional standard deviation.

5.3 Stochastic coupling between two stochastic dynamical systems

5.3.1 Method

In the second study, a DSM network was applied to predicting the conditional probability density for a time series generated from the stochastic coupling between the noisy logistic map and the kappa map (4.16). As before, the network contained $m = 1$ input unit, and ten nodes were employed in both the $\mathcal{S}$- and the $\mathcal{G}$-layer[4], $H = K = 10$, and the initialisation of the parameters

---

[4] A repetition of the simulations with $H = 20$ nodes in the $\mathcal{S}$-layer gave similar results as for $H = 10$. For $H = 5$, the prediction performance was slightly worse in that the spreading of the kernel centres over the relevant state-space domain, as shown in Fig.5.7, was less homogeneous and equi-distant as for $H = 10$.

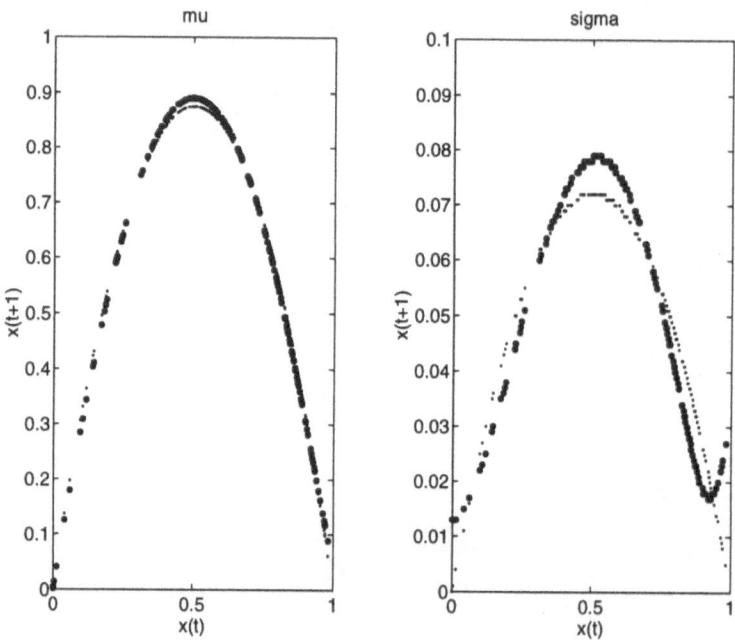

Fig. 5.5. Conditional mean and conditional standard deviation for the stochastic logistic map. The figures show a comparison between the predicted (*bold line*) and the actual (*narrow line*) value of the conditional mean $\langle x_{t+1}|x_t \rangle$ (left) and the conditional standard deviation $\sigma_{x_{t+1}|x_t}$ (right). The network employed was a DSM with two nodes in the $\mathcal{G}$-layer. The true functions were calculated from (4.11) and (4.14). Note that the deviations of the predictions for $\sigma_{x_{t+1}|x_t}$ at the interval margins result from the singularity of the conditional probability density $P(x_{t+1}|x_t)$ for $x_t \to 0, 1$.

was as described in Section 5.2.1. Two sets of $N_{\text{train}} = N_{\text{cross}} = 1000$ data points x_t were created for training and cross-validation, again with $C = 0.5 \approx \mu_{x_t}$ subtracted from the data. As in the previous case, the ratio of the learning rates has to be chosen with some care. Too large a value for γ_a causes, as before, an early decay of the priors of poorly-mapping network branches, which results in a loss of accuracy of the prediction[5]. A satisfactory choice was found to be $\gamma_w = 2$, $\gamma_a = \gamma_\beta = 0.01$. After about 7000 epochs, the cross-validation 'error' had become constant, and the resulting network configuration was taken for further evaluation.

---

[5] Recall that for predicting a state-space dependent conditional standard deviation of a sub-process, at least two network branches fitting that sub-process are required. Moreover note that due to its discontinuities, the true conditional probability density can only be predicted accurately with an infinite number of kernels. Hence every loss of a kernel further decreases the maximal obtainable prediction performance achievable with the given network.

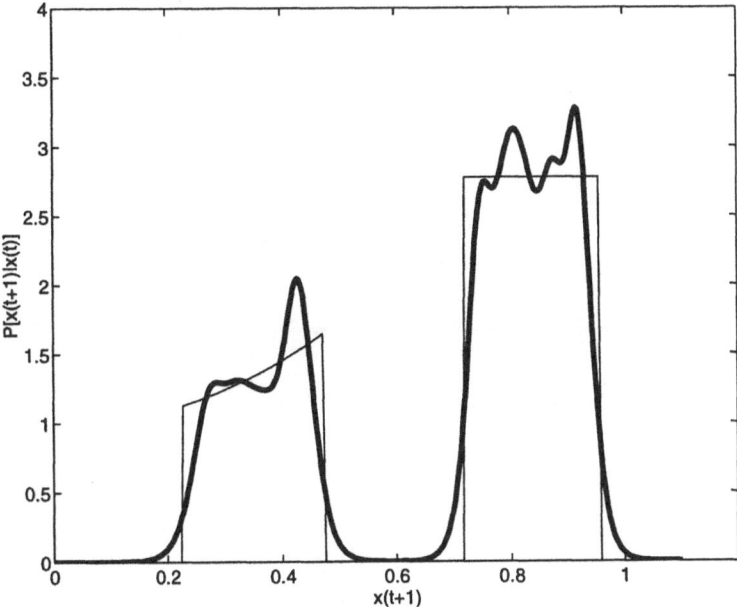

Fig. 5.6. Cross-section of the true and the predicted conditional probability for the logistic-kappa map. The figure shows a cross-section of the conditional probability density for $x_t = 0.6$, $P(x_{t+1}|x_t = 0.6)$, and compares the true function (narrow line, calculated from (4.22)) with that predicted by the DSM (bold line). Obviously, the network has captured the relevant features of the stochastic process and predicts a bimodal distribution. Note that this could not be achieved with the conventional approach of point-predictions, which aims at learning the conditional mean and thus predicts a value between the two modes, which actually never occurs.

5.3.2 Results

Figure 5.6 compares a cross-section of the predicted conditional density, $P(x_{t+1}|x_t, \mathbf{q})$, with the true density, $P_0(x_{t+1}|x_t)$, where the latter is calculated from (4.22). Apparently the network has captured the relevant features of the stochastic process and predicts a bimodal distribution. Recall that a conventional network for time series prediction, which only predicts the conditional mean $\langle x_{t+1}|x_t \rangle$, is completely inappropriate for the problem studied here since it would predict a value between the two clusters, which actually never occurs. In order to gain further insight into the network performance, Fig.5.7 shows a state-space plot of the kernel centres $\mu_k(x_t, \mathbf{w})$. It is seen that, similar to Fig.5.3, the different network branches focus on distinct regions of the state space, and that the kernels $\mu_k(x_t, \mathbf{w})$ spread out and become distributed over the whole relevant state-space domain. Summing over

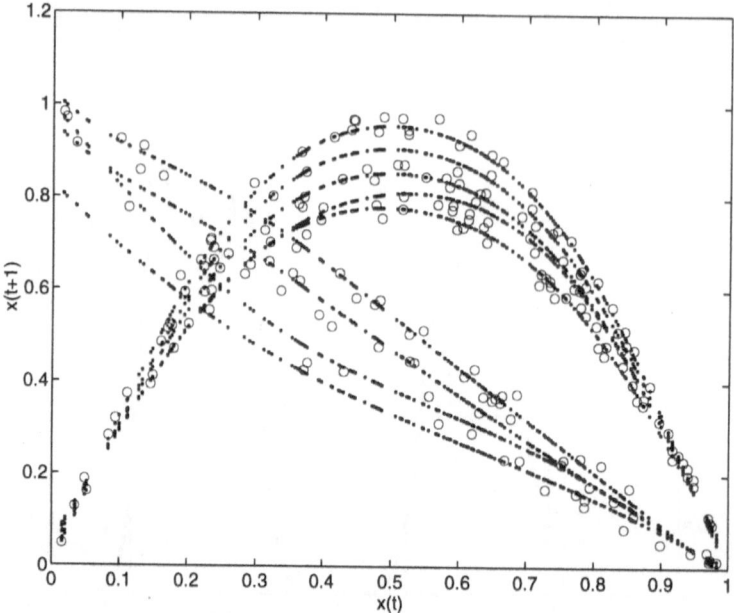

Fig. 5.7. State-space plot and kernel centres for the logistic-kappa map.
The circles show a state-space plot of the time series discussed in Section 4.2,
Equation 4.16. The lines represent the kernel centres $\mu_k(x_t; \mathbf{w})$ of a DSM network
after training. It can be seen how different network branches focus on different
regions of the state space, and how they spread out so as to cover the whole relevant
domain.

the output weights of branches fitting the kappa map and those learning the
logistic map gave the following ratio of the prior probabilities of the two
sub-processes:

$$\frac{P(\text{kappa}|\mathbf{q})}{P(\text{logistic}|\mathbf{q})} = \frac{\sum_{\text{kappa}} a_k}{\sum_{\text{logistic}} a_k} = \frac{0.330}{0.670}. \tag{5.4}$$

This is seen to be very close to the correct ratio of the priors, which from
page 61 is known to be $P(\text{kappa})/P(\text{logistic}) = 0.\bar{3}/0.\bar{6} = 0.5$.

5.3.3 Auto-pruning

In Fig.5.7 the different kernel centres $\mu_k(x_t, \mathbf{w})$ are seen to fit *either* the logistic map *or* the kappa map, i.e. one and only one of the two sub-processes of
(4.16). Counting the number of kernels reveals that the logistic map is fitted

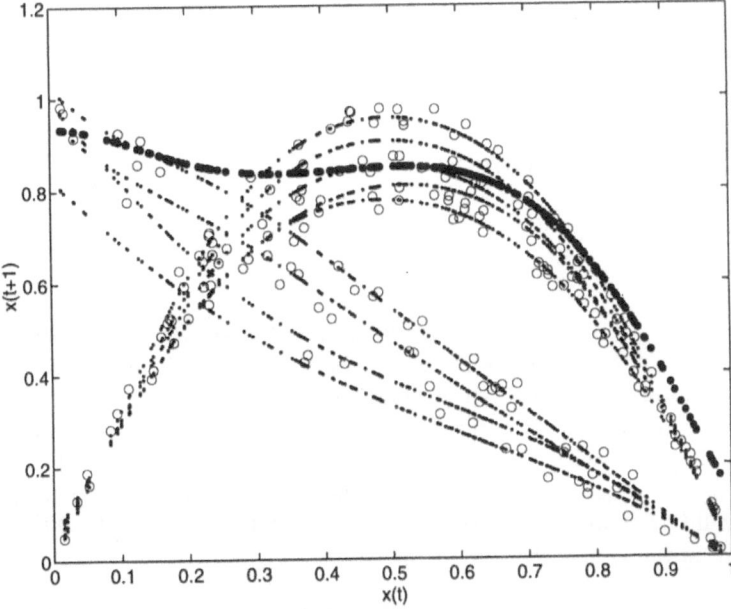

Fig. 5.8. Auto-pruning. The figure is basically identical to Fig.5.7, except that the bold line shows the centre of an additional kernel, $\mu_{10}(x_t; \mathbf{w})$, whose output weight a_{10} had decayed to zero during the training process. As discussed in the text, this is a consequence of the 'cross-over' mapping, that is, the fitting of parts of both sub-processes. This behaviour is inconsistent with the data, and the network switches the respective branch off.

by five centres, and the kappa map by four. The output weight of the tenth kernel, a_{10}, had decayed to zero during the training process, therefore the respective kernel centre is not plotted in the figure. The reason for switching off this kernel can be understood from Fig.5.8, which shows, by a bold line, the function implemented in the respective network branch, $\mu_{10}(x_t, \mathbf{w})$. It is seen that this function fits the kappa map in the left region of the state space, and then 'crosses over' to fit the logistic map in the right region. Since this 'crossing-over' leads to a change in the ratio $\sum_{\text{kappa}} a_k / \sum_{\text{logistic}} a_k$, the resulting network configuration cannot predict the true, $x-independent$ ratio of the priors, $P(\text{kappa})/P(\text{logistic}) = \text{const}$, unless a_{10} becomes zero. The observed decay of this output weight therefore proves to be an efficient auto-pruning scheme, which disposes of a branch that renders the network configuration inconsistent with the data. Note especially that this pruning scheme is of a fundamentally different nature than the one discussed in Section 5.1. The latter results from a wrong choice for the learning rates and leads to a reduction of network complexity *before* the pruned parts were given a chance to be optimised. The pruning observed here follows *after* the implementation of different mappings $\mu_k(x_t, \mathbf{w})$ in the different network branches, and switches off those that turn out to be inefficient.

5.4 Brownian motion in a double-well potential

5.4.1 Method

The application of a neural network to the double-well problem described in Section 4.3 was studied by others before, [45], [46], [66], and thus allows a comparison between the DSM model studied here and some of the alternative approaches described in Section 2.9. Equation (4.24) was numerically integrated with the leapfrog algorithm (stepsize $\delta = 0.1$). Two sets of $N_{train} = N_{cross} = 10,000$ data points were generated for training and cross-validation, and ten further independent sets of $N_{gen} = 10,000$ each for estimating the generalisation 'error'

$$E_{gen} = -\frac{1}{N_{gen}} \sum_{t=1}^{N_{gen}} \ln P(y_t | \mathbf{x}_t). \tag{5.5}$$

The data were normalised by subtracting the mean and deviding by the standard deviation, as in [45], [46]. The network architecture was chosen such that its complexity was similar to the CDEN network of Ormoneit and Neuneier et al. Recall from Section 2.9 that in the CDEN all kernel parameters are modelled as $\mathbf{x}$-dependent functions, given by the output of a seperate one-hidden-layer network. In [45], [46], Ormoneit and Neuneier et al. employed three networks for modelling the $\mathbf{x}$-dependent priors, widths and centres of $K = 3$ Gaussian kernels, thus modelling altogether *nine* $\mathbf{x}$-dependent functions $\{a_k(\mathbf{x}), \sigma_k(\mathbf{x}), \mu_k(\mathbf{x})\}_{k=1}^{3}$. In order to compare this with a DSM network of the same complexity, *nine* kernel nodes were employed in the $\mathcal{G}$-layer of the DSM, hence also modelling *nine* $\mathbf{x}$-dependent functions, $\{\mu_k(\mathbf{x})\}_{k=1}^{9}$.

Five previous particle positions at consecutive time steps (that is, a lag vector of dimension $m = 5$) were presented as inputs to the network[6], and the objective was to predict the conditional probability distribution for the position of the particle 25 time steps ahead, $P(x_{t+25} | \mathbf{x}_t) = P(x_{t+25} | x_t, x_{t-1}, \ldots, x_{t-4})$. Different learning rate combinations were tried in the simulations, and in each case the network was trained until E_{cross},

---

[6] This follows Weigend and Srivastava, [66], whereas Ormoneit and Neuneier et al., [45], [46], employed a 13-dimensional input vector containing the positions of the last ten steps, the last position itself, and exponentially smoothed averages of the previous movements and positions.

the 'error' on the cross-validation set, started to increase. The right graphs in Fig.5.9 show the prediction of the conditional probability density by the best network found[7]. This is to be compared with the graphs on the left of Fig.5.9, which show the empirical estimate of the true probability density, $P_0(x_{t+25}|x_t, \dot{x}_t)$, obtained by numerically integrating the equation of motion (4.24) 100,000 times the 25 time steps ahead, and then applying a Parzen estimator to the set of positions as described in Section 4.3.

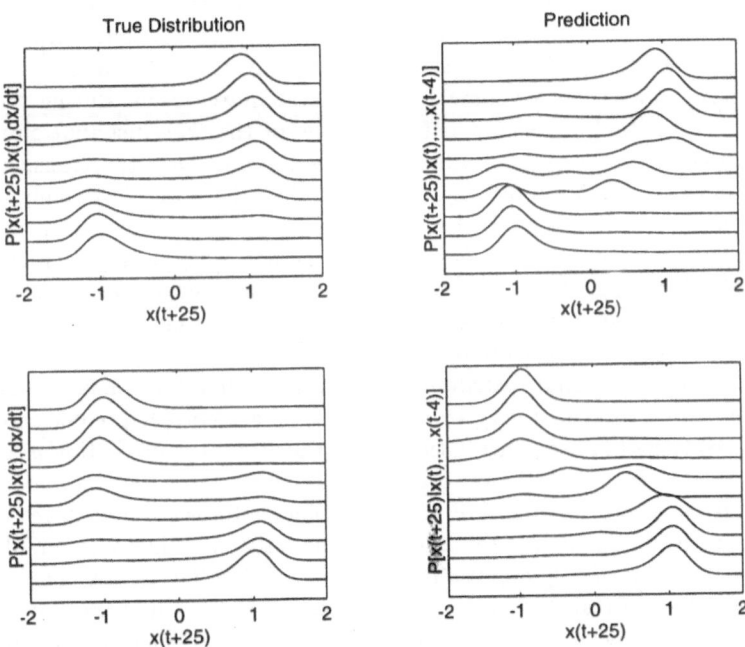

Fig. 5.9. Evolution of the true and the predicted conditional probability density during two phase transitions of the double-well problem. The graphs on the left show the evolution of the true conditional probability density $P(x_{t+25}|x_t, \dot{x}_t)$ during the phase transitions at (i) $t = 255 - 300$ and (ii) $t = 1510 - 1555$ in the time series segment plotted in Fig.4.4. The time difference between two contiguous graphs is $\Delta t = 5$, and the arrow of time points from the bottom to the top. The graphs on the right show the corresponding predictions $P(x_{t+25}|\mathbf{x}_t, \mathbf{q})$ with a DSM network. Note that the predictions are very accurate immediately before and after the transitions, and qualitatively correct, that is bimodal, during the transition.

---

[7] $\gamma_w = 2$, $\gamma_a = \gamma_\beta = 0.001$

5.4.2 Results

The left graphs in Fig.5.9 show the evolution of the (estimated) true conditional probability density along two 'phase transitions' of the system, taken between (i) times $t_1 = 255$ and $t_2 = 300$, and (ii) times $t_1 = 1510$ and $t_2 = 1555$ in the time series segment plotted in Fig.4.4 (test set). The estimates were taken in intervals of $\Delta t = 5$, and are plotted in the figures such that the arrow of time points from the bottom to the top. It can be seen that the distribution is unimodal before and after the 'phase transition', and undergoes a stage of bimodality during the transition. A closer inspection of the graphs also reveals some slight but systematic trends during the unimodal stages. In Fig.5.9, top left, the mode of the true distribution shows a slow drift towards smaller values both before and after the transition. In a similar way, Fig.5.9, bottom left reveals that the mode is moving slowly towards larger values before the transition occurs. Comparing these graphs with the conditional density predicted by the network, plotted in Fig.5.9 on the right, it is seen that the network captures the relevant characteristics of the dynamics and correctly models the evolution of the system from a unimodal distribution via a bimodal transition state to a new unimodal form. In fact, the conditional distribution before and after the 'phase transition' is found to be modelled very precisely, including a correct prediction of the drifts mentioned above. During the phase transition, the prediction is less accurate and only qualitatively correct. This deterioration can be understood from the time series segment of Fig.4.4, which contains six transitions within $t = 4000$ steps. The prior probability for a transition is thus very small, with an average value of only about one event in $t = 600 - 700$ steps. Consequently, there is a lack of training data in the corresponding region of the state space, leading inevitably to a degradation of the achievable prediction performance. However, in spite of this sparseness of data, the network succeeds in capturing the relevant feature of the transition: its bimodality.

5.4.3 Comparison with other approaches

The results of the previous simulations were compared[8] with the work of Ormoneit [46] and Neuneier et al. [45], who applied a variety of different models to the given problem. A summery of their results is listed in Tab.5.1, which shows the generalisation 'error' E_{gen} obtained on an independent test set. A brief overview of the different models employed was already given in Section 2.9. The second and third column of the table show two results obtained with the indirect approaches, where $P(y|\mathbf{x})$ is modelled by the quotient of

---

[8] Unfortunately, a comparison with the simulations of Weigend and Srivastava [66] is not possible since the authors do not report any quantitative results.

the outputs of two different networks via (2.40), $P(y|\mathbf{x}) = P(y,\mathbf{x})/P(\mathbf{x})$. The fourth and fifth column show the results obtained with the two *direct* approaches studied in [45] and [46], the CDEN and the DPMN (see Section 2.9). Finally, the last column gives the generalisation error of the DSM network studied in this book, where the numbers indicate the mean and standard deviation obtained on a set of ten different test sets with $N = 10,000$ data points each.

Network	Indir: Specht	Indir: best	CDEN	DPMN	DSM
Source	[45], [46]	[45], [46]	[45], [46]	[45], [46]	this work
E_{gen}	0.556	0.413	0.373	0.334	0.336 ± 0.009

Table 5.1. Comparison between the DSM and different alternative models. The table shows the results of network predictions obtained on the double-well problem of Section 4.3. The prediction performance of the DSM is compared with different alternative models, described in Section 2.9. The value E_{gen} gives the estimated generalisation 'error', as defined in (5.5). Smaller values thus indicate a better prediction performance.

From their simulations, Ormoneit and Neuneier conclude that a *direct* approach to the prediction of conditional probability densities is superior to the *indirect* methods. The results of the simulations carried out here support this conjecture and show that the DSM achieves a generalisation performance which is considerably better than that of the indirect approaches. A comparison with the alternative *direct* methods shows that the performance of the DSM is as good as that of the DPMN. The latter, however, suffers from certain restrictions, as discussed in Section 2.9 and pointed out in [45], [46]. The more interesting comparison is therefore between the DSM and the CDEN. The better performance of the DSM found here is not necessarily significant since the training simulations varied in some details[9] and were not carried out on identical training sets. However, the results do give strong support for the claim that the two principled approaches compared here, adopting a kernel expansion (2.31) with a few complex kernels (MDN, CDEN), or an expansion of the form (2.30) with many simple kernels (DSM, GM), do achieve a similar performance if networks of equivalent complexity are employed.

5.5 Conclusions

The study summarised in this chapter has empirically tested the network model and the training algorithm derived earlier in chapters 2 and 3 on

---

[9] In particular, Ormoneit and Neuneier conditioned the distribution on an input vector of dimension $m = 13$, whereas in the study reported here an input vector of dimension $m = 5$ was employed, thus possibly reducing the risk of overfitting.

different stochastic time series. The results suggest that the model succeeds in dealing with heteroscedasticity and multimodality.

An evaluation of the network performance on the first problem - a time series generated from the stochastic logistic map - demonstrates that the network predicts the (first two) conditional moments very accurately. The second and third time series were generated from the stochastic logistic-kappa map and Brownian dynamics in a double-well potential. A comparison between the predicted and the true conditional probability densities suggests that the network captures the relevant features and characteristics of the distributions, and that the deviations observed are mainly due to inadequacies inherent in the benchmark problems (discontinuities) or training data (limited amount of data in the relevant state-space region).

5.6 Discussion

Although the prediction results obtained on the three benchmark problems were very satisfactory, the computational costs were found to be discouragingly long. Typical training times for the simulations in Section 5.3 amounted to 5000-7000 adaptation steps, corresponding to 6-8 hours of CPU time[10] The simulations reported in Section 5.4 required typically about 2000 adaptation steps, equivalent to a CPU time of 2 days. Obviously, this raises serious doubts as to the applicability of the whole scheme to large-scale real-world applications.

A straightforward acceleration of the training process can be achieved by considering varying subsets of the training data. That is, for each epoch, the exemplars of the training set $\mathbf{D} = \{y_t, \mathbf{x}_t\}_{t=1}^{N}$ are accepted only with a probability $0 < P_{accept} \leq 1$, giving rise to the subset $\tilde{\mathbf{D}}_n = \{y_{t_i}, \mathbf{x}_{t_i}\}_{i=1}^{\tilde{N}} \subset \mathbf{D}$ with length $\tilde{N} \approx N P_{accept} < N$. In the nth adaptation step the network is thus trained on the modified error function

$$\tilde{E}_n \quad = \quad -\frac{1}{\tilde{N}} \sum_{(y_t, \mathbf{x}_t) \in \tilde{\mathbf{D}}_n} \ln P(y_t | \mathbf{x}_t, \mathbf{q}) \tag{5.6}$$

which is a noisy version of the correct error function

$$E \quad = \quad -\frac{1}{N} \sum_{(y_t, \mathbf{x}_t) \in \mathbf{D}} \ln P(y_t | \mathbf{x}_t, \mathbf{q}). \tag{5.7}$$

Since for each epoch n a different subset $\tilde{\mathbf{D}}_n$ is chosen, the noise can be expected to average out over the whole training process. The overall effect is

---

[10] The simulations were carried out on a Pentium 75 MHz PC.

thus a reduction of the effective epoch length from N to $\tilde{N}$. Moreover, the addition of noise can be expected to be helpful in escaping from local minima or flat regions of the error surface. In order to allow proper convergence of the algorithm, it is advisable to increase P_{accept} during the simulation to $P_{accept} = 1$. The simulations carried out in the present work were performed with a linear increase of P_{accept} with the epoch number n from 0.4 to 1.0. The results obtained were similar to those reported above in Section 5.3 and Section 5.4, with an acceleration of the training process by a factor of about two. However, even with this improvement the overall training times remained excessively long.

An alternative approach to be considered is the application of a second-order training scheme like conjugate gradients or the quasi-Newton method (reviewed in [7], Chapter 7). By including second-derivative information in the optimisation process, a considerable reduction of training time can be achieved if the error surface is convex. However, for highly *nonconvex* error surfaces, the application of a second-order scheme is rather dubious, since it is more likely to be trapped in local minima than simple gradient descent. For example, Gorse et al. [19] showed that for the XOR problem the probability of a suboptimal solution increases from 15% for gradient descent to 49% for conjugate gradients and 66% for the quasi-Newton method. For this reason, the approach adopted in the present work is first to seek a modification of the network architecture and the training scheme so as to *reduce* the *nonconvexity* of the error surface. This can be accomplished by a combination of two different training paradigms: the random vector functional link (RVFL) net approach, which constrains certain parameters to randomly chosen initial values, and the expectation maximisation (EM) algorithm, known from statistics. These concepts will be discussed in the two following chapters.

6. Random Vector Functional Link (RVFL) Networks

This chapter summarises the concept of the random vector functional link network (RVFL), that is, a multilayer perceptron (MLP) in which only the output weights are chosen as adaptable parameters, while the remaining parameters are constrained to random values independently selected in advance. It is shown that an RVFL is an efficient universal approximator that avoids the curse of dimensionality. The proof is based on an integral representation of the function to be approximated and subsequent evaluation of the integral by the Monte-Carlo approach. This is compared with the universal approximation capability of a standard MLP. The chapter terminates with a simple experimental illustration of the concept on a toy problem.

6.1 The RVFL theorem

Multilayer perceptrons (MLP) are universal approximators. However, due to the non-convexity of the error surface defined in the space of adaptable parameters, the training process can become rather awkward. It is well-known that local minima can lead to suboptimal network configurations, and 'flat' regions, that is regions corresponding to quasi-zero eigenvalues of the Hessian, normally result in extremely long convergence times. For this reason Pao, Park and Sobajic [48] studied a modification of the standard MLP, which they referred to as the *random vector functional link network* (RVFL). The architecture of an RVFL is identical to that of an MLP. Assume, for simplicity, that the function to be modelled is a scalar, then the output of the network is given by

$$f(\mathbf{x}) \quad = \quad \sum_{i=1}^{n} w_i g(\mathbf{u}_i^\dagger \mathbf{x} - b_i) \qquad (6.1)$$

where $g(.)$ denotes a nonlinear, usually sigmoidal, transfer function. The difference between the standard MLP and the RVFL is that in the former all network parameters, i.e. the output weights w_k, hidden weights $\mathbf{u}_i$ and hidden thresholds b_i, are optimised in the training process, while in the latter only the output weights w_k can be adapted. The parameters of the hidden layer, $\mathbf{u}_i$ and b_i, are selected randomly and independently in advance, and then remain constrained to these values during training. In this way the error function E becomes quadratic in the remaining adaptable parameters $\mathbf{w}$, which solves the problem of local minima and allows fast optimisation using conjugate gradients or matrix inversion techniques. That is,

$$E(\mathbf{w}) = E(\hat{\mathbf{w}}) + \frac{1}{2}(\mathbf{w} - \hat{\mathbf{w}})^\dagger \mathbf{H}(\mathbf{w} - \hat{\mathbf{w}}), \tag{6.2}$$

$$\nabla E(\mathbf{w}) = \mathbf{H}(\mathbf{w} - \hat{\mathbf{w}}) \tag{6.3}$$

where the Hessian $\mathbf{H}$ is semi positive definite. If $\mathbf{H}$ is also strictly positive definite, that is, if there exists a well-defined global minimum $\hat{\mathbf{w}}$, then this minimum can, in principle, be found in a single step by application of Newton's rule:

$$\hat{\mathbf{w}} = \mathbf{w} - \mathbf{H}^{-1}\nabla E(\mathbf{w}) = -\mathbf{H}^{-1}\nabla E(0) \tag{6.4}$$

The question of interest is, of course, what are the consequences of imposing the constraints on the hidden parameters? Igelnik and Pao [32] showed that, if the unknown function f satisfies certain regularity conditions (which are the same as for the standard universal approximation theorem, namely that f is bounded, continuous, and defined on a compact set), then an arbitrarily close approximation of f can be attained by an RVFL of sufficient complexity. This universal approximation capability is not, in itself, of much value, though, since it can also be achieved with fixed basis functions, like polynomials. The latter, however, are inflicted by the *curse of dimensionality*, with an approximation error that cannot be made smaller than of order $\mathcal{O}(1/n^{2/m})$, where m is the dimension of the input vector $\mathbf{x}$ and n the number of basis functions [3]. Obviously, in a high-dimensional setting this approximation rate of $2/m$ becomes vanishingly small. Multilayer perceptrons, on the other hand, have been found to have an approximation error of order $\mathcal{O}(\frac{1}{n})$ irrespective of the dimension of the input space m [3], [43]. The important finding of Igelnik and Pao [32] is that, provided the function to be approximated satisfies certain smoothness conditions[1], then the RVFL shows the same rate of approximation error convergence of $\mathcal{O}(\frac{1}{n})$ as an MLP. Hence an RVFL is, like an MLP, an *efficient* universal approximator that is free from the curse of dimensionality.

---

[1] The function needs to be Lipshitz continuous, as will be discussed in the next section. See (6.24) for a definition.

6.2 Proof of the RVFL theorem

The following section combines the ideas of [32] with the mathematical approach adopted in [43]. The main objective is to summarise and explain the basic principles of the RVFL scheme. Mathematical rigour is therefore sacrificed in part in order to focus on the main aspects in a simplified, less opaque language. The mathematically interested reader is referred to the original work by Igelnik and Pao [32] and Murata [43].

Let f, the function to be approximated, be a bounded and continuous function defined on a compact set $K \subset \mathbb{R}^m$ (these are the same conditions as required for the standard universal approximation theorem). Let g be a bounded, differentiable function on $\mathbb{R}$, whose derivative is square integrable (i.e. $\int_{\mathbb{R}} [g'(x)]^2 dx < \infty$)[2] . For convenience it will be further assumed that $|g(x)| \leq 1 \forall x$. Examples for g are sigmoidal functions, like $\tanh(x)$ and $\varsigma(x) = [1+e^{-x}]^{-1}$, which are usually employed as transfer functions in neural networks. Then the following integral representation for f can be shown to hold [32], [43]:

$$f(\mathbf{x}) \quad = \quad \int_{\mathbb{R}^{m+1}} T(\mathbf{u}, b) g(\mathbf{u}^\dagger \mathbf{x} - b) \, d\mathbf{u} \, db. \qquad (6.5)$$

Here $\mathbf{u} \in \mathbb{R}^m$, $b \in \mathbb{R}$, and $T : \mathbb{R}^{m+1} \to \mathbb{R}$ is given by an inverse transformation,

$$T(\mathbf{u}, b) \quad \propto \quad \int_{\mathbb{R}^m} \breve{g}(\mathbf{u}^\dagger \mathbf{x} - b) f(\mathbf{x}) dx. \qquad (6.6)$$

The so-called decomposing kernel $\breve{g}$ needs to satisfy certain *conjugacy* conditions listed in [43] to ensure that it is *conjugate* to the composing kernel g. However, for the purpose of this paragraph, these details are not important and will be omitted. Note that the above transformations have certain similarities with the Fourier transform, except that the Fourier transform and inverse Fourier transform give a one-to-one correspondence between functions defined on the same set $\mathbb{R}^m$, whereas (6.5) and (6.6) map between two different spaces of functions defined on $\mathbb{R}^m$ and $\mathbb{R}^{m+1}$. Hence (6.5) is more akin to a wavelet transform, allowing representation of f by different transforms $T_1(\mathbf{u}, b)$, $T_2(\mathbf{u}, b)$ with respect to different decomposing kernels $\breve{g}_1$, $\breve{g}_2$.

Let us define

---

[2] In [43], g itself is required to be square integrable, which excludes the sigmoidal function frequently employed in neural networks. A heuristic reasoning for the above generalisation is based on the fact that a bell-shaped function (which satisfies the stricter condition) can always be represented by a linear combination of two sigmoidals. A mathematical, though more cumbersome, proof can be found in [32]. Since the purpose of this section is to present ideas rather than maintain strict mathematical rigour, these subtleties will not be considered here.

$$V := [-\Omega, \Omega]^{m+1}, \qquad |V| := (2\Omega)^{m+1}, \tag{6.7}$$

where $\Omega \in \mathbb{R}$, and

$$f_\Omega(\mathbf{x}) := \int_V T(\mathbf{u}, b) g(\mathbf{u}^\dagger \mathbf{x} - b) \, d\mathbf{u} \, db$$

$$= \int_{-\Omega}^{\Omega} \cdots \int_{-\Omega}^{\Omega} T(u_1, \ldots, u_m, b) g(\mathbf{u}^\dagger \mathbf{x} - b) du_1 \ldots du_m db. \tag{6.8}$$

Obviously, from (6.5) and (6.8) it follows that

$$f(\mathbf{x}) = \lim_{\Omega \to \infty} f_\Omega(\mathbf{x}). \tag{6.9}$$

A two-stage approximation to the function f can now be attained. The first stage consists in approximating $f(\mathbf{x}) \approx f_\Omega(\mathbf{x})$, the second stage in obtaining an estimate of the integral (6.8) by using the Monte-Carlo method: $f_\Omega(\mathbf{x}) \approx \tilde{f}_n(\mathbf{x})$, where

$$\tilde{f}_n(\mathbf{x}) := \frac{|V|}{n} \sum_{i=1}^n T(\mathbf{u}_i, b_i) g(\mathbf{u}_i^\dagger \mathbf{x} - b_i) \tag{6.10}$$

and $\{(\mathbf{u}_1, b_1), \ldots, (\mathbf{u}_n, b_n)\}$ is a sample of size n drawn independently from the uniform distribution

$$P(\mathbf{u}_i, b_i) = \begin{cases} |V|^{-1} & \text{if } (\mathbf{u}_i, b_i) \in V \\ 0 & \text{otherwise} \end{cases} \tag{6.11}$$

If we define

$$w_i := \frac{|V|}{n} T(\mathbf{u}_i, b_i), \tag{6.12}$$

then

$$\tilde{f}_n(\mathbf{x}) = \sum_{i=1}^n w_i g(\mathbf{u}_i^\dagger \mathbf{x} - b_i) \tag{6.13}$$

has exactly the form of an RFVL network, where the hidden weights $\mathbf{u}_i$ and thresholds b_i are independently chosen subject to the probability distribution (6.11), and the output weights w_i are optimised in the training process. It will now be shown that

$$\langle \tilde{f}_n(\mathbf{x}) \rangle = f_\Omega(\mathbf{x}), \tag{6.14}$$

where $\langle . \rangle$ denotes the expectation value with respect to the probability distribution (6.11), and that the distance between f_Ω and $\tilde{f}_n$, defined as

$$d_K[f_\Omega, \tilde{f}_n] := \sqrt{\frac{1}{|K|} \left\langle \int_K [f_\Omega(\mathbf{x}) - \tilde{f}_n(\mathbf{x})]^2 d\mathbf{x} \right\rangle} \tag{6.15}$$

(where $|K|$ denotes the volume of the compact set K) scales as $d_K \propto \frac{1}{\sqrt{n}}$. With definition (6.10) one obtains

$$\left\langle \tilde{f}_n(\mathbf{x}) \right\rangle = \left\langle \sum_{i=1}^{n} \frac{T(\mathbf{u}_i, b_i)|V|}{n} g(\mathbf{u}_i^\dagger \mathbf{x} - b_i) \right\rangle \tag{6.16}$$

Since the $\mathbf{u}_i$ and b_i are drawn independently from the probability distribution (6.11), this gives

$$\left\langle \tilde{f}_n(\mathbf{x}) \right\rangle = n \left\langle \frac{T(\mathbf{u}, b)|V|}{n} g(\mathbf{u}^\dagger \mathbf{x} - b) \right\rangle = \left\langle T(\mathbf{u}, b)|V| g(\mathbf{u}^\dagger \mathbf{x} - b) \right\rangle$$

$$= \int T(\mathbf{u}, b)|V| g(\mathbf{u}^\dagger \mathbf{x} - b) P(\mathbf{u}, b)\, du\, db$$

$$= \int_V T(\mathbf{u}, b)|V| g(\mathbf{u}^\dagger \mathbf{x} - b) \frac{1}{|V|}\, du\, db$$

$$= \int_V T(\mathbf{u}, b) g(\mathbf{u}^\dagger \mathbf{x} - b)\, du\, db = f_\Omega(\mathbf{x}) \tag{6.17}$$

in which the last step follows from (6.8). In the same way, the variance is obtained as

$$\mathrm{Var}[\tilde{f}_n(\mathbf{x})] = \mathrm{Var}\left[\sum_i \frac{T(\mathbf{u}_i, b_i)|V|}{n} g(\mathbf{u}_i^\dagger \mathbf{x} - b_i) \right]$$

$$= n\,\mathrm{Var}\left[\frac{T(\mathbf{u}, b)|V|}{n} g(\mathbf{u}^\dagger \mathbf{x} - b) \right]$$

$$= \frac{1}{n}\mathrm{Var}\left[T(\mathbf{u}, b)|V| g(\mathbf{u}^\dagger \mathbf{x} - b) \right]$$

$$= \frac{1}{n}\left\{ \left\langle (T(\mathbf{u}, b)|V| g(\mathbf{u}^\dagger \mathbf{x} - b))^2 \right\rangle - \left\langle T(\mathbf{u}, b)|V| g(\mathbf{u}^\dagger \mathbf{x} - b) \right\rangle^2 \right\}$$

$$= \frac{1}{n}\left\{ \left\langle (T(\mathbf{u}, b)|V| g(\mathbf{u}^\dagger \mathbf{x} - b))^2 \right\rangle - \left[f_\Omega(\mathbf{x}) \right]^2 \right\}$$

where the last step follows from (6.17). The first term can be written as

$$\left\langle (T(\mathbf{u}, b)|V| g(\mathbf{u}^\dagger \mathbf{x} - b))^2 \right\rangle = \int \left(T(\mathbf{u}, b)|V| g(\mathbf{u}^\dagger \mathbf{x} - b) \right)^2 P(\mathbf{u}, b)\, du\, db$$

$$= \int_V \left(T(\mathbf{u}, b)|V| g(\mathbf{u}^\dagger \mathbf{x} - b) \right)^2 \frac{1}{|V|}\, du\, db$$

$$= |V| \int_V \left(T(\mathbf{u}, b) g(\mathbf{u}^\dagger \mathbf{x} - b) \right)^2 du\, db$$

Inserting this expression into the above equation for the variance yields

$$\mathrm{Var}[\tilde{f}_n(\mathbf{x})] = \frac{1}{n}\left\{ |V| \int_V \left(T(\mathbf{u}, b) g(\mathbf{u}^\dagger \mathbf{x} - b) \right)^2 du\, db - \left[f_\Omega(\mathbf{x}) \right]^2 \right\}. \tag{6.18}$$

Inserting (6.14) into the expression for the distance between f_Ω and $\tilde{f}_n$, Equation (6.15), gives

$$d_K^2[f_\Omega, \tilde{f}_n] = \frac{1}{|K|} \left\langle \int_K \left[f_\Omega(\mathbf{x}) - \tilde{f}_n(\mathbf{x})\right]^2 d\mathbf{x} \right\rangle = \frac{1}{|K|} \int_K \left\langle \left[f_\Omega(\mathbf{x}) - \tilde{f}_n(\mathbf{x})\right]^2 \right\rangle d\mathbf{x}$$

$$= \frac{1}{|K|} \int_K \left\langle \left[\left\langle \tilde{f}_n(\mathbf{x})\right\rangle - \tilde{f}_n(\mathbf{x})\right]^2 \right\rangle d\mathbf{x} = \frac{1}{|K|} \int_K \text{Var}[\tilde{f}_n(\mathbf{x})]d\mathbf{x} \quad (6.19)$$

We can now substitute the expression for the variance, equation (6.18), to obtain

$$d_K^2[f_\Omega, \tilde{f}_n] = \frac{1}{n|K|} \int_K d\mathbf{x} \left\{ |V| \int_V \left(T(\mathbf{u}, b)\, g(\mathbf{u}^\dagger \mathbf{x} - b)\right)^2 d\mathbf{u}\, db - \left[f_\Omega(\mathbf{x})\right]^2 \right\}$$
$$(6.20)$$

Since $|g(x)| \leq 1$, this can be bounded by

$$d_K^2[f_\Omega, \tilde{f}_n] \leq \frac{1}{n|K|} \int_K d\mathbf{x} \left\{ |V| \int_V \left(T(\mathbf{u}, b)\, g(\mathbf{u}^\dagger \mathbf{x} - b)\right)^2 d\mathbf{u}\, db \right\}$$
$$\leq \frac{|V|}{n|K|} \int_K d\mathbf{x} \int_V T^2(\mathbf{u}, b)\, d\mathbf{u}\, db = \frac{|V||K|}{n|K|} \int_V T^2(\mathbf{u}, b)\, d\mathbf{u}\, db$$

and hence

$$d_K[f_\Omega, \tilde{f}_n] \leq \frac{C}{\sqrt{n}} \quad (6.21)$$

where

$$C := \sqrt{|V| \int_V T^2(\mathbf{u}, b)\, d\mathbf{u}\, db} = \sqrt{(2\Omega)^{m+1} \int_V T^2(\mathbf{u}, b)\, d\mathbf{u}\, db} \quad (6.22)$$

Let us now consider the distance between $\tilde{f}_n$ and the actual function f, which can be bounded by

$$d_K[f, \tilde{f}_n] \leq \sup_{\mathbf{x} \in K} |f(\mathbf{x}) - f_\Omega(\mathbf{x})| + d_K[f_\Omega, \tilde{f}_n]. \quad (6.23)$$

Because of (6.9), the first term can be made arbitrarily small by choosing large enough values for the parameter Ω. In this case, however, the second term becomes very large and tends to infinity as $\Omega \to \infty$ (as seen from (6.21)). It is therefore necessary to restrict the class of continuous functions f to the class of smoother functions satisfying the *Lipshitz* condition

$$\exists \kappa > 0: \quad |f(\mathbf{x}) - f(\mathbf{y})| \leq \kappa \sum_{i=1}^m |x_i - y_i| \quad \forall \mathbf{x}, \mathbf{y} \in K. \quad (6.24)$$

Then the first term on the right of (6.23) becomes negligibly small for finite Ω (see [32]) so that (6.21) can be re-written as

$$d_K[f, \tilde{f}_n] \leq \frac{C}{\sqrt{n}}; \quad C = \sqrt{(2\Omega)^{m+1} \int_V T^2(\mathbf{u}, b) \, d\mathbf{u} \, db}; \quad \Omega < \infty$$

$$(6.25)$$

In practical applications the parameter Ω, which determines the distribution of the random parameters $\mathbf{u}_i$ and b_i, needs to be optimised in the training process. This will be discussed in more detail in Chapter 8.

6.3 Comparison with the multilayer perceptron

In order to provide additional insight into the universal approximation capability of the RVFL, it is useful to look over the corresponding proof for the universal approximation capability of the standard multilayer perceptron (MLP), as given e.g. by Murata [43]. First define

$$\mathcal{N} := \int_{\mathbb{R}^{m+1}} |T(\mathbf{u}, b)| \, d\mathbf{u} \, db, \qquad (6.26)$$

where the integral is assumed to be finite[3], and rewrite the integral representation (6.5) in the form

$$f(\mathbf{x}) = \int_{\mathbb{R}^{m+1}} \text{sign}[T(\mathbf{u}, b)] |T(\mathbf{u}, b)| \frac{\mathcal{N}}{\mathcal{N}} g(\mathbf{u}^\dagger \mathbf{x} - b) \, d\mathbf{u} \, db. \qquad (6.27)$$

With the definition of the probability distribution

$$P(\mathbf{u}, b) := \frac{1}{\mathcal{N}} |T(\mathbf{u}, b)|, \qquad (6.28)$$

equation (6.27) leads to

$$f(\mathbf{x}) = \mathcal{N} \int_{\mathbb{R}^{m+1}} \text{sign}[T(\mathbf{u}, b)] \, g(\mathbf{u}^\dagger \mathbf{x} - b) \, P(\mathbf{u}, b) \, d\mathbf{u} \, db. \qquad (6.29)$$

Similar to (6.10), this integral can be approximated with the Monte-Carlo method $f(\mathbf{x}) \approx \tilde{f}_n(\mathbf{x})$, where

$$\tilde{f}_n(\mathbf{x}) := \frac{\mathcal{N}}{n} \sum_{i=1}^{n} \text{sign}[T(\mathbf{u}_i, b_i)] g(\mathbf{u}_i^\dagger \mathbf{x} - b_i) \qquad (6.30)$$

and the $\{(\mathbf{u}_1, b_1), \ldots, (\mathbf{u}_n, b_n)\}$ are independently chosen subject to the probability distribution (6.28). Define

$$w_i := \frac{\mathcal{N}}{n} \text{sign}[T(\mathbf{u}_i, b_i)], \qquad (6.31)$$

---

[3] See [43] for a discussion of the divergent case.

then

$$\tilde{f}_n(\mathbf{x}) \;=\; \sum_{i=1}^{n} w_i\, g(\mathbf{u}_i^\dagger \mathbf{x} - b_i) \tag{6.32}$$

has a form similar to (6.13), where the w_i can be interpreted as output weights, the $\mathbf{u}_i$ as hidden weights, and the b_i as hidden thresholds. However, whereas in (6.13) the $\mathbf{u}_i$ and b_i are drawn *randomly* from the domain V (RVFL), the probability distribution considered here includes information about the unknown function f via $T(\mathbf{u}, b)$, as seen from (6.6) and (6.28), and thus needs to be learned in the training process (MLP). In other words, whereas the RVFL uses a simple mechanism for random selection of the parameters $\mathbf{u}_i$, b_i (uniform distribution) and includes all information about the unknown function f in the output weights to be learned, w_i, the MLP includes further information about f in the mechanism of random selection. This corresponds to the variance reduction method of *importance sampling* (see [32], [51], pp.316-317) and allows a closer approximation of the integral (6.5). In fact, repeating the steps (6.16), (6.18), and (6.19) for the distribution defined in (6.28) leads to (see [43])

$$d_K[f, \tilde{f}_n] \;\leq\; \frac{C_{MLP}}{\sqrt{n}}; \qquad C_{MLP} = \int_{\mathbb{R}^{m+1}} |T(\mathbf{u}, b)|\, d\mathbf{u}\, db. \tag{6.33}$$

Since it is assumed that the integral on the right is finite, it is valid for sufficiently large Ω to make the approximation

$$C_{MLP} \approx \int_V |T(\mathbf{u}, b)|\, d\mathbf{u}\, db = \frac{|V|}{|V|} \int_V |T(\mathbf{u}, b)|\, d\mathbf{u}\, db = |V| \Big\langle |T(\mathbf{u}, b)| \Big\rangle \tag{6.34}$$

where the expectation value $\langle . \rangle$ is taken with respect to the uniform distribution defined in (6.11). Compare this with the constant C defined in (6.21), which will be denoted C_{RVFL} henceforth,

$$C_{RVFL}^2 = |V| \int_V T^2(\mathbf{u}, b)\, d\mathbf{u}\, db = |V|^2 \frac{1}{|V|} \int_V T^2(\mathbf{u}, b)\, d\mathbf{u}\, db = |V|^2 \Big\langle T^2(\mathbf{u}, b) \Big\rangle \tag{6.35}$$

(where again the expectaion value is taken with respect to the uniform distribution (6.11)). Combining (6.34) and (6.35) gives

$$C_{RVFL}^2 - C_{MLP}^2 = |V|^2 \left(\Big\langle T^2(\mathbf{u}, b) \Big\rangle - \Big\langle |T(\mathbf{u}, b)| \Big\rangle^2 \right) = |V|^2 \mathrm{Var}|T(\mathbf{u}, b)| \;\geq\; 0 \tag{6.36}$$

and hence

$$C_{RVFL} \;\geq\; C_{MLP}. \tag{6.37}$$

Consequently, for a given value of n, the MLP gives a closer approximation to f than the RVFL, as expected. However, the functional dependence of the approximation error on the complexity of the model is in both cases of the form $\propto 1/\sqrt{n}$, so both models are *efficient* universal approximators that avoid an exponential increase in the number of hidden units.

6.4 A simple illustration

It was proved in the previous section that the RVFL network, as distinct from a network with fixed basis functions, is an efficient universal approximator that avoids the curse of dimensionality. The purpose of this section is to give a simple, plausible illustration for this fact. Consider the problem of determining the volume of an m-dimensional (hyper-)sphere by numerical integration. First, define the binary indicator function

$$I(\mathbf{x}) = \begin{cases} 1 & \text{if} \quad ||\mathbf{x}|| \leq R \\ 0 & \text{if} \quad ||\mathbf{x}|| > R \end{cases} \tag{6.38}$$

where R is the radius of the sphere. The true volume is known (see e.g. [7], pp. 28-29) to be

$$V_{Sphere} \quad = \quad \int_{\mathbb{R}^m} I(\mathbf{x}) d^m \mathbf{x} \quad = \quad \frac{2\pi^{m/2} R^m}{m\Gamma(m/2)}. \tag{6.39}$$

A standard numerical integration approximates the integral by a simple Riemann sum. That is, first introduce a hypercube $K = [-R, R]^m$ of side $L = 2R$ and divide each side into k equi-distant fragments of length $l = \frac{L}{k}$. This leads to a division of the whole space into a grid of $n = k^m$ equally-sized cells or sub-cubes with associated volume

$$V_1 = \dots V_n = l^m = \left(\frac{L}{k}\right)^m = \frac{L^m}{n} = \frac{(2R)^m}{n}. \tag{6.40}$$

Next, define $\mathbf{x}_i$ to be the centre of the ith sub-cube and approximate the integral (6.38) by

$$V_{Sphere}^{Grid} \quad = \quad \sum_{i=1}^{n} I(\mathbf{x}_i) V_i \quad = \quad \frac{(2R)^m}{n} \sum_{i=1}^{n} I(\mathbf{x}_i). \tag{6.41}$$

An alternative method is the Monte-Carlo technique. Here we introduce the uniform distribution

$$P(\mathbf{x}_i) \quad = \quad \begin{cases} |K|^{-1} = (2R)^{-m} & \text{if} \quad \mathbf{x}_i \in K \\ 0 & \text{otherwise} \end{cases} \tag{6.42}$$

The Monte-Carlo approximation to (6.39) is given by

$$V_{Sphere}^{MC} \quad = \quad \frac{|K|}{n} \sum_{i=1}^{n} I(\mathbf{x}_i) \quad = \quad \frac{(2R)^m}{n} \sum_{i=1}^{n} I(\mathbf{x}_i), \tag{6.43}$$

where the $\mathbf{x}_i$ are selected independently from the hypercube according to the distribution (6.42). Note that the essential difference between equations

(6.41) and (6.43), which formally look identical, is the different choice of the vectors $\mathbf{x}_i$.

The first method can be likened to a neural network with fixed basis functions, with each cell or sub-cube corresponding to one particular basis function. A sufficient approximation accuracy requires cells with small side $l = L/k$, and hence large values of k. On the other hand, the total number of cells, n, is given by $n = k^m$. This leads to an exponential increase in the number of cells with the input dimension m.

The second method is akin to the RVFL approach. As the $\mathbf{x}_i$ are chosen at random rather than systematically, the integration space K can be explored with a much smaller number of 'basis functions' n. The approximation error can be estimated by

$$\epsilon \approx \frac{C}{\sqrt{n}} \tag{6.44}$$

where

$$C = (2R)^m \sqrt{\langle I^2 \rangle - \langle I \rangle^2}, \quad \langle I^2 \rangle = \frac{1}{n} \sum_{i=1}^{n} I^2(\mathbf{x}_i), \quad \langle I \rangle = \frac{1}{n} \sum_{i=1}^{n} I(\mathbf{x}_i) \tag{6.45}$$

(see e.g. [51], p.305). Its functional dependence on n is thus independent of m (i.e. m does not occur in the exponent of n, but only enters indirectly via C), which avoids the curse of dimensionality.

Table 6.1 contrasts the relative approximation error $\varepsilon = \frac{100|V_{predict} - V_{Sphere}|}{V_{Sphere}}$ for the two methods, where $V_{predict}$ is the predicted volume of the sphere, given either by (6.41) or by (6.43), and V_{Sphere} is the true volume (6.39). The computations were repeated for different dimensions of the hyper-sphere, m, and different numbers of 'basis functions', n. In each case, the Monte-Carlo scheme was repeated ten times (starting from different random number generator seeds), and the mean error was determined. It can be seen that the errors obtained with the Monte-Carlo method are significantly smaller than for the grid method, especially in the case of the large dimension $m = 10$.

6.5 Summary

The proof of the universal approximation capability for both the MLP and the RVFL is based on an integral representation of the function to be approximated and subsequent evaluation of the integral by the Monte-Carlo approach. In both cases the approximation error is of the order $\frac{C}{\sqrt{n}}$, where n is the number of nodes in the hidden layer, and C a constant independent of n. Hence it follows that both the MLP and the RVFL are free of the *curse*

dimension	number of kernels	Grid: relative error	MC: relative error
3	27	34.4	12.8
3	125	23.8	7.8
3	1000	5.4	2.1
10	1024	100.0	38.5
10	59049	36.7	4.9

Table 6.1. Simple Monte Carlo. The table shows, for the simple case of determining the volume of an m-dimensional sphere, the dependence of the relative error, $\varepsilon = \frac{100|V_{predict} - V_{Sphere}|}{V_{Sphere}}$ (where $V_{predict}$ is the predicted volume and V_{Sphere} the correct one), on the number of 'kernels' or 'basis functions' for the grid and the Monte Carlo (MC) method.

of dimensionality, as opposed to a network with fixed basis functions. The difference between the RVFL and the MLP is that the former approximates the integral with a *simple* Monte-Carlo method, whereas the latter follows the more sophisticated scheme of *importance sampling* for variance reduction. The advantage of the MLP is therefore the possibility of a closer approximation of f with the same model complexity. Its disadvantage, however, is the non-convexity of the optimisation scheme, which renders the training process slow and susceptible to local minima.

This chapter has been concerned with learning in a conventional neural network for point predictions. The following chapter will show how the RVFL approach can be applied to GM networks for predicting conditional probabilities. The first section will review the EM algorithm, a powerful parameter adaptation scheme known from statistics. It will be shown, however, that its direct application to the GM model is not immediately possible, and that only in combination with the RVFL approach can a significant speed-up of the training scheme be achieved.

7. Improved Training Scheme Combining the Expectation Maximisation (EM) Algorithm with the RVFL Approach

This chapter reviews the Expectation Maximisation (EM) algorithm, and points out its principled advantage over a conventional gradient descent scheme. It is shown, however, that its application to the GM network is not directly feasible since the M-step of this algorithm is intractable with respect to one class of network parameters. A simple simulation demonstrates that as a consequence of this bottleneck effect, a gain in training speed can only be achieved at the price of a concomitant deterioration in the generalisation performance. It is therefore suggested to combine the EM algorithm with the RVFL concept discussed in the previous chapter. The parameter adaptation rules for the resulting GM-RVFL network are derived, and questions of numerical stability of the algorithm discussed. The superiority of this scheme over training a GM model with standard gradient descent will be demonstrated in the next chapter.

7.1 Review of the Expectation Maximisation (EM) algorithm

The basic idea of the EM (Expectation Maximisation) algorithm is that the original optimisation problem could be alleviated considerably if a set of further so-called hidden variables Λ were known. Recall, from Section 3.1, the definition of the maximum-likelihood error function

$$E(\mathbf{q}) \quad = \quad -\ln P(\mathbf{D}|\mathbf{q}), \tag{7.1}$$

where, for simplifying the notation, the factor $\frac{1}{N}$ has been omitted. Let us further define

$$\Psi(\mathbf{q}, \Lambda) \quad := \quad -\ln P(\mathbf{D}, \Lambda|\mathbf{q}) \tag{7.2}$$

$$U(\mathbf{q}|\mathbf{q}') \quad := \quad \Big\langle \Psi(\mathbf{q}, \Lambda) \Big\rangle_{\Lambda|\mathbf{D},\mathbf{q}'} \tag{7.3}$$

$$S(\mathbf{q}|\mathbf{q}') \quad := \quad \Big\langle \ln(P(\Lambda|\mathbf{D}, \mathbf{q}) \Big\rangle_{\Lambda|\mathbf{D},\mathbf{q}'} \tag{7.4}$$

where $\left\langle X \right\rangle_{\Lambda|\mathbf{D},\mathbf{q}'} := \int X(\Lambda) P(\Lambda|\mathbf{D},\mathbf{q}') d\Lambda.$

THEOREM
If the probability model is chosen such that

$$\int P(\mathbf{D}, \Lambda|\mathbf{q}') d\Lambda \;=\; P(\mathbf{D}|\mathbf{q}'), \tag{7.5}$$

then

$$E(\mathbf{q}) \;=\; U(\mathbf{q}|\mathbf{q}') + S(\mathbf{q}|\mathbf{q}'). \tag{7.6}$$

PROOF
Condition (7.5) implies that

$$\int P(\Lambda|\mathbf{D},\mathbf{q}') d\Lambda \;=\; \frac{1}{P(\mathbf{D}|\mathbf{q}')} \int P(\Lambda,\mathbf{D}|\mathbf{q}') d\Lambda \;=\; \frac{P(\mathbf{D}|\mathbf{q}')}{P(\mathbf{D}|\mathbf{q}')} = 1$$

and therefore

$$\begin{aligned}
\left\langle \ln P(\mathbf{D}|\mathbf{q}) \right\rangle_{\Lambda|\mathbf{D},\mathbf{q}'} &= \int \ln P(\mathbf{D}|\mathbf{q}) P(\Lambda|\mathbf{D},\mathbf{q}') d\Lambda \\
&= \ln P(\mathbf{D}|\mathbf{q}) \int P(\Lambda|\mathbf{D},\mathbf{q}') d\Lambda \\
&= \ln P(\mathbf{D}|\mathbf{q}).
\end{aligned}$$

Then apply Bayes' rule and (7.1), (7.3), (7.4) to obtain

$$\begin{aligned}
P(\Lambda,\mathbf{D}|\mathbf{q}) &= P(\Lambda|\mathbf{D},\mathbf{q}) P(\mathbf{D}|\mathbf{q}) \\
\Rightarrow \quad \ln P(\Lambda,\mathbf{D}|\mathbf{q}) &= \ln P(\Lambda|\mathbf{D},\mathbf{q}) + \ln P(\mathbf{D}|\mathbf{q}) \\
\Rightarrow \quad \left\langle \ln P(\Lambda,\mathbf{D}|\mathbf{q}) \right\rangle_{\Lambda|\mathbf{D},\mathbf{q}'} &= \left\langle \ln P(\Lambda|\mathbf{D},\mathbf{q}) \right\rangle_{\Lambda|\mathbf{D},\mathbf{q}'} + \ln P(\mathbf{D}|\mathbf{q}) \\
\Rightarrow \quad -U(\mathbf{q}|\mathbf{q}') &= S(\mathbf{q}|\mathbf{q}') - E(\mathbf{q}), \tag{7.7}
\end{aligned}$$

which leads to (7.6).
◇

Now from Kullback's theorem (3.3) it is known that $S(\mathbf{q}|\mathbf{q}')$ has its maximum at $\mathbf{q}'$, i.e. $S(\mathbf{q}|\mathbf{q}') \leq S(\mathbf{q}'|\mathbf{q}')$ and, assuming that $S(\mathbf{q}|\mathbf{q}')$ is differentiable, $[\nabla_{\mathbf{q}} S(\mathbf{q}|\mathbf{q}')]_{\mathbf{q}=\mathbf{q}'} = 0$. Together with Equation (7.6) this proves the following theorem:

THEOREM

$$\begin{aligned}
U(\mathbf{q}|\mathbf{q}') \leq U(\mathbf{q}'|\mathbf{q}') \quad &\Longrightarrow \quad E(\mathbf{q}) \leq E(\mathbf{q}') \tag{7.8} \\
[\nabla_{\mathbf{q}} U(\mathbf{q}|\mathbf{q}')]_{\mathbf{q}=\mathbf{q}'} \quad &= \quad [\nabla_{\mathbf{q}} E(\mathbf{q})]_{\mathbf{q}=\mathbf{q}'} \tag{7.9}
\end{aligned}$$

Equation (7.8) states that a parameter adaptation $\mathbf{q}' \to \mathbf{q}$ which decreases U also decreases the original maximum-likelihood error function E. Equation (7.9) implies that the critical points (that is, points with zero gradient) of E and U are the same. Thus (7.8) and (7.9) suggest the following training scheme known as the *EM algorithm*:

1. Given $\mathbf{q}'$, find $\mathbf{q}$ that minimises $U(\mathbf{q}|\mathbf{q}')$ (M-step).

2. If $\mathbf{q} \neq \mathbf{q}'$, define $\mathbf{q}' := \mathbf{q}$, recalculate $U(\ldots|\mathbf{q}')$, and go to 1 (E-step).

Because of (7.8) and (7.9), the sequence of $\mathbf{q}$ converges to a local minimum or, less likely, to a saddle point[1] of E. The advantage over a conventional gradient descent scheme, as derived in Chapter 3 and applied in Chapter 5, is that for an appropriate choice of the hidden variables Λ the minimisation of U can be much easier than the minimisation of the original 'error' function E. This will be discussed in more detail shortly. In order to apply the EM algorithm to the GM model, let us introduce the K by N matrix of binary hidden variables $\lambda_k(t)$,

$$\Lambda := \left(\lambda(1), \ldots, \lambda(N)\right), \quad \lambda(t) := \begin{pmatrix} \lambda_1(t) \\ \vdots \\ \lambda_K(t) \end{pmatrix}, \quad \lambda_k(t) \in \{0, 1\}, \quad \|\lambda(t)\| = 1 \; \forall t,$$

$$(7.10)$$

where $\lambda_k(t) = 1$ indicates that the tth exemplar $(\mathbf{x}_t, y_t)$ has been generated from the kth component of the mixture distribution (2.30), and the last equation on the right implies that at any particular time t one and only one $\lambda_k(t)$ is set to 1, the probability for which was already defined in (3.31),

$$\pi_k(t) := P\left(\lambda_k(t) = 1 | \mathbf{D}, \mathbf{q}'\right) = P\left(\lambda_k(t) = 1 | y_t, \mathbf{x}_t, \mathbf{q}'\right) := P(k | y_t, \mathbf{x}_t, \mathbf{q}').$$

$$(7.11)$$

This probability can be calculated by application of Bayes' rule

$$\pi_k(t) = P(k | y_t, \mathbf{x}_t, \mathbf{q}') = \frac{P(y_t | \mathbf{x}_t, k, \mathbf{q}') P(k | \mathbf{x}_t, \mathbf{q}')}{P(y_t | \mathbf{x}_t, \mathbf{q}')} = \frac{P(y_t | \mathbf{x}_t, k, \mathbf{q}') P(k)}{P(y_t | \mathbf{x}_t, \mathbf{q}')}$$

$$(7.12)$$

Let us further define

$$P(\mathbf{D}, \Lambda | \mathbf{q}) := \prod_{t=1}^{N} \prod_{k=1}^{K} \left(P(k) P(y_t | \mathbf{x}_t, \mathbf{q}, k)\right)^{\lambda_k(t)}. \qquad (7.13)$$

In the appendix on page 118 it is shown that this choice of the probability model satisfies condition (7.5) and thus allows the application of the EM

---

[1] A convergence to a saddle point is in principle possible since the inverse of (7.8) is not true. If $U(\mathbf{q}|\mathbf{q}') > U(\mathbf{q}'|\mathbf{q}')$ but $|U(\mathbf{q}|\mathbf{q}') - U(\mathbf{q}'|\mathbf{q}')| < |S(\mathbf{q}|\mathbf{q}') - S(\mathbf{q}'|\mathbf{q}')|$, then, as seen from (7.6), $E(\mathbf{q}) < E(\mathbf{q}')$ in spite of an increase in U.

algorithm. Now recall from Section 2.6 and Section 2.7 that the GM model is given by

$$P(y_t|\mathbf{x}_t, k, \mathbf{q}) = \mathcal{G}_{\beta_k}(y_t - \mu_k(\mathbf{x}_t; \mathbf{w})) = \sqrt{\frac{\beta_k}{2\pi}} \exp\left(-\frac{\beta_k}{2}\left(y_t - \mu_k(\mathbf{x}_t; \mathbf{w})\right)^2\right),$$

$$P(k) = a_k.$$

Inserting this into (7.13) yields

$$P(\mathbf{D}, \mathbf{\Lambda}|\mathbf{q}) = \prod_{t=1}^{N} \prod_{k=1}^{K} \left[a_k \sqrt{\frac{\beta_k}{2\pi}} \exp\left(-\frac{\beta_k}{2}\left(y_t - \mu_k(\mathbf{x}_t; \mathbf{w})\right)^2\right)\right]^{\lambda_k(t)} \tag{7.14}$$

and, with (7.2),

$$\Psi(\mathbf{q}, \mathbf{\Lambda}) = -\ln P(\mathbf{D}, \mathbf{\Lambda}|\mathbf{q}) \tag{7.15}$$

$$= \sum_{t=1}^{N} \sum_{k=1}^{K} \lambda_k(t) \left[\frac{\beta_k}{2}\left(y_t - \mu_k(\mathbf{x}_t; \mathbf{w})\right)^2 - \ln a_k - \frac{1}{2}\ln\left(\frac{\beta_k}{2\pi}\right)\right]$$

Since $\lambda_k(t)$ is binary and thus

$$\left\langle \lambda_k(t) \right\rangle_{\mathbf{\Lambda}|\mathbf{D}, \mathbf{q}'} = 1 \cdot P(\lambda_k(t)=1|\mathbf{D}, \mathbf{q}') + 0 \cdot P(\lambda_k(t)=0|\mathbf{D}, \mathbf{q}')$$

$$= P(\lambda_k(t)=1|\mathbf{D}, \mathbf{q}') = \pi_k(t) \tag{7.16}$$

equations (7.3) and (7.15) lead to

$$U(\mathbf{q}|\mathbf{q}') = \left\langle \Psi(\mathbf{q}, \mathbf{\Lambda}) \right\rangle_{\mathbf{\Lambda}|\mathbf{D}, \mathbf{q}'} \tag{7.17}$$

$$= \sum_{t=1}^{N} \sum_{k=1}^{K} \pi_k(t) \left[\frac{\beta_k}{2}\left(y_t - \mu_k(\mathbf{x}_t; \mathbf{w})\right)^2 - \ln a_k - \frac{1}{2}\ln\left(\frac{\beta_k}{2\pi}\right)\right]$$

The minimisation of U with respect to $\{\beta_k\}$ and $\{a_k\}$ can be done immediately. Taking the corresponding derivatives of U gives

$$\frac{\partial U}{\partial \beta_k} = \sum_{t=1}^{N} \frac{\pi_k(t)}{2} \left[\left(y_t - \mu_k(\mathbf{x}_t; \mathbf{w})\right)^2 - \frac{1}{\beta_k}\right] \tag{7.18}$$

$$\frac{\partial U}{\partial a_k} = -\frac{1}{a_k} \sum_{t=1}^{N} \pi_k(t) + \frac{1}{a_K} \sum_{t=1}^{N} \pi_K(t) \tag{7.19}$$

The last term on the right of (7.19) results from the constraint $\sum_{k=1}^{K} a_k = 1$. Without loss of generality the first $(K-1)$ variables a_k are chosen as free parameters. The value of a_K is then given by $a_K = 1 - \sum_{k=1}^{K-1} a_k$, whence $\frac{\partial a_K}{\partial a_k} = -1$. Setting the expression in (7.19) to zero leads to

$$\frac{1}{a_k}\left[\sum_{t=1}^{N}\pi_k(t)\right] = \frac{1}{a_K}\left[\sum_{t=1}^{N}\pi_K(t)\right] \tag{7.20}$$

Since this holds for all k, the expression must be constant,

$$\frac{1}{a_k}\left[\sum_{t=1}^{N}\pi_k(t)\right] = C, \tag{7.21}$$

where C follows from the constraint $\sum_{k=1}^{K}a_k = 1$,

$$C = \sum_{k=1}^{K}\left[\sum_{t=1}^{N}\pi_K(t)\right] = \sum_{t=1}^{N}\sum_{k=1}^{K}\pi_K(t) = N. \tag{7.22}$$

This leads to

$$a_k = \frac{1}{N}\sum_{t=1}^{N}\pi_k(t). \tag{7.23}$$

Setting the derivative in (7.18) to zero leads to the following update rule for the kernel widths:

$$\frac{1}{\beta_k} = \frac{\sum_{t=1}^{N}\pi_k(t)\left(y_t - \mu_k(\mathbf{x}_t;\mathbf{w})\right)^2}{\sum_{t=1}^{N}\pi_k(t)} \tag{7.24}$$

Let us compare these update rules with those obtained from the standard maximum likelihood approach, discussed in Chapter 3. Formally, the expressions (7.23) and (7.24) can be obtained by setting the terms on the right of (3.36) and (3.50) to zero. The difference, however, is that in this case the ensuing equations cannot be solved directly, since the posterior probabilities $\pi_k(t) = P(k|y_t, \mathbf{x}_t, k, \mathbf{q})$ themselves are dependent on the parameters $\mathbf{q}$. As discussed before, this leads to a complex nonlinear optimisation problem. When applying the EM algorithm, on the other hand, the posteriors depend only on the *old* parameters $\mathbf{q}'$, $\pi_k(t) = P(k|y_t, \mathbf{x}_t, k, \mathbf{q}')$, which are constant. Consequently, the complexity of the optimisation problem has been considerably reduced, allowing us to find a solution to the update equations immediately. This comparison suggests that the main advantage of the EM algorithm is the simplification of the optimisation problem by replacing a complex non-convex 'error' surface, $E(\mathbf{q})$, by a simple convex one, $U(\mathbf{q}|\mathbf{q}')$. The disadvantage, though, is that for $\mathbf{q}'$ far away from the optimal parameters, $\mathbf{q}_{\text{opt}}$, the 'error' functions $U(\mathbf{q}|\mathbf{q}')$ and $E(\mathbf{q})$ deviate strongly, necessitating an iteration of the whole process. However, when the required number of iterations is sufficiently small, the considerable acceleration of the minimisation process itself can lead to a net speed-up of the overall training process. This was indeed observed in the simulations, as will be reported in Chapter 8.

Let us finally turn to the remaining network parameters, $\mathbf{w}$. Since U as a function of $\mathbf{w}$ is still *non-convex*, the optimisation with respect to these parameters cannot be performed in a single step. Rather, U can only be decreased by following the gradient, which from (7.17) is given by

$$-\nabla_\mathbf{w} U \;\; = \;\; \sum_{t=1}^{N}\sum_{k=1}^{K} \pi_k(t)\beta_k \Big[y_t - \mu_k(\mathbf{x}_t;\mathbf{w}) \Big] \nabla_\mathbf{w}\mu_k(\mathbf{x}_t;\mathbf{w}) \qquad (7.25)$$

where $\nabla_\mathbf{w}\mu_k(\mathbf{x}_t;\mathbf{w})$ is calculated by backpropagation (recall that the $\mu_k(\mathbf{x}_t;\mathbf{w})$, which determine the centres of the Gaussian kernels in the $\mathcal{G}$-layer, are given by the outputs of a one-hidden-layer network, as seen from Fig.2.1c). For reasons discussed earlier in Section 3.2.5, the training process is improved by leaving the factor β_k out, which corresponds to the shape modification of the error surface given by (3.69). Moreover, a momentum term is applied in practice, leading to the update equation

$$\Delta\mathbf{w} \;\; = \;\; \gamma_\mathbf{w} \sum_{t=1}^{N}\sum_{k=1}^{K} \pi_k(t) \Big[y_t - \mu_k(\mathbf{x}_t;\mathbf{w}) \Big] \nabla_\mathbf{w}\mu_k(\mathbf{x}_t;\mathbf{w}) + \eta[\Delta\mathbf{w}]_\text{old} \quad (7.26)$$

in which $\gamma_\mathbf{w}$ is the learning rate, $\eta \in [0,1[$ the momentum parameter, and $[\Delta\mathbf{w}]_\text{old}$ the weight change obtained from the previous epoch. However, even with this improvement the adaptation of the weights $\mathbf{w}$ constitutes an iterative process that is slow compared to the fast adaptation rules (7.23) and (7.24). It can therefore be anticipated to result in a bottleneck effect which slows down the overall convergence speed of the algorithm. This is confirmed in the following simulations[2].

7.2 Simulation: Application of the GM network trained with the EM algorithm

7.2.1 Method

In order to compare the EM algorithm with the gradient descent scheme discussed in chapters 3 and 5, a GM network of the same architecture as in Section 5.3, $m = 1$, $H = 10$, $K = 10$, was applied to the time series generated from the logistic-kappa map, described in Section 4.2. The simulations closely followed those of Section 5.3. Two sets of $N_{train} = N_{cross} = 1000$ data points, for training ($\mathbf{D}_{train}$) and cross-validation ($\mathbf{D}_{cross}$), and a third set of $N_{gen} = 2000$ data points, for testing the generalisation performance ($\mathbf{D}_{gen}$), were

---

[2] The simulation results presented in the following section were first published in [29].

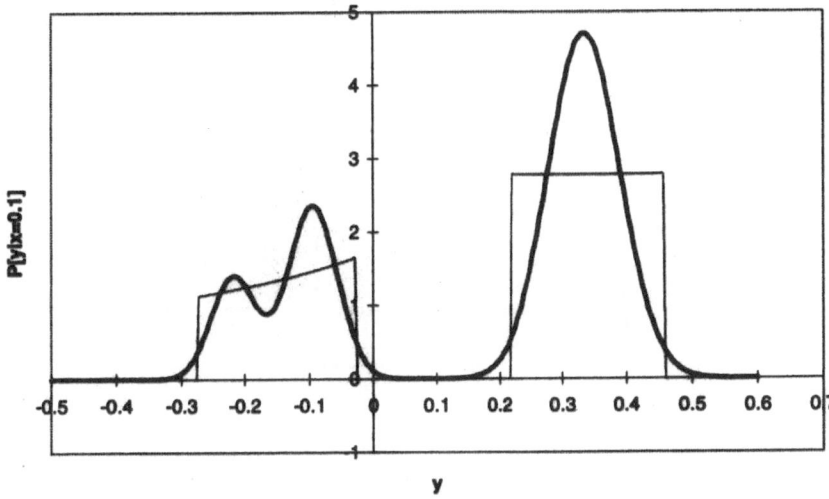

Fig. 7.1. Cross-section of the predicted conditional probability density for the logistic-kappa map problem, training without retardation. The figure shows a cross-section of the conditional probability density for $x_t = 0.6$, $P(x_{t+1}|x_t = 0.6)$, and compares the true function (narrow line, calculated from (4.22)) with that predicted by the GM network (bold line). Training was performed with the EM algorithm, (7.23), (7.24), and (7.26). The prediction result is suboptimal, as discussed in the text. Reprinted from [31], with permission from Elsevier Science.

generated from (4.16), with a value of $0.5 (\approx$ mean) subtracted from the data. The network parameters were initialised as described in Section 5.2.1. The output weights a_k and the inverse kernel widths β_k were adapted according to (7.23) and (7.24). For the remaining weights $\mathbf{w}$, the iterative scheme given by Equation (7.26) was applied, with a learning rate chosen small enough to prevent instabilities, $\gamma_{\mathbf{w}} = 1$, and a momentum parameter of $\eta = 0.9$. Training was performed until the cross-validation 'error' E_{cross} became constant or started to increase.

7.2.2 Results

The average convergence time took about 1000 parameter adaptation steps, which is significantly less than the 6000-8000 adaptation steps found in the earlier maximum-likelihood based simulations of Section 5.3. However, the

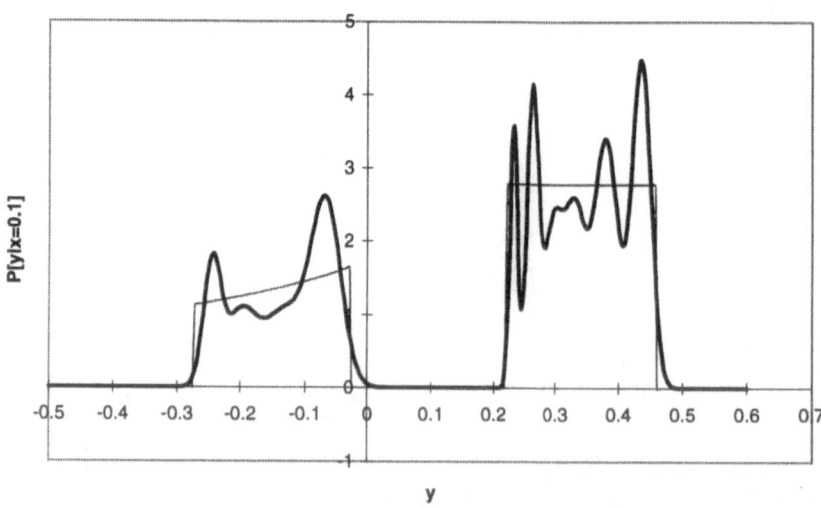

Fig. 7.2.   Cross-section of the predicted conditional probability density for the logistic-kappa map problem, training with retardation. As in Fig.7.1, the figure shows a cross-section of the conditional probability density $P(x_{t+1}|x_t = 0.6)$, and compares the true function (narrow line) with that predicted by the GM network (bold line). Training was performed with the EM algorithm, (7.23), (7.24), (7.26), combined with the retardation scheme (7.27). A comparison with Fig.7.2 reveals that the prediction is considerably improved by the retardation. Reprinted from [31], with permission from Elsevier Science.

prediction performance degraded significantly from $E_{gen} = -1.02$ to values around $E_{gen} = -0.9$. The reason for this deterioration can be understood from Fig.7.1 and Fig.7.3, which show a typical network performance after training. Figure 7.1 depicts a cross-section of the conditional probability density, $P(y(t)|\mathbf{x}(t) = 0.6)$, and compares the network prediction (bold line) with the correct function (narrow line). Apparently the network captures the multimodality of the distribution, but fails to predict its fine-grained characteristics[3]. Figure 7.3 shows a state-space plot of the time series and the kernel centres $\mu_k(x_t; \mathbf{w})$. Only three kernels, indicated by bold lines, cover relevant regions in the state space, while the other seven kernels (indicated by narrow lines) have gone astray. In fact, their corresponding output weights $a_k = P(k)$ were found to have decayed to zero. This suggests that as a consequence of

---

[3] This includes the sharp edges and the state-space dependence of the cluster widths. Figure 7.3 shows that only *one* kernel fits the parabola-shaped part of the distribution. Since its width parameter β_k is indepent of x, the state-space dependence of the respective cluster width cannot be predicted.

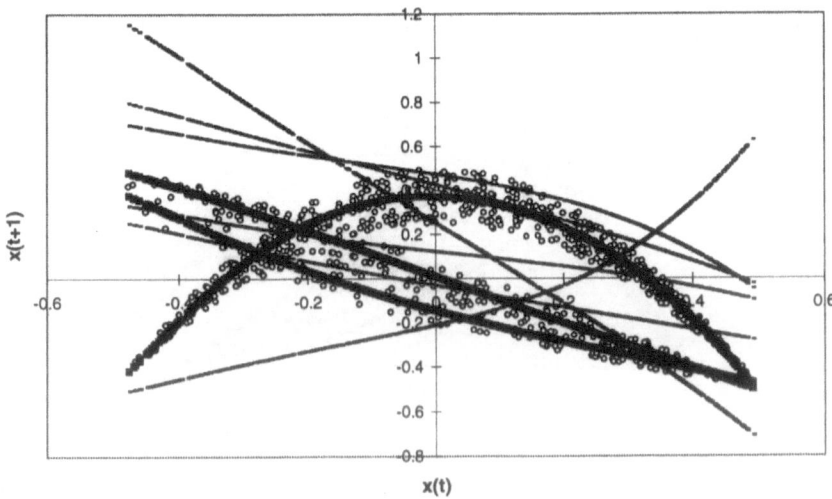

Fig. 7.3. State-space plot and kernel centres for the logistic-kappa map, training without retardation. The circles show a state-space plot of the time series generated from Equation (4.16). The lines show the kernel centres $\mu_k(x_t; \mathbf{w})$ implemented in a GM network after training with the EM algorithm, (7.23), (7.24), and (7.26). Only three kernels, indicated by the bold lines, cover relevant regions in state space. The narrow lines show the centres of kernels whose output weights had decayed to zero. Reprinted from [31], with permission from Elsevier Science.

the fast adaptation rule (7.23), kernels with poor mappings $\mu_k(x_t; \mathbf{w})$ are 'switched off' before they can be significantly improved. Moreover, the fast adaptation of the kernels widths (7.24) can result in kernels focusing on irrelevant parts of the state space, occluding the relevant region and thus leading to serious overfitting. The adaptation of the priors and kernel widths was therefore retarded by introducing a first-order relaxation scheme:

$$X_{new} = X_{old} + \frac{1}{\tau}(X_{EM} - X_{old}), \qquad (7.27)$$

where 'X' stands for a_k or β_k, respectively, 'old' for the old parameter values, 'EM' for the values resulting from (7.23) and (7.24), and X_{new} for the new values eventually adopted. The relaxation time τ determines the mean time it takes for the system to converge to the EM-values (note that for $\tau = 1$, (7.23) and (7.24) are regained). The simulations yielded a decrease in E_{gen} with increasing τ, and for $\tau = 100$ the same prediction performance of $E_{gen} = -1.02$ was obtained as in the earlier simulations of Section 5.3 (shown in Fig.7.2 and Fig.7.4). However, the training time had increased to over 6000

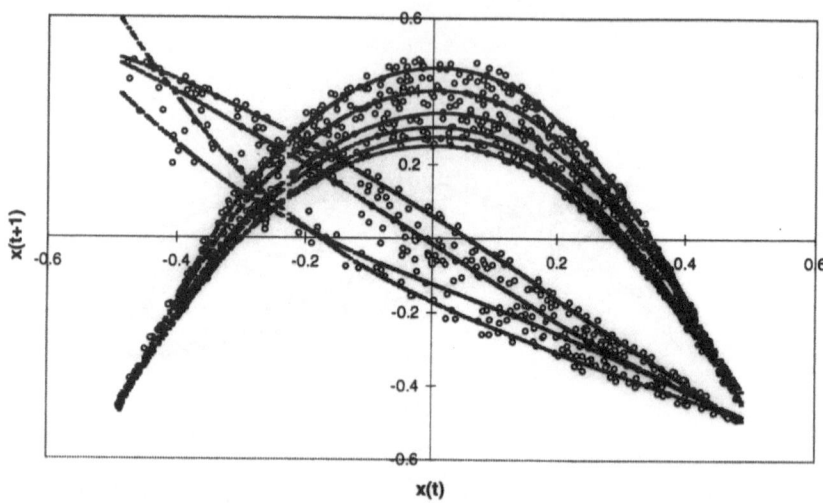

Fig. 7.4. State-space plot and kernel centres for the logistic-kappa map, training with retardation. Similar to Fig.7.3, the figure shows a state-space plot of the time series (4.16) and the kernel centres implemented in the network. Training was performed with the retarded EM algorithm, (7.23), (7.24), (7.26), and retardation scheme (7.27). As a consequence of the retardation, no prior decays to zero, allowing all kernels to fit some relevant region of the state space. Reprinted from [31], with permission from Elsevier Science.

epochs, hence no improvement over the maximum-likelihood scheme had been achieved.

7.2.3 Discussion

The simulations suggest that speed of convergence and prediction performance are strongly correlated, with a fast convergence making the algorithm susceptible to local minima and a degradation in the generalisation performance. The adoption of a retarding relaxation scheme could alleviate these problems, though at the price of increased training times. The reason for this impact on the network predictions by a speed-up of the training process can be understood when inspecting the learning rules (7.23), (7.24), and (7.26). Only a subset of the parameters, namely the output weights or prior probabilities $\{a_k\}$ and the kernel widths $\{\sigma_k = \beta_k^{-1/2}\}$, follow a complete minimisation step (M-step) of the EM-algorithm. The M-step for the remaining parameters, namely the weights $\mathbf{w}$ feeding into and out of the $\mathcal{S}$-layer, cannot be carried out. The reason is that the objective function U is a complex non-convex function of $\mathbf{w}$, so the best we can do is to decrease U by following the gradient rather than minimise U in one step. The training algorithm is therefore only of a GEM type (*Generalised Expectation Maximisation*, see [14]), which is known to converge more slowly than a proper EM scheme [34]. Moreover, it is the difference in convergence speed for the different parameter classes which caused the observed impact on the prediction performance and called for the retarding relaxation scheme.

7.3 Combining EM and RVFL

The bottleneck effect caused by the non-convexity of U in the parameters $\mathbf{w}$ motivated Neuneier et al. [45] and Ormoneit [46] to do away with these parameters altogether and develop the DPMN network, which was briefly discussed in Section 2.9.4. However, as pointed out earlier, this approach does not ensure universal approximation capability, and will therefore not be further considered here.

In order to carry out the adaptation of the weights $\mathbf{w}$ in a single step, like the adaptation of the β_k's and a_k's, the architecture of the GM network needs to be modified, though. The objective is to make $U(\mathbf{w})$ in (7.25) quadratic in $\mathbf{w}$. This will ensure that U is a *convex* function of $\mathbf{w}$, whose single (global) minimum can, in principle, easily be found by inverting the Hessian. In their 'hierarchical mixture of experts' (HME) approach, briefly summarised in Section 2.9.5, Jordan and Jacobs [34] introduced a hierarchical structure of linear networks, with $\mathbf{x}$-dependent priors $a_k = a_k(\mathbf{x}; \mathbf{v})$ ('gating network') softly switching between different input regimes. Obviously, this guarantees that U is quadratic in $\mathbf{w}$. However, a drawback of this approach is that due to the normalisation (2.48), U is not quadratic in $\mathbf{v}$, the parameters that determine the outputs of the gating network. These parameters hence still need to be adapted in an iterative way over several training passes, which slows down the overall training process.

The approach adopted in this book, therefore, is to keep the priors $\mathbf{x}$-independent as before, and draw on the random vector functional link (RVFL) approach, presented in the previous chapter, to ensure the convexity of U in all parameters. Recall from Chapter 6 that an RVFL network is a multilayer perceptron (MLP) where only the output weights are adjusted in the training process. The remaining weights are drawn from some random distribution and remain fixed during the training process. Since in this way the output of the network becomes a linear function of $\mathbf{w}$ (but not of $\mathbf{x}$!), the error function becomes quadratic in $\mathbf{w}$. Moreover, recall from Chapter 6 that an

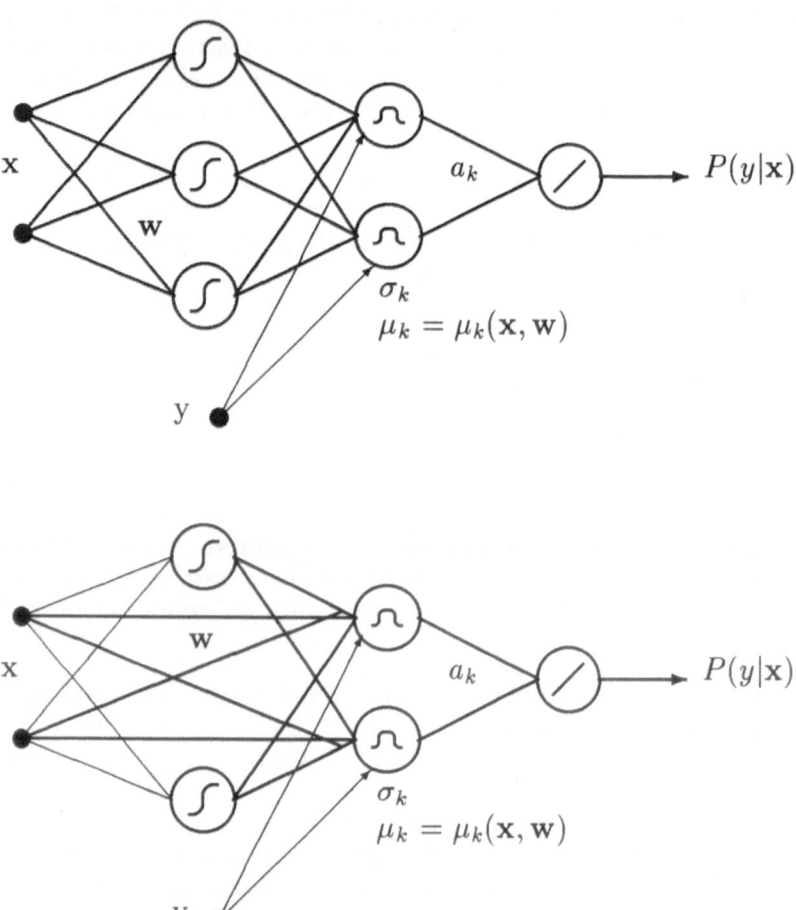

Fig. 7.5. GM and GM-RVFL network architectures. The top figure shows the architecture of a GM network, which contains sigmoidal units in the first hidden layer ($\mathcal{S}$-layer) and Gaussian units in the second hidden layer ($\mathcal{G}$-layer). The widths of the Gaussian bumps, σ_k, are treated as adaptable parameters, and the output weights a_k are positive and normalised. Arrows symbolise weights with constant value 1, and bold lines repesent wires with adaptable weights. All units are connected to a bias element of constant activity 1 (not shown). The GM-RVFL network, shown in the bottom figure, is obtained from the GM network by selecting the weights between the input and the $\mathcal{S}$-layer, shown by the narrow lines, randomly and independently in advance and keeping them fixed during training. The centres μ_k of the Gaussian kernels thus become linear in the weights, which allows application of the EM algorithm. Note that direct weights between the input and the $\mathcal{G}$-layer have been added in order to extract the linearly predictable part of the function to be learned, and free up the nonlinear resources, i.e. the random part of the network, to be employed only where really needed.

RVFL network has the same universal approximation capabilities as an MLP, and that the approximation error as a function of H, the number of hidden nodes, is of the same order as for an MLP. This encourages the following modification of the GM network, henceforth referred to as the GM-RVFL network and depicted in Fig.7.5. The weights feeding into the nodes of the $\mathcal{S}$-layer are constrained to random values, which are independently drawn from a normal distribution $N(0, \sigma_{rand})$. (The choice for σ_{rand} will be discussed shortly.) Additional direct connections between the input layer and the $\mathcal{G}$-layer can be introduced, in order to extract the linearly predictable part of the function to be learned and free up the nonlinear resources, i.e. the random part of the network, to be employed only where really needed. In this way most of the network parameters (output weights a_k, kernel widths σ_k, and all the weights feeding into the $\mathcal{G}$-nodes) are still adapted in the training process, thus giving a better ratio between the number of adjusted to the number of random parameters than in the original RVFL.

Let us denote the weights feeding into the kth $\mathcal{G}$-layer node by $\mathbf{w}_k \in \mathbb{R}^W$, where $W = H + m + 1$ or $W = H + 1$, depending on whether or not direct connections between the input and the $\mathcal{G}$-layer are employed (the '+1' stems from the bias). Let $\mathbf{g}(\mathbf{x}_t) \in \mathbb{R}^W$ contain the activities of all the nodes in the input (if direct weights are included), bias[4] and $\mathcal{S}$-layer upon presenting pattern $\mathbf{x}_t$. Then

$$\mu_k(\mathbf{x}_t; \mathbf{w}) = \mu(\mathbf{x}_t; \mathbf{w}_k) := \mathbf{w}_k^\dagger \mathbf{g}(\mathbf{x}_t), \qquad (7.28)$$

and U becomes quadratic in $\mathbf{w} = (\mathbf{w}_1^\dagger, \ldots, \mathbf{w}_K^\dagger)^\dagger \in \mathbb{R}^{WK}$. From (7.25) we obtain

$$-\nabla_{\mathbf{w}_k} U = \beta_k \sum_{t=1}^{N} \pi_k(t) \Big[y_t - \mu(\mathbf{x}_t; \mathbf{w}_k) \Big] \nabla_{\mathbf{w}_k} \mu(\mathbf{x}_t; \mathbf{w}_k)$$

$$= \beta_k \sum_{t=1}^{N} \pi_k(t) \Big[y_t - \mathbf{w}_k^\dagger \mathbf{g}(\mathbf{x}_t) \Big] \mathbf{g}(\mathbf{x}_t)$$

$$= \beta_k \left(\sum_{t=1}^{N} \pi_k(t) y_t \mathbf{g}(\mathbf{x}_t) - \sum_{t=1}^{N} \pi_k(t) \mathbf{g}(\mathbf{x}_t) [\mathbf{g}(\mathbf{x}_t)]^\dagger \mathbf{w}_k \right) \quad (7.29)$$

Introducing the definitions

$$\mathbf{y} := (y_1, \ldots, y_N)^\dagger \qquad (7.30)$$

$$\mathbf{G} := \Big(\mathbf{g}(\mathbf{x}_1), \ldots, \mathbf{g}(\mathbf{x}_N) \Big) \qquad \text{W by N matrix} \qquad (7.31)$$

$$\mathbf{\Pi}_k : \quad \text{diagonal N by N matrix with} \quad (\mathbf{\Pi}_k)_{tt'} := \pi_k(t)\delta_{tt'} \quad (7.32)$$

$$\mathbf{I} : \quad \text{unit matrix,} \quad \mathbf{I}_{ij} := \delta_{ij} \qquad (7.33)$$

---

[4] The simplest way to consider a bias is to introduce an additional layer with one unit of constant activity *one*, and connect it to all other units.

the expression for the gradient (7.29) can be rewritten as

$$\nabla_{\mathbf{w}_k} U \quad = \quad \beta_k \left[\left(\mathbf{G} \mathbf{\Pi}_k \mathbf{G}^\dagger \right) \mathbf{w}_k - \mathbf{G} \mathbf{\Pi}_k \mathbf{y} \right]. \qquad (7.34)$$

Setting the gradient to zero,

$$\nabla_{\mathbf{w}} U \quad = \quad (\nabla_{\mathbf{w}_1}^\dagger U, \ldots, \nabla_{\mathbf{w}_K}^\dagger U)^\dagger = 0 \qquad (7.35)$$

yields

$$\left(\mathbf{G} \mathbf{\Pi}_k \mathbf{G}^\dagger \right) \mathbf{w}_k \quad = \quad \mathbf{G} \mathbf{\Pi}_k \mathbf{y} \qquad (7.36)$$

so that by inverting the matrix on the left the M-step with respect to $\mathbf{w}$ can in principle be carried out. In practice, however, the matrix $\mathbf{G} \mathbf{\Pi}_k \mathbf{G}^\dagger$ may be *singular* or *ill-conditioned*[5]. In the first case, there is no unique solution to (7.36). In the second case, a unique solution exists, but it will most likely be found to be an extremely poor approximation to the correct value due to considerable numerical corruptions caused by roundoff errors. These scenarios need therefore to be treated with special care, as will be discussed in the following section.

7.4 Preventing numerical instability

For the following discussion, some well-known facts from linear algebra are briefly reviewed. Let $\mathbf{H}$ be a symmetric, positive semi-definite n by n matrix with eigenvectors $\mathbf{u}_i$ and eigenvalues ε_i,

$$\mathbf{H}\mathbf{u}_i \quad = \quad \varepsilon_i \mathbf{u}_i, \qquad (7.37)$$

let $\mathbf{U}$ be an n by n matrix composed of the eigenvectors,

$$\mathbf{U} \quad := \quad (\mathbf{u}_1, \ldots, \mathbf{u}_n) \qquad (7.38)$$

and let $\mathbf{D}$ be a diagonal n by n matrix containing the eigenvalues in the diagonal,

$$\mathbf{D} \quad := \quad \begin{pmatrix} \varepsilon_1 & & 0 \\ & \ddots & \\ 0 & & \varepsilon_n \end{pmatrix} \qquad (7.39)$$

Then

---

[5] A matrix is *ill-conditioned* if its eigenvalues vary over several orders of magnitude such that the ratio $\varepsilon_{min}/\varepsilon_{max}$ between its smallest and largest eigenvalues is smaller than the floating point precision of the computer. Ill-conditioning usually arises when the matrix is analytically singular, but zero eigenvalues actually differ from zero due to numerical roundoff errors.

$$\mathbf{HU} = (\mathbf{Hu}_1, \ldots, \mathbf{Hu}_n) = (\varepsilon_1 \mathbf{u}_1, \ldots, \varepsilon_n \mathbf{u}_n)$$

$$= (\mathbf{u}_1, \ldots, \mathbf{u}_n) \begin{pmatrix} \varepsilon_1 & & 0 \\ & \ddots & \\ 0 & & \varepsilon_n \end{pmatrix} = \mathbf{UD} \qquad (7.40)$$

and

$$\mathbf{U}^{-1}\mathbf{HU} = \mathbf{U}^{-1}\mathbf{UD} = \mathbf{D}. \qquad (7.41)$$

Taking the transpose on both sides of the above equation and considering that $\mathbf{H}^\dagger = \mathbf{H}$ (because $\mathbf{H}$ is symmetric) and $\mathbf{D}^\dagger = \mathbf{D}$ (because $\mathbf{D}$ is diagonal) gives

$$\mathbf{U}^\dagger \mathbf{H}^\dagger (\mathbf{U}^{-1})^\dagger = \mathbf{U}^\dagger \mathbf{H}(\mathbf{U}^{-1})^\dagger = \mathbf{D}^\dagger = \mathbf{D} = \mathbf{U}^{-1}\mathbf{HU} \qquad (7.42)$$

and consequently

$$\mathbf{U}^\dagger = \mathbf{U}^{-1}. \qquad (7.43)$$

Thus $\mathbf{U}$ is orthogonal, with

$$\mathbf{UU}^\dagger = \mathbf{U}^\dagger \mathbf{U} = \mathbf{I}, \qquad \mathbf{u}_i^\dagger \mathbf{u}_j = \delta_{ij}. \qquad (7.44)$$

We can therefore write, using (7.41) and (7.44),

$$\mathbf{H} = \mathbf{UDU}^\dagger \qquad (7.45)$$
$$\mathbf{D} = \mathbf{U}^\dagger \mathbf{HU}. \qquad (7.46)$$

Moreover, if the inverse of $\mathbf{H}$ exists, then

$$\mathbf{H}^{-1} = \mathbf{UD}^{-1}\mathbf{U}^\dagger. \qquad (7.47)$$

The decomposition (7.45) can in practice be carried out with the method of *singular value decomposition* (SVD). A powerful subroutine implementing this algorithm, which has been used in the simulations reported in this book, is listed in [51]. Let us now introduce the following definitions:

$$\mathcal{R}(\mathbf{H}) := \{\mathbf{u}_i | \varepsilon_i \neq 0\} \qquad (7.48)$$
$$\mathcal{N}(\mathbf{H}) := \{\mathbf{u}_i | \varepsilon_i = 0\} \qquad (7.49)$$
$$\mathrm{Range}(\mathbf{H}) := \langle \mathcal{R}(\mathbf{H}) \rangle \qquad (7.50)$$
$$\mathrm{Nullspace}(\mathbf{H}) := \langle \mathcal{N}(\mathbf{H}) \rangle \qquad (7.51)$$

where $\langle \ldots \rangle$ denotes the linear hull of the argument. Obviously, the following relations hold: (i) Range($\mathbf{H}$) and Nullspace($\mathbf{H}$) are orthogonal, (ii) dimRange($\mathbf{H}$)+ dimNullspace($\mathbf{H}$) $= n$, (iii) $\mathbf{Hw} = 0$ iff $\mathbf{w} \in$ Nullspace($\mathbf{H}$). Now consider the optimisation problem

$$\mathbf{Hw} = \mathbf{b} \qquad (7.52)$$

which is of the form of Equation (7.36) if we define

$$\mathbf{H} := \mathbf{G}\Pi_k\mathbf{G}^\dagger, \qquad \mathbf{b} := \mathbf{G}\Pi_k\mathbf{y}, \qquad \mathbf{w}_k := \mathbf{w}. \qquad (7.53)$$

For $\mathbf{b} \neq 0$, a solution exists if and only if $\mathbf{b} \in \text{Range}(\mathbf{H})$[6]. This can be assumed to hold here: since for the GM-RVFL network $U(\mathbf{w})$ of Equation (7.17) is quadratic in $\mathbf{w}$ with a positive semi-definite Hessian, it must have a minimum, that is a solution to (7.36) must exist. However, for singular $\mathbf{H}$ this solution is not unique, since any vector in Nullspace$(\mathbf{H})$ can be added to $\mathbf{w}$ in any linear combination. In order to remove this ambiguity, one usually considers the modified equation

$$\lim_{\lambda \to 0}(\mathbf{H} + \lambda\mathbf{I})\mathbf{w} = \mathbf{b}, \qquad (7.54)$$

where λ is a small positive constant and $\mathbf{I}$ the n by n unit matrix. The additional term $\lambda\mathbf{I}$ can be envisaged as arising from an additional 'penalty term' $\frac{\lambda}{2}\mathbf{w}^\dagger\mathbf{w}$ added to the standard error function, which favours, among the set of all degenerate solutions to (7.52), the one with the smallest modulus. Let us define the modified diagonal matrix $\tilde{\mathbf{D}}$ with

$$[\tilde{\mathbf{D}}]_{ii} = \begin{cases} \varepsilon_i & \text{if} \quad \varepsilon_i \neq 0 \\ \lambda & \text{if} \quad \varepsilon_i = 0 \end{cases} \qquad (7.55)$$

Note that $\tilde{\mathbf{D}}$ is invertible, with

$$[\mathbf{D}\tilde{\mathbf{D}}^{-1}]_{ii} = \begin{cases} \varepsilon_i\varepsilon_i^{-1} = 1 & \text{if} \quad \varepsilon_i \neq 0 \\ 0\lambda^{-1} = 0 & \text{if} \quad \varepsilon_i = 0 \end{cases} \qquad (7.56)$$

Thus $\mathbf{U}\mathbf{D}\tilde{\mathbf{D}}^{-1}\mathbf{U}^\dagger$ is a projector onto Range$(\mathbf{H})$,

$$\mathbf{U}\mathbf{D}\tilde{\mathbf{D}}^{-1}\mathbf{U}^\dagger\mathbf{w} = \sum_{i|\varepsilon_i \neq 0} \mathbf{u}_i\mathbf{u}_i^\dagger\mathbf{w} \in \text{Range}(\mathbf{H}). \qquad (7.57)$$

From (7.44) and (7.45) we obtain

$$\mathbf{H} + \lambda\mathbf{I} = \mathbf{U}(\mathbf{D} + \lambda\mathbf{I})\mathbf{U}^\dagger = \mathbf{U} \begin{pmatrix} \varepsilon_1 + \lambda & & 0 \\ & \ddots & \\ 0 & & \varepsilon_n + \lambda \end{pmatrix} \mathbf{U}^\dagger \quad (7.58)$$

$$\Rightarrow (\mathbf{H} + \lambda\mathbf{I})^{-1} = \mathbf{U} \begin{pmatrix} \frac{1}{\varepsilon_1 + \lambda} & & 0 \\ & \ddots & \\ 0 & & \frac{1}{\varepsilon_n + \lambda} \end{pmatrix} \mathbf{U}^\dagger \qquad (7.59)$$

---

[6] This can be shown by *reductio ad absurdum*. Assume that $\exists \mathbf{w}$ such that $\mathbf{H}\mathbf{w} = \mathbf{b}$ for $\mathbf{b} \notin \text{Range}(\mathbf{H})$. The latter implies that $\mathbf{b}$ has a non-zero component in Nullspace$(\mathbf{H})$, call it $\mathbf{b}_0$, so $\mathbf{b} = \mathbf{b}_0 + \mathbf{b}_1$, where $\mathbf{b}_1 \in \text{Range}(\mathbf{H})$ and $||\mathbf{b}_0|| > 0$. Consequently, $\mathbf{H}\mathbf{w} = \mathbf{b}_0 + \mathbf{b}_1$, which when multiplied by $\mathbf{b}_0^\dagger$ from the left gives $\mathbf{b}_0^\dagger\mathbf{H}\mathbf{w} = \mathbf{b}_0^\dagger\mathbf{b}_0 > 0$. However, since $\mathbf{b}_0 \in \text{Nullspace}(\mathbf{H})$, $\mathbf{b}_0^\dagger\mathbf{H} = 0$, which leads to $0 > 0$. Obviously, this is wrong, hence $\mathbf{b} \in \text{Range}(\mathbf{H})$ must hold.

For nonzero ε_i the constant λ can be neglected due to the limit $\lambda \to 0$. With (7.55) this gives

$$(\mathbf{H} + \lambda\mathbf{I})^{-1} = \mathbf{U}\tilde{\mathbf{D}}^{-1}\mathbf{U}^\dagger. \tag{7.60}$$

Thus $\mathbf{w}_0 := \mathbf{U}\tilde{\mathbf{D}}^{-1}\mathbf{U}^\dagger\mathbf{b}$ solves (7.54) and is therefore, for $\mathbf{b} \in \text{Range}(\mathbf{H})$, the solution to (7.52) with the smallest modulus. In fact, the above statement can be generalised to any arbitrary λ, although for finite λ this does not allow an interpretation in terms of (7.54) any more (since in this case the step from (7.59) to (7.60) is not valid).

THEOREM
Let $\mathbf{b} \in \text{Range}(\mathbf{H})$, and let $\mathbf{U}$ and $\tilde{\mathbf{D}}$ be n by n matrices defined in (7.38) and (7.55), respectively. Then for any nonzero value of λ, $\mathbf{w}_0 := (\mathbf{U}\tilde{\mathbf{D}}^{-1}\mathbf{U}^\dagger)\mathbf{b}$ is the solution to (7.52) with the minimum modulus.

PROOF
Inserting the expression for $\mathbf{w}_0$ into (7.52) and making use of $\mathbf{H} = \mathbf{U}\mathbf{D}\mathbf{U}^\dagger$ (7.45), the fact that $\mathbf{U}$ is orthogonal (7.44), and equation (7.57) gives

$$\begin{aligned}
\mathbf{H}\mathbf{w}_0 &= \mathbf{H}(\mathbf{U}\tilde{\mathbf{D}}^{-1}\mathbf{U}^\dagger)\mathbf{b} = \mathbf{U}\mathbf{D}\mathbf{U}^\dagger(\mathbf{U}\tilde{\mathbf{D}}^{-1}\mathbf{U}^\dagger)\mathbf{b} \\
&= \mathbf{U}\mathbf{D}\tilde{\mathbf{D}}^{-1}\mathbf{U}^\dagger\mathbf{b} = \sum_{i|\varepsilon_i\neq 0} \mathbf{u}_i\mathbf{u}_i^\dagger\mathbf{b}
\end{aligned}$$

Since $\mathbf{b} \in \text{Range}(\mathbf{H})$, any projection into Nullspace$(\mathbf{H})$ is zero, $\sum_{i|\varepsilon_i=0} \mathbf{u}_i\mathbf{u}_i^\dagger\mathbf{b} = 0$. This sum can therefore be added to the expression on the right of the above equation, leading to

$$\mathbf{H}\mathbf{w}_0 = \sum_i \mathbf{u}_i\mathbf{u}_i^\dagger\mathbf{b} = \mathbf{U}\mathbf{U}^\dagger\mathbf{b} = \mathbf{b},$$

where the last step follows from the fact that $\mathbf{U}$ is orthogonal. This proves that $\mathbf{w}_0$ is a solution to (7.52). In order to show that it is the solution with the smallest modulus, consider an alternative solution $\mathbf{w} = \mathbf{w}_0 + \mathbf{v}$, where $\mathbf{v} \in \text{Nullspace}(\mathbf{H})$. Now

$$\begin{aligned}
|\mathbf{w}| &= |\mathbf{w}_0 + \mathbf{v}| = |\mathbf{U}\tilde{\mathbf{D}}^{-1}\mathbf{U}^\dagger\mathbf{b} + \mathbf{v}| \\
&= |\mathbf{U}\tilde{\mathbf{D}}^{-1}\mathbf{U}^\dagger\mathbf{b} + \mathbf{U}\mathbf{U}^\dagger\mathbf{v}| = |\tilde{\mathbf{D}}^{-1}\mathbf{U}^\dagger\mathbf{b} + \mathbf{U}^\dagger\mathbf{v}|,
\end{aligned}$$

where the last step follows from the unitarity of the matrix $\mathbf{U}$. Since $\mathbf{v} \in$ Nullspace$(\mathbf{H})$, $\mathbf{U}^\dagger\mathbf{v} = (\mathbf{u}_1, \ldots, \mathbf{u}_n)^\dagger\mathbf{v}$ has nonzero component only for those i with $\varepsilon_i = 0$, whereas $\mathbf{U}^\dagger\mathbf{b}$ has nonzero components only for i with $\varepsilon_i \neq 0$ (since $\mathbf{b} \in \text{Range}(\mathbf{H})$). Consequently, the two vectors on the right-hand side of the above equation are orthogonal and

$$|\mathbf{w}| = |\tilde{\mathbf{D}}^{-1}\mathbf{U}^\dagger\mathbf{b} + \mathbf{U}^\dagger\mathbf{v}| = \sqrt{|\tilde{\mathbf{D}}^{-1}\mathbf{U}^\dagger\mathbf{b}|^2 + |\mathbf{U}^\dagger\mathbf{v}|^2}$$
$$\geq |\tilde{\mathbf{D}}^{-1}\mathbf{U}^\dagger\mathbf{b}| = |\mathbf{U}\tilde{\mathbf{D}}^{-1}\mathbf{U}^\dagger\mathbf{b}| = \mathbf{w}_0$$

◇

The significance of this result becomes clear when a numerical implementation of the algorithm is performed. When calculating $\mathbf{w}_0 := (\mathbf{U}\tilde{\mathbf{D}}^{-1}\mathbf{U}^\dagger)\mathbf{b}$, the components of $\mathbf{U}^\dagger\mathbf{b}$ corresponding to $\mathbf{u}_i \in$ Nullspace($\mathbf{H}$) are analytically zero, since $\mathbf{b} \in$ Range($\mathbf{H}$): $\mathbf{u}_i^\dagger\mathbf{b} = 0$ if $\varepsilon_i = 0$. Numerically, however, this will normally not be the case due to roundoff errors in the singular value decomposition of the matrix $\mathbf{H}$. The component $\mathbf{u}_i^\dagger\mathbf{b}$ can therefore be expected to be small but nonzero. Since the corresponding element in $\tilde{\mathbf{D}}^{-1}$ is $[\tilde{\mathbf{D}}^{-1}]_{ii} = \frac{1}{\lambda}$, the results of the product $[\tilde{\mathbf{D}}^{-1}]_{ii}\mathbf{u}_i^\dagger\mathbf{b}$ can become large when $\lambda \to 0$. The vector $\tilde{\mathbf{D}}^{-1}\mathbf{U}^\dagger\mathbf{b}$ can therefore have components in Nullspace($\mathbf{H}$) that are orders of magnitude larger than those in Range($\mathbf{H}$). The following multiplication from the left with $\mathbf{U}$ leads to a mixing of these components and further compounds the roundoff problem. Consequently, the calculated result for $\mathbf{w}_0$ is most likely to be so corrupted by numerical roundoff errors that it will only be a poor approximation to the correct value. However, the above scenario can simply be staved off by setting $\lambda = \infty$. In this case the diagonal elements in $\tilde{\mathbf{D}}^{-1}$ corresponding to $\varepsilon_i = 0$ are zero, $[\tilde{\mathbf{D}}^{-1}]_{ii} = 0$, and the troublesome nullspace-components of $\mathbf{U}^\dagger\mathbf{b}$ are exactly switched off.

Now it is not only the eigenvectors $\mathbf{u}_i$ that are prone to roundoff errors in the singular value decomposition $\mathbf{H}$: the same holds, of course, for the eigenvalues ε_i. This implies that although $\mathbf{H}$ may be analytically singular, numerically this is in general not the case and eigenvalues that are analytically zero will most likely be small, but nonzero. If this is not dealt with in some way, one will be faced with the same scenario as described above: a small, but nonzero nullspace component $\mathbf{u}_i^\dagger\mathbf{b}$ will be divided by a small, but nonzero eigenvalue ε_i. This quotient of two numerical zeroes can lead again to large nullspace components, which can cause considerable numerical instabilities. Consequently, it is best to identify small eigenvalues of $\mathbf{H}$ as numerical zeroes and set

$$[\tilde{\mathbf{D}}]_{ii} = \begin{cases} \varepsilon_i & \text{if } \varepsilon_i > \theta \\ \infty & \text{if } \varepsilon_i \leq \theta \end{cases} \tag{7.61}$$

where θ is some appropriate constant. This is the reason for the at first sight paradoxical suggestion to replace small eigenvalues $[\mathbf{D}]_{ii} = \varepsilon_i$ by infinity. Since the corresponding elements in the inverse $\mathbf{D}^{-1}$ are set to zero and their corresponding contributions therefore 'cut off',

$$[\tilde{\mathbf{D}}^{-1}]_{ii} = \begin{cases} \frac{1}{\varepsilon_i} & \text{if } \varepsilon_i > \theta \\ 0 & \text{if } \varepsilon_i \leq \theta \end{cases} \tag{7.62}$$

this method will be referred to as *eigenvalue cutoff* henceforth.

The question now is how to choose the constant θ. The best way would be to inspect the spectrum of eigenvalues explicitly and then decide on which values are likely to be numerical zeroes and therefore to be cut off. A simpler way, suggested in [51], is to consider the ratio $\varepsilon_i/\varepsilon_{max}$, where $\varepsilon_{max} = \max\{\varepsilon_i\}$, and discard all contributions with a ratio smaller than the machine's floating point precision. For single-value precision, a value of 10^{-6} is recommended in [51], that is

$$[\tilde{\mathbf{D}}^{-1}]_{ii} = \begin{cases} \frac{1}{\varepsilon_i} & \text{if} \quad \frac{\varepsilon_i}{\varepsilon_{max}} > 10^{-6} \\ 0 & \text{if} \quad \frac{\varepsilon_i}{\varepsilon_{max}} \leq 10^{-6} \end{cases} \tag{7.63}$$

7.5 Regularisation

Consider again (7.54), which always has a unique solution. However, as discussed on page 116, the choice of $\lambda \to 0$ does not help us to solve the numerical problem mentioned above: a nullspace-component $\mathbf{u}_i^\dagger \mathbf{b}$, which due to roundoff errors deviates from zero, is divided by λ and gives rise to a large contribution. This problem can be averted by choosing a larger value for λ, leading to

$$(\mathbf{H} + \lambda \mathbf{I})\mathbf{w} \quad = \quad \mathbf{b}, \qquad \lambda > 0. \tag{7.64}$$

However, the step from (7.59) to (7.60) cannot be done in this case, and the theorem of page 115 does not apply. In fact, the solution to (7.64) deviates from the one to (7.52) in that the resulting $\mathbf{w}$ is systematically shifted towards 0. The reason is that (7.64) results from adding a penalty term of the form $\frac{\lambda}{2}\mathbf{w}^\dagger\mathbf{w}$ to the error function, which penalises large vector lengths. Such a regularisation term, equivalent to a weight decay term in backpropagation-like training schemes, is usually applied to prevent overfitting and over-complexity of the model[7]. This can be understood from the effective number of model parameters γ, which, as will be discussed in Chapter 10 (see also [37]), is given by the eigenvalues ε_i of the Hessian,

$$\gamma \quad = \quad \sum_i \frac{\varepsilon_i}{\varepsilon_i + \lambda}, \tag{7.65}$$

and which decreases as λ increases. Interestingly, the scheme given by Equation (7.62) similarly reduces the number of adaptable network parameters, and the application of larger values for θ - beyond what is required for numerical stabilisation - can be regarded as an alternative regularisation method for reducing over-complexity. A comparison of these schemes, henceforth referred to as *diagonal incrementation* and *eigenvalue cutoff*, can be found in Chapter 14.

---

[7] This method is known as *ridge regression* in statistics.

7.6 Summary

This chapter has reviewed the expectation maximisation (EM) algorithm. It has been demonstrated that the direct application of this scheme to the proposed network architecture is not immediately possible since one class of parameters is refractory to the M-step of the algorithm. This limitation can be overcome by imposing constraints on a subset of the network parameters such that the EM error surface becomes convex in the space of the unconstrained (adaptable) parameters. The constrained parameters are selected randomly and independently in advance and are then frozen, that is, not further updated during the training process. This scheme is thus equivalent to the RVFL concept and should, as discussed in Chapter 6, not cause any major limitations on the universal approximation capability of the model.

An empirical corroboration of this concept will be given in the next chapter.

7.7 Appendix

In order to prove the relation

$$\sum_{\Lambda} P(\mathbf{D}, \Lambda | \mathbf{q}) \quad = \quad P(\mathbf{D} | \mathbf{q}) \tag{7.66}$$

let us introduce the definition $z_k(t) := P(k)P(y_t | \mathbf{x}_t, \mathbf{q}, k)$. Then from (7.13) we have

$$P(\mathbf{D}, \Lambda | \mathbf{q}) \quad := \quad \prod_{t=1}^{N} \prod_{k=1}^{K} [z_k(t)]^{\lambda_k(t)} \tag{7.67}$$

and (2.30) and (3.14) lead to

$$P(\mathbf{D} | \mathbf{q}) \quad = \quad \prod_{t=1}^{N} \sum_{k=1}^{K} z_k(t). \tag{7.68}$$

The relation to be proved thus reads

$$\sum_{\Lambda} \prod_{t=1}^{N} \prod_{k=1}^{K} [z_k(t)]^{\lambda_k(t)} \quad = \quad \prod_{t=1}^{N} \sum_{k=1}^{K} z_k(t). \tag{7.69}$$

Recall that at any particular time t one and only one of the $\lambda_k(t)$ is different from zero, so we can write $\lambda_k(t) = \delta_{k,k_t}$ with $k_t \in \{1, \ldots, K\}$. Thus

$$\sum_{\Lambda} \prod_{t=1}^{N} \prod_{k=1}^{K} [z_k(t)]^{\lambda_k(t)} = \sum_{k_1=1}^{K} \cdots \sum_{k_t=1}^{K} \cdots \sum_{k_N=1}^{K} \prod_{t=1}^{N} \prod_{k=1}^{K} [z_k(t)]^{\delta_{k,k_t}}$$

$$= \sum_{k_1=1}^{K} \cdots \sum_{k_t=1}^{K} \cdots \sum_{k_N=1}^{K} \prod_{t=1}^{N} z_{k_t}(t). \qquad (7.70)$$

The proof is now given by induction. For $N = 1$ Equation (7.69) is obviously regained. Now assume that (7.69) holds for $N - 1$. Then

$$\sum_{\Lambda} \prod_{t=1}^{N} \prod_{k=1}^{K} [z_k(t)]^{\lambda_k(t)} = \sum_{k_1=1}^{K} \cdots \sum_{k_t=1}^{K} \cdots \sum_{k_N=1}^{K} \prod_{t=1}^{N} z_{k_t}(t)$$

$$= \sum_{k_1=1}^{K} \cdots \sum_{k_t=1}^{K} \cdots \sum_{k_N=1}^{K} \left(\prod_{t=1}^{N-1} z_{k_t}(t) \right) z_{k_N}(N)$$

$$= \left(\sum_{k_1=1}^{K} \cdots \sum_{k_t=1}^{K} \cdots \sum_{k_{N-1}=1}^{K} \prod_{t=1}^{N-1} z_{k_t}(t) \right) \sum_{k_N=1}^{K} z_{k_N}(N)$$

$$= \left(\prod_{t=1}^{N-1} \sum_{k_t=1}^{K} z_{k_t}(t) \right) \sum_{k_N=1}^{K} z_{k_N}(N) = \prod_{t=1}^{N} \sum_{k_t=1}^{K} z_{k_t}(t)$$

where in the last line (7.70) and the fact that (7.69) holds for $N - 1$ have been used. Q.e.d.

8. Empirical Demonstration: Combining EM and RVFL

The GM-RVFL network and the EM training scheme derived in the previous chapter are applied to the stochastic time series of sections 4.2 and 4.3. The prediction performance is found to depend critically on the distribution width for the random weights, with too small a value making the network incapable of learning the non-linearities of the time series, and too large a value degrading the performance due to excessive non-linearity and overfitting. However, the training process is accelerated by about two orders of magnitude, which allows the training of a whole ensemble of networks at the same computational costs as required otherwise for training a single model. In this way, 'good' values for the distribution width can easily be obtained by a discrete random search. Combining the best models in a committee leads to an improvement of the generalisation performance over that obtained with an individual fully-adaptable model. For the double-well time series of Section 4.3, a committee of GM-RVFL networks is found to outperform all alternative models otherwise applied to this problem.

The results presented in this chapter were published, in part, in [31] and [28].

8.1 Method

The objective of the study reported in this chapter was to apply the GM-RVFL network to the benchmark problems of Chapter 4 and to compare the required training times and the achieved prediction performance with those obtained in the earlier studies of Chapter 5 and Section 7.2. Training was performed with the EM algorithm, that is, the priors a_k were adapted according to (7.23), the inverse kernel widths β_k followed (7.24), and the remaining weights $\mathbf{w}$ were updated by (7.36). Obviously, application of (7.24) runs into trouble when a prior a_k has decayed to a very small value. In this case the sum over the posteriors, $\sum_{t=1}^{N} \pi_k(t)$, that is, the denominator in (7.24), will most likely also be very small, and the update scheme for the kernel widths becomes unstable. For this reason, a pruning scheme was applied, which switched a network branch off when a_k had fallen below a certain threshold:

$$a_k < \epsilon \quad \Rightarrow \quad a_k := 0. \tag{8.1}$$

In the simulations, a value of $\epsilon = 10^{-6}$ was used.

For adapting the remaining weights $\mathbf{w}$, it is clear that a direct application of (7.36) is, in general, not possible since the matrix on the left of (7.36) is likely to be singular or ill-conditioned. For this reason, the method of eigenvalue cutoff (7.63), explained in Section 7.4, was applied. Since the invocation of this scheme requires explicit knowledge of the eigenvalues, the matrix on the left of (7.36) was decomposed with *singular value decomposition* ([51], Section 2.6). The practical computation was carried out with the subroutine listed in [51], pp.67-70. All the network architectures employed for this study contained direct weights between the input and the $\mathcal{G}$-layer. For certain problems these direct weights were found to lead to a significant improvement of the network performance, as seen, for instance, from Fig. 16.1. For the time series generated from the logistic-kappa map, (4.16), however, no improvement was observed, and in later simulations, reported in Chapter 12 and Chapter 14, these direct connections were omitted.

It may be recalled that the following definitions for the training, cross-validation and generalisation 'error' are used:

$$E_{train} \quad := \quad -\frac{1}{N_{train}} \ln P(\mathbf{D}_{train}|\mathbf{q}) \tag{8.2}$$

$$E_{cross} \quad := \quad -\frac{1}{N_{cross}} \ln P(\mathbf{D}_{cross}|\mathbf{q}) \tag{8.3}$$

$$E_{gen} \quad := \quad -\frac{1}{N_{gen}} \ln P(\mathbf{D}_{gen}|\mathbf{q}) \tag{8.4}$$

in which convention (3.14) is applied.

8.2 Application of the GM-RVFL network to predicting the stochastic logistic-kappa map

8.2.1 Training a single model

As a first test, the GM-RVFL network was applied to the time series generated from the stochastic logistic-kappa map (4.16), described in Section 4.2.

Problem	Network	Training Method	Iterations	CPU Time
Section 4.2	GM	GEM gradient descent	5000 – 7000	6 – 8 h
Section 4.2	GM	GEM conjugate gradients	1000 – 1500	$2\frac{1}{2} - 3$ h
Section 4.2	GM-RVFL	EM	30-50	3 – 5 min
Section 4.3	GM	GEM gradient descent	2000	48 h
Section 4.3	GM-RVFL	EM	10	50 min

Table 8.1. Comparison of typical training times. The table shows typical training times for training a single network. 'Iterations' refers to the number of adaptation steps for the respective algorithm. The GM-RVFL network was trained with the EM algorithm, (7.23), (7.24), (7.36), where the update equation for the weights $\mathbf{w}$, (7.36), was solved with singular value decomposition. The GM network was trained according to the GEM scheme (7.23), (7.24), (7.26). Here, the adaptation of $\mathbf{w}$ either followed a single gradient descent step, with an appropriate retardation (7.27) for the adaptation of the a_k and β_k, or a limited number (≤ 50) of conjugate gradient steps (each line search counting as one iteration). 'CPU time' refers to simulations carried out on a 75 MHz Pentium PC. The network architectures were as follows: First problem: $m = 1$ input, $H = 10$ $\mathcal{S}$-nodes, $K = 10$ $\mathcal{G}$-nodes. Second problem: $m = 5$ inputs, $H = 15$ $\mathcal{S}$-nodes, $K = 9$ $\mathcal{G}$-nodes.

The architecture of the network was the same as in the previous studies (Section 5.3 and Section 7.2), with $m = 1$ input node, $H = 10$ nodes in the $\mathcal{S}$-layer, and $K = 10$ units in the $\mathcal{G}$-layer. The initialisation of the network parameters followed Section 5.2

Figure 8.3 shows the state-space plot of the kernel centres $\mu(x_t; \mathbf{w}_k)$ after a successful training simulation. A comparison with Fig.5.7 and Fig.7.4 suggests that indeed the same prediction performance results as for the DSM and the GM network. Yet this has been achieved after less than 50 EM-steps, corresponding to a reduction in CPU-time by two orders of magnitude to about 2% of the time required for training the GM network, as seen from Tab.8.1. A considerable acceleration of the training process can thus be achieved.

The bottleneck, as one might expect, is the dependence of the prediction performance on an appropriate choice of the random weights, as mentioned earlier, on page 93. The simulations revealed that an $\mathcal{S}$-layer of $H = 10$ units is sufficiently large for the particular choice of the random values (depending on the choice of the random number generator seed) to have only a minor influence on the outcome of the training process. It is the distribution width σ_{rand}, however, that does have an important influence and determines the generalisation performance of the model. Recall that the nonlinear complexity of a function implemented in a network depends on the size of the weights feeding into the sigmoidal units. Too small a value for σ_{rand} leads, generi-

Network	Training method	Prediction	E_{gen}
GM	GEM (7.23), (7.24), (7.26)	single predictor (best result obtained)	-0.91
GM	GEM, retardation, $\tau = 100$ (7.23), (7.24), (7.26), (7.27)	single predictor (best result obtained)	-1.020
GM-RVFL	EM (7.23), (7.24), (7.36)	single predictor (average over committee)	-1.035
GM-RVFL	EM (7.23), (7.24), (7.36)	committee prediction	-1.047

Table 8.2. Comparison between the GM and the GM-RVFL model. The table compares the generalisation performance between the GM-RVFL and the GM networks on the first time series prediction problem (Section 4.16). The last column shows the value for the generalisation 'error' E_{gen}. Smaller values indicate a better generalisation performance.

cally, to a restriction to the linear range of the transfer function, and the first hidden layer basically performs just a *linear* transformation of the input data. Consequently, as shown in Fig.8.1, left graph, and Fig.8.2, the nonlinearities of the underlying processes cannot be learned. Conversely, if σ_{rand} is too large and most of the inputs to the hidden units are in the saturation range, the model becomes *too nonlinear*. The consequence is overfitting and a serious impact on the generalisation performance. This is shown in Fig.8.1, right graph, and Fig.8.4, which also give a cautioning example that *cross-validation* and *early stopping* alone cannot remedy this defect. Although the graph on the right of Fig.8.1 suggests that overfitting is clearly indicated, the generalisation performance in the early-stopping point itself is sub-optimal.

In order to understand the failure of *early stopping* as a guard against overfitting, let us briefly recapitulate its basic principle in standard network training. Starting from a configuration of small randomly chosen initial weights, a network at the early stage of a training process is basically a *linear* model. As training continues and the weight values increase, the network becomes increasingly more *nonlinear*. Training is stopped when the nonlinear complexity reaches a level where the network starts to overfit. Now recall that the degree of nonlinearity of a network depends basically on the weights *feeding into* the sigmoidal, i.e. first-hidden-layer neurons. In the GM-RVFL, these weights are constrained to their initial values. This implies that the nonlinear complexity is an *a priori* choice rather than an entity that evolves dynamically during training. Consequently, the efficiency of *early stopping* is dramatically reduced.

However, this deficiency can be made up for by the fact that we gain two orders of magnitude in training time, which will allow us to train a whole ensemble of networks with different a priori choices for σ_{rand}. The GM-RVFL approach therefore offers a principled alternative to conventional network training. The latter focuses on optimising the training process for a

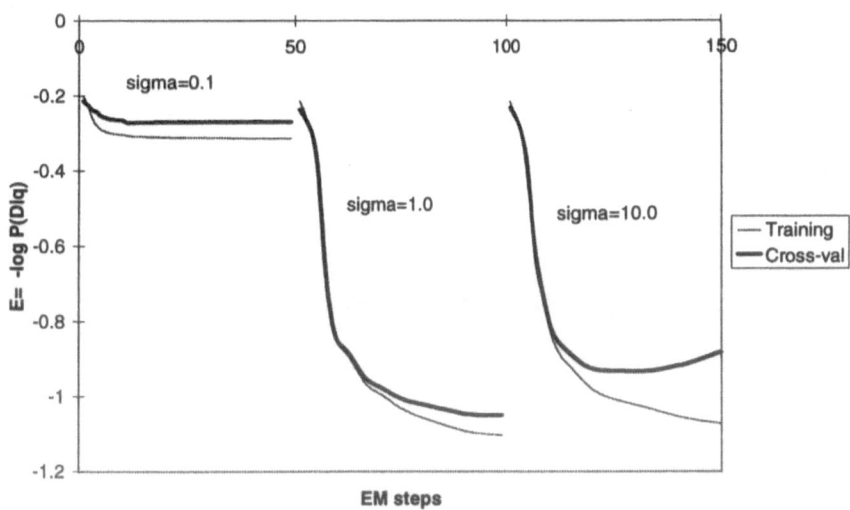

Fig. 8.1. Training and cross-validation error for the GM-RVFL network.
The graphs show the training (narrow line) and cross-validation (bold line) 'error'
for a GM-RVFL network with three different values for σ_{rand}. Each training simulation was carried out over 50 EM steps. **Left:** $\sigma_{rand} = 0.1$. The network is effectively
a linear model, which cannot learn the nonlinearities of the underlying processes.
Right: $\sigma_{rand} = 10.0$. The nonlinear complexity is too large, resulting in serious
overfitting. Note that even in the cross-validation point the network performance is
suboptimal. *Early stopping* alone is therefore not sufficient to prevent overfitting.
Middle: $\sigma_{rand} = 1.0$. The nonlinear complexity is chosen appropriately, allowing
the network to learn the underlying processes without overfitting. Reprinted from
[31], with permission from Elsevier Science.

single model with respect to its generalisation performance. As discussed in
Section 5.6, the resulting training times become rather long. Adopting the
RVFL approach, on the other hand, allows creation of an ensemble of predictors at the same computational costs as required for training an individual
GM network. Since this will allow an exploration of a much broader range of
the configuration space, it may be hoped to eventually come across predictors
that are better than the single GM model. Combining them in a *committee*
can be expected to result in a further improvement of the generalisation
performance.

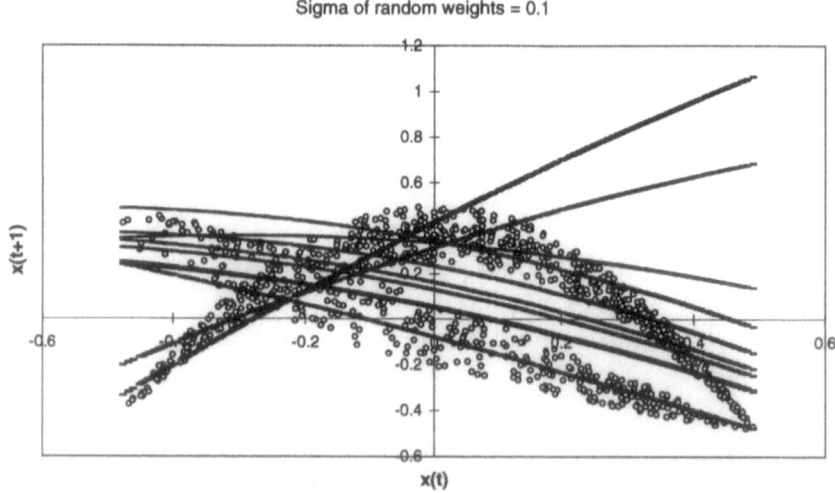

Fig. 8.2. State-space plots for a GM-RVFL with too small a value for sigma. A GM-RVFL network with $K = 10$ kernels was trained on a time series generated from (4.16) (circles). The lines show state-space plots of the kernel centres, $\mu(x_t|w_k)$, at the minimum of the cross-validation error (see Fig. 8.1). The distribution width for the random weights has been chosen too small: $\sigma_{rand} = 0.1$. The network is effectively a linear model that cannot learn the nonlinearities of the underlying subprocesses. Reprinted from [31], with permission from Elsevier Science.

8.2.2 Training an ensemble of models

Figures 8.5 and 8.6 show the results of training a committee of 120 GM-RVFL networks with six different values for σ_{rand}, $\ln \sigma_{rand} = -2, -1, 0, 1, 2, 3$, (or, equivalently, $\sigma_{rand} = $ 0.14, 0.37, 1.0, 2.72, 7.39, 20.09). In each case, the simulation was repeated with 20 different initialisations, using the same set of random number generator seeds. Each network was trained over 35 and 50 EM cycles[1], and the generalisation 'error' E_{gen} was determined on an independent test set of 2000 data points. The values of $E_{gen}(\sigma_{rand})$ belonging to networks initialised with the same random number generator seed were plotted against the corresponding value for $\ln \sigma_{rand} = 0$, $E_{gen}(\ln \sigma_{rand} = 0)$.

---

[1] Ideally, the method of cross-validation should be used. At the time of these simulations, this had not been implemented in the program. The fixed training times chosen were motivated from inspection of the training curves for a single model, like those shown in Fig.8.1.

Sigma of random weights = 1.0

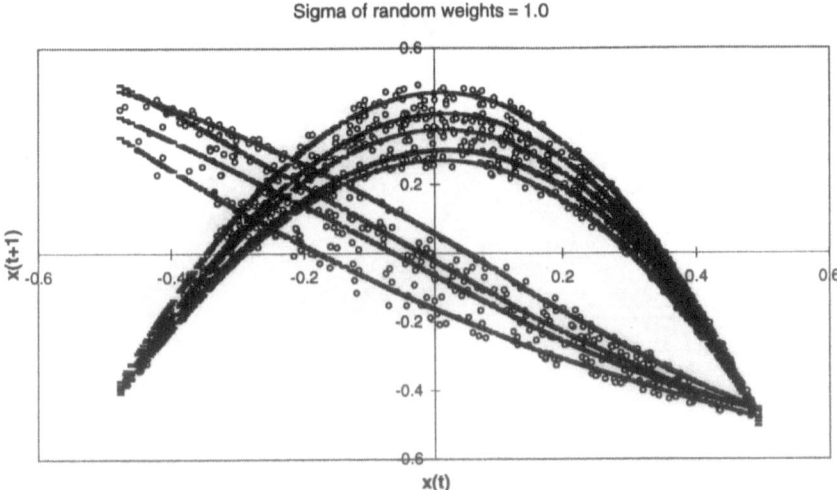

Fig. 8.3. State-space plots for a GM-RVFL with a good choice for sigma.
A GM-RVFL network with $K = 10$ kernels was trained on a time series generated from (4.16) (circles). The lines show state-space plots of the kernel centres,
$\mu(x_t|\mathbf{w}_k)$, at the minimum of the cross-validation error (see Fig. 8.1). The distribution width for the random weights has been chosen appropriately, $\sigma_{rand} = 1.0$, which
allows the network to learn the nonlinearities of the underlying subprocesses, but is
small enough to make the interpolants smooth and prevent overfitting. Reprinted
from [31], with permission from Elsevier Science.

Entries below the dashed line show a performance better than that obtained
with $\ln \sigma_{rand} = 0$, for entries above the dashed line the performance is worse
than for $\ln \sigma_{rand} = 0$. Figures 8.5 and 8.6 suggest that the best results can
be obtained with $\ln \sigma_{rand} = -1$. Increasing this value gradually degrades
the generalisation performance due to excessive nonlinear complexity and
overfitting (as discussed above). A decrease in $\ln \sigma_{rand}$, on the other hand,
leads to a sharp increase in the generalisation 'error' E_{gen}, since the network
then becomes effectively a *linear* model that can no longer capture the non-
linearities of the processes.

In order to select a committee of 'best predictors', a cross-validation set
$\mathbf{D}_{cross}$ of 1000 data points was used, and an arbitrary threshold of $E_{cross}^{cutoff} =
-1.02$ was defined (which was the best value of E_{gen} found for the GM
models). A committee of predictors was set up, including all networks with

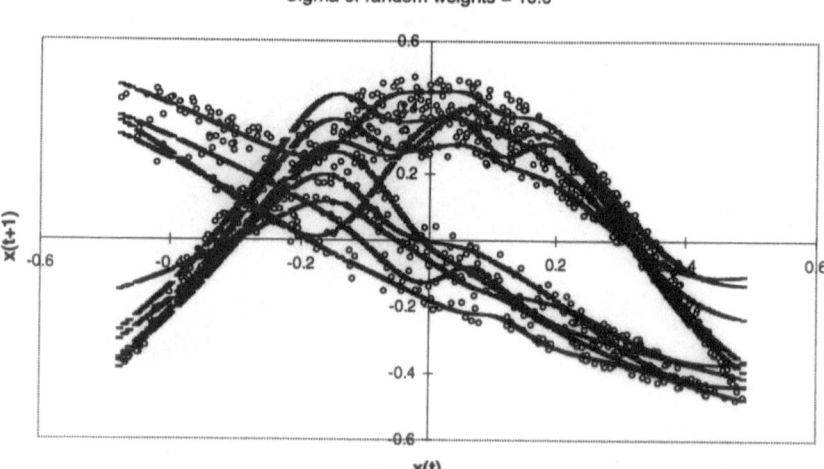

Fig. 8.4. State-space plots for a GM-RVFL with too large a value for sigma. A GM-RVFL network with $K = 10$ kernels was trained on a time series generated from (4.16) (circles). The lines show state-space plots of the kernel centres, $\mu(x_t|\mathbf{w}_k)$, at the minimum of the cross-validation error (see Fig. 8.1). The distribution width for the random weights has been chosen too large, $\sigma_{rand} = 10.0$, resulting in serious overfitting. Reprinted from [31], with permission from Elsevier Science.

$E_{cross} < E_{cross}^{cutoff}$. Thirteen networks satisfied this condition[2], four of which had been initialised with $\ln \sigma_{rand} = 0$, and the remaining nine with $\ln \sigma_{rand} = -1$. The resulting generalisation performance for the committee prediction,

$$P_{Com}(y|\mathbf{x}) = \frac{1}{N_{Com}} \sum_{i=1}^{N_{Com}} P(y|\mathbf{x}, \mathbf{q}_i), \tag{8.5}$$

where $N_{Com} = 13$, is shown in Tab.8.2. A comparison with the results obtained from the conventional approach of training a fully-adaptable GM network suggests that indeed a (slight) performance improvement over the conventional single-predictor scheme has been achieved.

---

[2] Networks initialised with the same random number seed, but trained over a different number of EM-steps were *not* considered as different variants. Instead, the better alternative of the two options '35 EM-steps' and '50 EM-steps' was chosen.

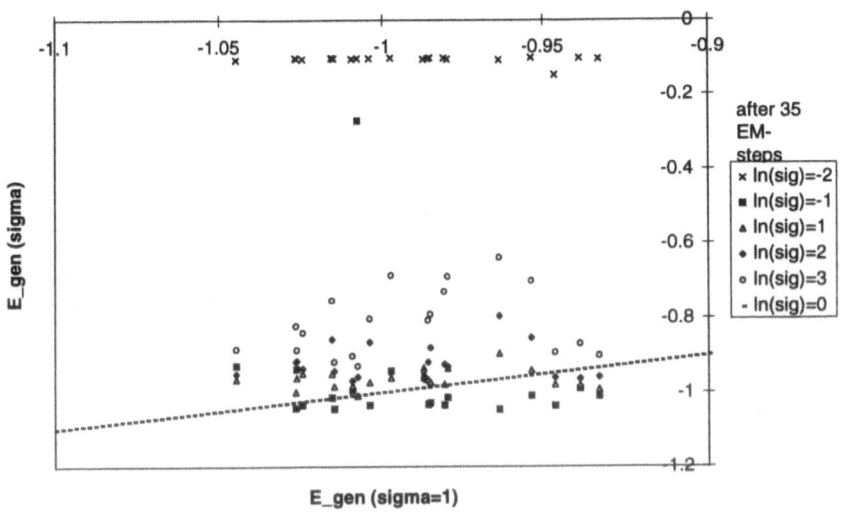

Fig. 8.5. Dependence of the generalisation performance on the random weights. The scatter plot shows, for the time series generated from (4.16), the dependence of the generalisation performance on σ_{rand}, the standard deviation of the distribution from which the random weights in the GM-RVFL are drawn. For each of the six values $\ln \sigma_{rand} =$ -2 (crosses), -1(squares), 0 (dashed line), 1 (triangulars), 2 (diamonds) and 3 (circles) (corresponding to $\sigma = 0.14, 0.37, 1.0, 2.72, 7.39, 20.09$, respectively), 20 training simulations over 35 EM steps were carried out, starting from different random number generator seeds. The generalisation 'error' E_{gen} was measured on an independent test set of 2000 data points. Values of $E_{gen}(\sigma_{rand})$ obtained from the same random number generator seed were plotted against the corresponding value for $\ln \sigma_{rand} = 0$, $E_{gen}(\ln \sigma_{rand} = 0)$. The dashed line indicates the performance for $\ln \sigma_{rand} = 0$. Choosing a smaller value of $\ln \sigma_{rand} = -1$ results in some improvement, whereas the generalisation ability deteriorates steadily as σ_{rand} increases. However, too small a value for σ_{rand}, $\ln \sigma_{rand} = -2$, leads to a dramatic increase in the generalisation error, which is due to a loss of nonlinear modelling capability. Reprinted from [31], with permission from Elsevier Science.

8.3 Application of the GM-RVFL network to the double-well problem

Experimental setup. In a second study, the GM-RVFL network was applied to the double-well problem of Section 4.3. Four different architectures were chosen, namely (i) 3 inputs, 10 tanh-units, 9 kernels, (ii) 5 inputs, 10 tanh-units, 9 kernels, (iii) 5 inputs, 10 tanh-units, 5 kernels, (iv) 10 inputs, 15 tanh-units, 9 kernels. Each architecture was combined with three distribu-

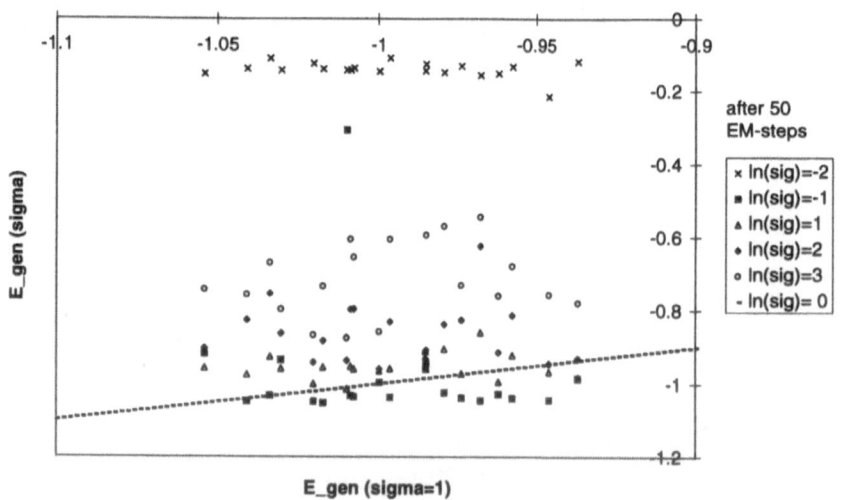

Fig. 8.6. Dependence of the generalisation performance on the random weights. The figure is similar to Fig. 8.5, except that training was performed over 50 rather than 35 EM steps. A comparison with Fig. 8.5 reveals that, as a consequence of overfitting, the poor performance of networks initialised with a large value of σ_{rand} deteriorates when the training time is increased, whereas for small values of σ_{rand} the performance remains constant or even slightly improves. Reprinted from [31], with permission from Elsevier Science.

tion widths $\ln \sigma_{rand} \in \{-1, 1, 3\}$. For each combination, three networks were created from different random number generator seeds. The overall size of the network ensemble thus obtained was $N_{Com} = 4 \cdot 3 \cdot 3 = 36$. The networks were trained on a training set of $N_{train} = 10,000$ data points, generated from (4.24) in the way described in Section 5.4. From this ensemble, a committee of 'best predictors' was selected on the basis of the cross-validation 'error' E_{cross}, estimated on a cross-validation set of the same size as the training set, $N_{cross} = 10,000$.

8.3.1 Committee selection

Figure 8.7 shows the resulting prediction performance of the networks on the training and cross-validation set after (i) 10 EM-steps and (ii) 25 EM-steps. Obviously, the longer training time leads to a slight degradation in the generalisation performance due to overfitting. Therefore, only the networks

Network Model	Prediction Method	E_{gen}	Source
1) Specht	single predictor	0.556	[45], [46]
2) CDEN	single predictor	0.373	[45], [46]
3) DPMN	single predictor	0.334	[45], [46]
4) DSM	single predictor	0.336 ± 0.009	Section 5.4
5) GM-RVFL	single predictor, average over the whole ensemble	0.337 ± 0.027	this study
6) GM-RVFL	single predictor, average over the committee of best models	0.307 ± 0.009	this study
7) GM-RVFL	committee prediction	0.276 ± 0.009	this study

Table 8.3. Comparison between the GM-RVFL and different alternative models. The prediction performance of the GM-RVFL network is compared with different alternative models applied to the double-well problem of Section 4.3. The value E_{gen} gives the estimated generalisation 'error', defined in (8.4). Smaller values indicate a better prediction performance. For the GM-RVFL network, an ensemble of $N_{total} = 36$ models were trained, as described in the text. From this, a committee of $N_{Com} = 12$ networks, having the smallest cross-validation 'error', was selected. The value for E_{gen} in the fifth row was obtained by applying all $N_{total} = 36$ models to a test set of $N_{gen} = 20,000$ data points. The indicated uncertainty is the standard deviation for a single network. The values in the last two rows represent the mean and the standard deviation of the generalisation 'error' obtained by applying the committee of the $N_{Com} = 13$ 'best' networks to *ten* different test sets of $N_{gen} = 10,000$ data points each.

trained over 10 EM-steps were further considered, from which a committee of 'best predictors' was selected by applying a (somewhat arbitrary) threshold of $E_{cross}^{cutoff} = 0.32$ for the cross-validation 'error' E_{cross} (i.e. only networks with $E_{cross} \leq E_{cross}^{cutoff} = 0.32$ 'qualified', leading to a total committee size of $N_{Com} = 13$.).

8.3.2 Prediction

The prediction performance achieved with the committee can be assessed by inspection of Fig. 8.8. The graphs at the top left of the figure show the evolution of the true conditional probability density $P(x_{t+25}|x_t, \dot{x}_t)$ during the phase transitions at $t = 1510 - 1555$ in the time series segment of Fig.4.4. (See Section 4.3 for an explanation of how these graphs were obtained.) This is compared with two neural network predictions. The graphs on the right show the results obtained with an individual GM network, trained with the maximum likelihood scheme of Chapter 3 in a similar way as described in Section 5.4. The graphs at the bottom right depict the predictions by the above committee of 13 GM-RVFL networks. The comparison reveals that in the latter case overfitting is greatly reduced, and that a considerably closer approximation to the true distributions is obtained than with the (single) GM network.

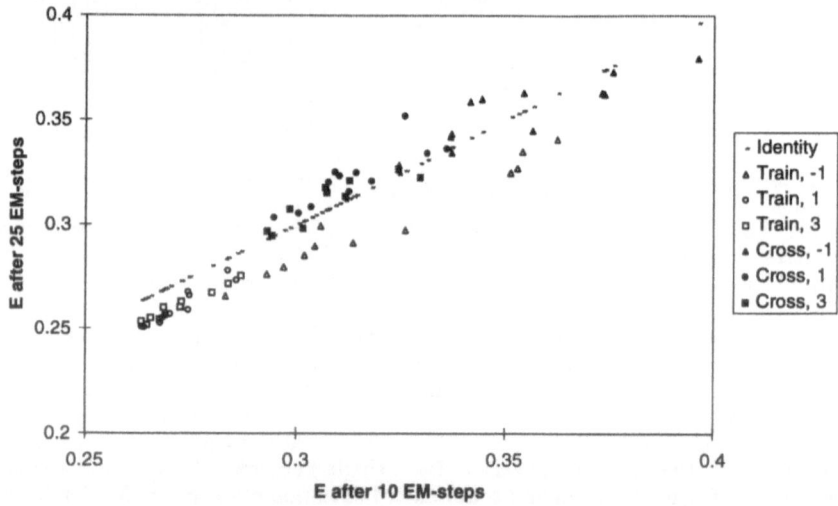

Fig. 8.7. Predictions with an ensemble of GM-RVFL networks trained on the Brownian-motion problem. The scatter plot shows the training and test set 'errors', E_{train} (empty symbols) and E_{test} (shaded symbols), for an ensemble of $N_{Com} = 36$ GM-RVFL networks trained on the time series generated from (4.24). The abscissa gives the value after 10 EM-steps, the ordinate represents the value after 25 EM-steps. Symbols *over* the dashed line indicate an increase of the 'error' with longer training time, symbols *under* the dashed line a decrease of the 'error' with longer training time. While the performance on the training set trivially improves with longer training time, the average performance on the test set deteriorates due to overfitting. Different symbols represent different values for the width of the random-weight distribution, σ_{rand}. Triangles: $\ln \sigma = -1$, circles: $\ln \sigma = 1$, squares: $\ln \sigma = 3$. Different from the first problem [logistic-kappa map, (4.16)], larger values for σ_{rand} are found to give a better performance. Reprinted from [31], with permission from Elsevier Science.

8.3.3 Comparison with other approaches

In order to assess the prediction performance of the GM-RVFL model quantitatively, the generalisation 'error' was measured on ten independent test sets $\mathbf{D}_{gen}^1, \ldots, \mathbf{D}_{gen}^{10}$ of $N_{gen} = 10,000$ data points each. The results were then compared with those listed in Tab.5.1, which for the reader's convenience are shown here, in Tab.8.3, again. It is seen, from row 5 of Tab.8.3, that an individual GM-RVFL network typically achieves prediction results similar to those obtained with the best alternative models listed in Tab.8.3. However, recall that due to the considerable acceleration of the training process, as

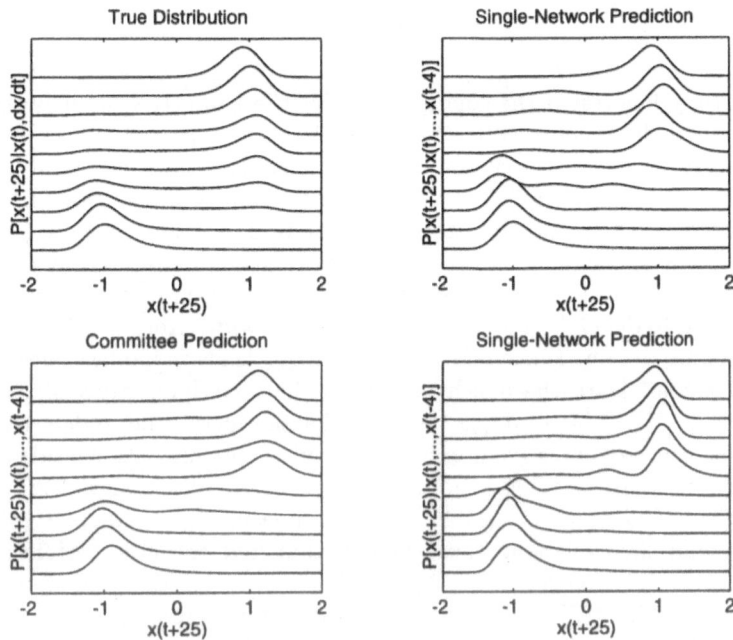

Fig. 8.8. Evolution of the true and the predicted conditional probability density during a phase transition of the double-well problem. The graphs in the figure on the top left show the evolution of the true conditional probability density $P(x_{t+25}|x_t, \dot{x}_t)$ during the phase transitions at $t = 1510 - 1555$ in the time series segment of Fig.4.4. The time difference between two contiguous graphs is $\Delta t = 5$, and the arrow of time points from the bottom to the top. The remaining figures show different network predictions. The graphs on the right show two predictions by an individual GM network, trained with the maximum likelihood scheme described in Chapter 3. The graphs at the bottom left show the prediction by a committee of GM-RVFL networks. Obviously, the latter show considerably fewer signs of overfitting, and give a closer approximation to the true distributions than obtained with an individual GM network. Reprinted from [28] with permission from the Academy of Sciences of the Czech Republic.

demonstrated in Tab.8.1, a whole ensemble of GM-RVFL networks can be trained at the same computational costs as required for training an individual, fully adaptable model. This allows a wider exploration of the model and configuration space, which eventually finds predictors that are better than the best conventional models (row 6 of Tab.8.3). Combining these best GM-RVFL networks in a committee significantly improves the generalisation performance (row 7 of Tab.8.3).

8.4 Discussion

This study has tested the concept developed in Chapter 7, namely the combination of the random vector functional link net approach (RVFL) with the fast Expectation Maximisation (EM) training algorithm. The resulting GM-RVFL network was applied to two stochastic time series, generated from the logistic-kappa map and the Brownian dynamics in the double-well potential. In both cases, a significant acceleration of the training process by about two orders of magnitudes was achieved. The generalisation performance, though, was found to depend on σ_{rand}, the distribution width from which the random weights are drawn. Too small a value for σ_{rand} leads to a quasi-linear model, which cannot learn the non-linearities of the underlying processes, whereas for too large a value of σ_{rand} the generalisation performance deteriorates as a consequence of excessive non-linearity and overfitting. The considerable acceleration of the training process, however, easily allows an optimisation of this parameter by carrying out several simulations with different values for σ_{rand}, and then selecting those models that give the best performance on the cross-validation set. For a 'good' choice of σ_{rand}, the prediction performance of a GM-RVFL network was typically found to be as good as that of a fully-adaptable GM network. By combining the best GM-RVFL models in a committee, a further improvement of the generalisation performance could be achieved, which for the double-well problem was found to be significant. In fact, a committee of GM-RVFL networks outperformed all competing single-model predictors. However, several cautioning remarks are to be made here.

1. Choice of the distribution widths for the random weights
In the present study, all the random weights in the GM-RVFL network were drawn from the same distribution, whose width parameter was optimised by a discrete random search: different values for σ_{rand} were tried out, and the best ones, selected on the basis of the cross-validation performance, were chosen. However, in general it cannot be assumed that an optimal model can be found by a simultaneous scaling of all random weights. This is especially the case if there are several inputs of different nature and different relevance level. In this case it would be advantageous to introduce several *weight groups* such that weights exiting different input nodes are drawn from different distributions. This will allow different scales and different *relevance levels* of various inputs to be taken into account. A random variation of the width parameters, however, is then no longer an efficient procedure because of the larger dimension of the space spanned by these parameters. A more systematic method of adaptation will be discussed in Chapter 15 and empirically tested in Chapter 16.

2. Cross-validation
In the simulations reported here, the method of cross-validation was applied

for regularisation[3] and model selection. When the data set is sparse, this is not possible as in this case all the available data are needed for training. An alternative method, which does not require to hold out data for monitoring overfitting, is the Bayesian evidence approach. This scheme was introduced to the neural network community by MacKay, who developed it under the assumption of uniform Gaussian noise on the target variable [37]. A generalisation of this scheme to the modelling of arbitrary conditional probability densities will be derived in Chapter 10. This is followed by an empirical corroboration in Chapter 12.

3. Selection of committees

The selection of network committees followed a rather *ad hoc* procedure: an arbitrary threshold was defined, and all networks with a cross-validation 'error' smaller than this value became part of the committee. It can be assumed that a more sophisticated weighting scheme for networks will lead to improved results. This will be discussed in Chapter 13.

4. Improvement of the generalisation performance by applying network committees

The simulation results suggest that the application of network committees leads to an improvement in the generalisation performance. However, whereas for the second time series (Section 8.3) a considerable improvement was observed, the improvement obtained in the first simulation (Section 8.2) was rather small. This raises the question about the conditions on which the degree of the improvement depends. This aspect will be discussed in Chapter 13.

5. Size of the training set

One reason why the decrease in the generalisation 'error' by applying a network committee was only so small for the first application (Section 8.2) can be the rather large amount of training data (compared to the number of parameters of the data-generating process). In this case, overfitting is not so much of a problem in the first place and, consequently, there is not much to be gained by using committees. This suggests repeating the simulations with a smaller training set. Results can be found in chapters 12 and 14.

---

[3] Although no proper *'early stopping'* was applied in the simulations, note that the fixed number of EM training steps was chosen on the basis of the cross-validation performance.

9. A simple Bayesian regularisation scheme

A regularisation scheme is derived from a simple Bayesian approach, where the maximum likelihood estimate of the network parameters is replaced by the mode of their posterior distribution. Conjugate priors for the various network parameters are introduced, which give rise to regularisation terms that can be viewed as a generalisation of simple weight decay. It is shown how the posterior mode can be found with a slightly modified version of the EM algorithm.

9.1 A Bayesian approach to regularisation

So far the network parameters $\mathbf{q}$ have been adapted to maximise the likelihood $P(\mathbf{D}|\mathbf{q})$, as discussed in Section 3.1. For sparse training sets this approach can lead to suboptimal results, where the network overfits the training data and is poor at generalising to new data not seen before. A standard approach to alleviate this problem is to introduce a regularisation or penalty term, which in a Bayesian framework can be interpreted as a prior distribution of the parameters. Consider the simple case of modelling y as a function of $\mathbf{x}$, $f(\mathbf{x}; \mathbf{w})$, where $\mathbf{w}$ is an M-dimensional vector of all the network weights. The latter are adapted in an iterative scheme so as to minimise the cost or error function

$$E^*(\mathbf{w}; \beta, \alpha) \quad = \quad E(\mathbf{w}; \beta) + R(\mathbf{w}; \alpha) \tag{9.1}$$

which is composed of the data misfit term

$$E(\mathbf{w}; \beta) \quad := \quad \frac{\beta}{2} \sum_{t=1}^{N} [y_t - f(\mathbf{x}_t; \mathbf{w})]^2 + C_E \tag{9.2}$$

and a regularisation or weight-decay term

$$R(\mathbf{w}; \alpha) \quad := \quad \frac{1}{2} \sum_{k=1}^{K} \alpha_k \mathbf{w}_k^\dagger \mathbf{w}_k + C_R \tag{9.3}$$

Here K denotes the number of weight groups $\mathbf{w} = (\mathbf{w}_1, \ldots, \mathbf{w}_K)$ in the network, $\boldsymbol{\alpha} = (\alpha_1, \ldots, \alpha_K)$ are the associated weight decay parameters, C_E and C_R symbolise constants, and the dagger indicates matrix transpose. This optimisation problem can be cast into a probabilistic framework by assuming independent and identically distributed Gaussian noise of variance $\frac{1}{\beta}$ on the targets y_t,

$$P(y_t | \mathbf{x}_t, \mathbf{w}, \beta) = \sqrt{\frac{\beta}{2\pi}} \exp\left[-\frac{\beta}{2}\left(y_t - f(\mathbf{x}_t; \mathbf{w})\right)^2\right] \qquad (9.4)$$

and a multivariate Gaussian prior distribution of the weights $\mathbf{w}$,

$$P(\mathbf{w}|\boldsymbol{\alpha}) = \prod_{k=1}^{K} P(\mathbf{w}_k | \alpha_k) = \prod_{k=1}^{K} \left(\frac{\alpha_k}{2\pi}\right)^{M_k/2} \exp\left(-\frac{\alpha_k}{2}\mathbf{w}_k^{\dagger}\mathbf{w}_k\right) \qquad (9.5)$$

in which $\mathbf{w}_k$ denotes an M_k-dimensional vector of all weights in the kth weight group. Equation (9.5) is related to the regularisation term of Equation (9.3) via[1]

$$R(\mathbf{w}; \boldsymbol{\alpha}) = -\ln P(\mathbf{w}|\boldsymbol{\alpha}) \qquad (9.6)$$

The likelihood of the training data $\mathbf{D}$ is given by

$$P(\mathbf{D}|\mathbf{w}, \beta) = \prod_{t=1}^{N} P(y_t; \mathbf{x}_t | \mathbf{w}, \beta) = \prod_{t=1}^{N} P(y_t | \mathbf{x}_t, \mathbf{w}, \beta) \prod_{t=1}^{N} P(\mathbf{x}_t | \mathbf{w}, \beta)$$

$$= C_L \prod_{t=1}^{N} P(y_t | \mathbf{x}_t, \mathbf{w}, \beta) \qquad (9.7)$$

where the first step follows from the assumed independence of the individual exemplars $(\mathbf{x}_t, y_t)$, and the last step is a consequence of the fact that for supervised training the distribution of the input vectors $\mathbf{x}_t$ is not modelled by the network and therefore independent of its parameters. Taking the negative logarithm of the likelihood $P(\mathbf{D}|\mathbf{w}, \beta)$ and making use of equations (9.2), (9.4), and (9.7), we get[2]

$$E(\mathbf{w}; \beta) = -\ln P(\mathbf{D}|\mathbf{w}, \beta) \qquad (9.8)$$

Applying Bayes' rule

$$P(\mathbf{w}|\mathbf{D}, \beta, \boldsymbol{\alpha}) \propto P(\mathbf{D}|\mathbf{w}, \beta) P(\mathbf{w}|\boldsymbol{\alpha}) \qquad (9.9)$$

taking logarithms on both sides and applying equations (9.1), (9.6), and (9.8) shows that

---

[1] Define the constant C_R in Equation (9.3) by $C_R := -\sum_k \frac{M_k}{2} \ln \frac{\alpha_k}{2\pi}$.
[2] Define the constant C_E in Equation (9.2) by $C_E := -\frac{N}{2} \ln \frac{\beta}{2\pi} - \ln C_L$, where C_L is the constant of Equation (9.7).

$$E^*(\mathbf{w}; \beta, \alpha) \quad = \quad -\ln P(\mathbf{w}|\mathbf{D}, \beta, \alpha) + C \tag{9.10}$$

which implies that the minimum of the error function $E^*(\mathbf{w}; \beta, \alpha)$ with respect to the network weights $\mathbf{w}$ for fixed β and α is identical to the mode of the posterior distribution $P(\mathbf{w}|\mathbf{D}, \beta, \alpha)$. Note that for a constant prior $P(\mathbf{w}|\alpha) = const$ the regularisation term also becomes constant, $R(\mathbf{w}; \alpha) = const$, so that the minima of $E^*(\mathbf{w}; \beta, \alpha)$ and $E(\mathbf{w}; \beta)$ with respect to $\mathbf{w}$ are identical. This reproduces the results of the unregularised maximum likelihood scheme.

9.2 A simple example: repeated coin flips

Consider as a further example the situation of a repeated coin flip. The model has a single parameter p, which represents the probability for the coin landing heads, and the outcome of the experiment is a binary time series, where 1 indicates that the coin landed head, and 0 that it landed tail. Now assume that after N coin flips we have observed m heads and $N - m$ tails. Then the likelihood for the data is given by

$$P(\mathbf{D}|p) \quad = \quad \binom{N}{m} p^m (1 - p)^{N-m} \tag{9.11}$$

Taking logarithms and multiplying both sides by (-1) gives for the negative log-likelihood or error function:

$$E \;=\; -\ln P(\mathbf{D}|p) \;=\; -m \ln p - (N - m) \ln(1 - p) + C \tag{9.12}$$

Maximising the likelihood (9.11) is equivalent to minimising the error function (9.12):

$$\frac{dE}{dp} = -\frac{m}{p} + \frac{N - m}{1 - p} = \frac{Np - m}{p(1 - p)} = 0 \tag{9.13}$$

which leads to the maximum likelihood estimate

$$\hat{p} \quad = \quad \frac{m}{N} \tag{9.14}$$

This is a reasonable result if N is large, since according to the law of large numbers the observed frequency m/N will almost surely be very close to the true value of p. But suppose N is small, say $N = 3$, and that in a sequence of this length we do not observe a single head. Is it reasonable to set the probability p to zero according to (9.14)? Probably not, especially if we do not have any reason to suspect that the coin is highly biased. We therefore have to include our prior assumption that we do not favour model parameters with value 0. We can do this by taking a non-uniform prior distribution for p, for example the beta distribution

$$P(p) \quad \propto \quad p^{a-1}(1-p)^{b-1} \tag{9.15}$$

where a and b are the parameters of the distribution. Applying Bayes rule and making use of equations (9.11) and (9.15) gives

$$P(p|\mathbf{D}) \propto P(\mathbf{D}|p)P(p) \propto p^m(1-p)^{N-m}p^{a-1}(1-p)^{b-1} = p^{\tilde{a}-1}(1-p)^{\tilde{b}-1} \tag{9.16}$$

where the definitions

$$\tilde{a} = a + m, \qquad \tilde{b} = b + N - m \tag{9.17}$$

have been introduced. It is seen that the posterior distribution (9.16) has the same functional form as the prior distribution (9.15). A prior which satisfies this property is called a *conjugate* prior (see below). For the mode of the posterior distribution, the so-called *maximum a posteriori* (MAP) estimate, we obtain (by setting the derivative of $\ln P(p|\mathbf{D})$ to zero):

$$\frac{d}{dp}\ln P(p|\mathbf{D}) = \frac{\tilde{a}-1}{p} - \frac{\tilde{b}-1}{1-p} = \frac{\tilde{a}-1-(\tilde{a}+\tilde{b}-2)p}{p(1-p)} = 0$$

$$\Rightarrow \quad \hat{p} = \frac{\tilde{a}-1}{\tilde{a}+\tilde{b}-2} = \frac{m+a-1}{N+a+b-2} \tag{9.18}$$

Comparing equations (9.18) and (9.14), we see that for $a = b = 1$ the mode of the posterior and the maximum likelihood estimator are identical. This is not surprising since in this case the prior is uniform, as seen from Equation (9.15). For $a \neq 1$ or $b \neq 1$, however, the maximum a posteriori parameter and the maximum likelihood estimate are different. In particular, if $a > 1$ and $b > 1$, the estimate of $\hat{p}$ can never become zero, as we wished to implement. A comparison between equations (9.18) and (9.14) suggests that the effect of the beta prior is equivalent to adding pseudo-counts to the observed data, where the actual number of heads is augmented by $(a-1)$ additional pseudo-heads out of a total of $(a+b-2)$ pseudo-observations.

The above examples can easily be generalised to the prediction of arbitrary conditional probability densities with the GM (or GM-RVFL) network, as will be shown in the following sections.

9.3 A conjugate prior

We first have to choose a prior on the network parameters, $P(\mathbf{q})$. A common choice for the weights $\mathbf{w}$ is a Gaussian distribution, which corresponds to a linear weight decay term in network training, as shown in Section 9.1. Since there are K independent weight groups it is reasonable to take a product of K independent Gaussian distributions with different variances $\sigma_k^2 := \alpha_k^{-1}$,

$$P(\mathbf{w}|\boldsymbol{\alpha}) \;=\; P(\mathbf{w}_1,\ldots,\mathbf{w}_K|\alpha_1,\ldots,\alpha_K) \;:=\; \prod_{k=1}^{K} P(\mathbf{w}_k|\alpha_k)$$

$$:=\; \prod_{k=1}^{K} \left(\frac{\alpha_k}{2\pi}\right)^{\frac{W}{2}} \exp\left(-\frac{\alpha_k}{2}\mathbf{w}_k^{\dagger}\mathbf{w}_k\right) \tag{9.19}$$

For the remaining parameters we choose *conjugate* priors, that is, prior distributions that belong to the same function family as the ensuing posterior distributions; see [13], Chapter 9 for a detailed exposition of this subject.

A conjugate prior for the precisions is a gamma distribution ([13], Chapter 4)

$$P(\boldsymbol{\beta}|\rho,\xi) \;=\; P(\beta_1,\ldots,\beta_K|\rho,\xi) \;:=\; \left(\frac{\xi^{\rho}}{\Gamma(\rho)}\right)^{K} \prod_{k=1}^{K} \beta_k^{\rho-1} \exp(-\xi\beta_k) \tag{9.20}$$

where $\xi,\rho > 0$. A proper conjugate prior for the mixture weightings $\mathbf{a}$ is a Dirichlet density ([13], Chapter 5),

$$P(\mathbf{a}|\nu) \;=\; P(a_1,\ldots,a_{K-1}|\nu) \;:=\; \frac{\Gamma(K\nu)}{[\Gamma(\nu)]^{K}} \prod_{k=1}^{K} a_k^{(\nu-1)}, \tag{9.21}$$

where $\nu > 0$ and $a_K = 1 - \sum_{k=1}^{K-1} a_k$. For the overall prior, let us take a product of these three terms, that is a product of a Gaussian, a gamma, and a Dirichlet density,

$$P(\mathbf{q}|\mathbf{r}) \;=\; P(\mathbf{w},\boldsymbol{\beta},\mathbf{a}|\boldsymbol{\alpha},\rho,\xi,\nu) \;:=\; P(\mathbf{w}|\boldsymbol{\alpha})P(\boldsymbol{\beta}|\rho,\xi)P(\mathbf{a}|\nu) \tag{9.22}$$

where the three terms on the right are defined in (9.19)-(9.21), $\mathbf{q} = (\mathbf{w},\boldsymbol{\beta},\mathbf{a})$ denotes the vector of all network parameters, and $\mathbf{r} := (\boldsymbol{\alpha},\rho,\xi,\nu)$ represents the set of all *hyperparameters*, that is, the parameters that determine the functional form of the prior[3]. In generalisation of the simple weight decay scheme of Section 9.1, equations (9.3) and (9.6), let us introduce the *penalty* or *regularisation term*

$$R(\mathbf{q}|\mathbf{r}) \;:=\; -\ln P(\mathbf{q}|\mathbf{r}) \tag{9.23}$$

which becomes

$$R(\mathbf{q}|\mathbf{r}) = \sum_{k=1}^{K} \left[\frac{\alpha_k}{2}\mathbf{w}_k^{\dagger}\mathbf{w}_k - \frac{W}{2}\ln\left(\frac{\alpha_k}{2\pi}\right) + \xi\beta_k - (\rho-1)\ln\beta_k - (\nu-1)\ln a_k\right]$$
$$+ K\ln\Gamma(\rho) - K\rho\ln\xi + K\ln\Gamma(\nu) - \ln\Gamma(K\nu) \tag{9.24}$$

---

[3] Here and in what follows, the term *'prior'* is used both for the prior distribution on the network parameters as well as for the output weights a_k. It should always be clear from the context which of these alternatives is meant, so confusions are unlikely to occur.

Now apply Bayes rule

$$P(\mathbf{q}|\mathbf{D},\mathbf{r}) \quad \propto \quad P(\mathbf{D}|\mathbf{q},\mathbf{r})P(\mathbf{q}|\mathbf{r}) \tag{9.25}$$

and define in the same way as in equations (9.1) and (9.8):

$$E(\mathbf{q}) \quad := \quad -\ln P(\mathbf{D}|\mathbf{q}) \tag{9.26}$$

$$E^*(\mathbf{q}|\mathbf{r}) \quad := \quad E(\mathbf{q}|\mathbf{r}) + R(\mathbf{q}|\mathbf{r}) \tag{9.27}$$

E is usually called the *likelihood error function*, and E^* is the *total error function*. As before in Section 9.1, it is seen that finding the mode $\hat{\mathbf{q}}$ of $P(\mathbf{q}|\mathbf{D},\mathbf{r})$ is equivalent to finding the minimum of $E^*(\mathbf{q}|\mathbf{r})$,

$$\hat{\mathbf{q}} \quad := \quad \text{argmax}_{\mathbf{q}}\{P(\mathbf{q}|\mathbf{D},\mathbf{r})\} \quad = \quad \text{argmin}_{\mathbf{q}}\{E^*(\mathbf{q}|\mathbf{r})\}. \tag{9.28}$$

This can be accomplished with a modified version of the EM algorithm.

9.4 EM algorithm with regularisation

In Section 7.1, the standard EM algorithm for finding the *maximum likelihood estimate* of a probability model was reviewed. It is now shown that this algorithm can easily be generalised to include a regularisation term. Recall the definitions (7.2) - (7.4),

$$\Psi(\mathbf{q},\mathbf{\Lambda}) \quad := \quad -\ln P(\mathbf{D},\mathbf{\Lambda}|\mathbf{q}) \tag{9.29}$$

$$U(\mathbf{q}|\mathbf{q}') \quad := \quad \left\langle \Psi(\mathbf{q},\mathbf{\Lambda}) \right\rangle_{\mathbf{\Lambda}|\mathbf{D},\mathbf{q}'} \tag{9.30}$$

$$S(\mathbf{q}|\mathbf{q}') \quad := \quad \left\langle \ln(P(\mathbf{\Lambda}|\mathbf{D},\mathbf{q})) \right\rangle_{\mathbf{\Lambda}|\mathbf{D},\mathbf{q}'} \tag{9.31}$$

where $\left\langle X \right\rangle_{\mathbf{\Lambda}|\mathbf{D},\mathbf{q}'} := \int X(\mathbf{\Lambda})P(\mathbf{\Lambda}|\mathbf{D},\mathbf{q}')d\mathbf{\Lambda}$.

THEOREM
If the condition for the application of the EM algorithm, (7.5), is satisfied, then

$$U(\mathbf{q}|\mathbf{q}') + R(\mathbf{q}|\mathbf{r}) \leq U(\mathbf{q}'|\mathbf{q}') + R(\mathbf{q}'|\mathbf{r}) \implies E^*(\mathbf{q}|\mathbf{r}) \leq E^*(\mathbf{q}'|\mathbf{r}) \tag{9.32}$$

$$\left[\nabla_{\mathbf{q}}U(\mathbf{q}|\mathbf{q}')\right]_{\mathbf{q}=\mathbf{q}'} + \left[\nabla_{\mathbf{q}}R(\mathbf{q}|\mathbf{r})\right]_{\mathbf{q}=\mathbf{q}'} = \left[\nabla_{\mathbf{q}}E^*(\mathbf{q}|\mathbf{r})\right]_{\mathbf{q}=\mathbf{q}'} \tag{9.33}$$

PROOF
From (9.27) and (7.6) we have

$$E^*(\mathbf{q}|\mathbf{r}) \quad = \quad U(\mathbf{q}|\mathbf{q}') + R(\mathbf{q}|\mathbf{r}) + S(\mathbf{q}|\mathbf{q}'). \tag{9.34}$$

Equations (9.32) and (9.33) follow then immediately from the fact that $S(\mathbf{q}|\mathbf{q}')$ has its maximum at $\mathbf{q}'$, i.e. $S(\mathbf{q}|\mathbf{q}') \leq S(\mathbf{q}'|\mathbf{q}')$ and $[\nabla_{\mathbf{q}}S(\mathbf{q}|\mathbf{q}')]_{\mathbf{q}=\mathbf{q}'} = 0$.

◇

Equation (9.32) implies that a parameter adaptation $\mathbf{q}' \to \mathbf{q}$ which decreases $U + R$ also decreases the total error function E^*. Equation (9.33) states that the critical points (which means, points with zero gradient) of E^* and $U + R$ are the same. Thus (9.32) and (9.33) lead to the following *regularised EM algorithm* for finding the mode of $P(\mathbf{q}|\mathbf{D}, \mathbf{r})$ or, equivalently, the minimum of $E^*(\mathbf{q})$:

1. Given $\mathbf{q}'$, find $\mathbf{q}$ that minimises $U(\mathbf{q}|\mathbf{q}') + R(\mathbf{q}|\mathbf{r})$ (M-step).

2. If $\mathbf{q} \neq \mathbf{q}'$, define $\mathbf{q}' := \mathbf{q}$, recalculate $U(\ldots|\mathbf{q}')$, and goto 1 (E-step).

9.5 The posterior mode

The mode of the posterior distribution can now easily be obtained by applying the regularised version of the EM algorithm, described in Section 9.4. Setting $\nabla_{\mathbf{q}}E^* = 0$, we obtain from (9.33) the following condition for the mode:

$$\frac{\partial U}{\partial a_k} = -\frac{\partial R}{\partial a_k}, \qquad \frac{\partial U}{\partial \beta_k} = -\frac{\partial R}{\partial \beta_k}, \qquad \nabla_{\mathbf{w}_k}U = -\nabla_{\mathbf{w}_k}R \qquad (9.35)$$

The derivatives of U with respect to the network parameters have already been calculated in sections 7.1 and 7.3. Inserting (7.18), (7.19), and (7.34) into (9.35) gives

$$\sum_{t=1}^{N} \frac{\pi_k(t)}{2} \left[\left(y_t - \mu(\mathbf{x}_t; \mathbf{w}) \right)^2 - \frac{1}{\beta_k} \right] = -\frac{\partial R}{\partial \beta_k} \qquad (9.36)$$

$$-\frac{1}{a_k} \sum_{t=1}^{N} \pi_k(t) + \frac{1}{a_K} \sum_{t=1}^{N} \pi_K(t) = -\frac{\partial R}{\partial a_k} \qquad (9.37)$$

$$\beta_k \left[(\mathbf{G}\mathbf{\Pi}_k\mathbf{G}^\dagger) \mathbf{w}_k - \mathbf{G}\mathbf{\Pi}_k\mathbf{y} \right] = -\nabla_{\mathbf{w}_k}R \qquad (9.38)$$

The derivatives of R follow from (9.24):

$$\nabla_{\mathbf{w}_k}R = \alpha_k\mathbf{w}_k \qquad (9.39)$$

$$\frac{\partial R}{\partial \beta_k} = \xi - \frac{\rho - 1}{\beta_k} \qquad (9.40)$$

$$\frac{\partial R}{\partial a_k} = -\frac{\nu - 1}{a_k} + \frac{\nu - 1}{a_K} \qquad (9.41)$$

Recall that the last equation follows from the fact that $\frac{\partial a_K}{\partial_k} = -1$ due to the normalisation constraint $a_K = 1 - \sum_{k=1}^{K-1} a_k$. Inserting (9.39) into (9.38) yields

$$\left(\mathbf{G}\Pi_k\mathbf{G}^\dagger + \frac{\alpha_k}{\beta_k}\mathbf{I} \right) \hat{\mathbf{w}}_k = \mathbf{G}\Pi_k\mathbf{y}. \tag{9.42}$$

Note that, by comparison with Equation (7.64), the quotient $\lambda_k := \frac{\alpha_k}{\beta_k}$ corresponds to a weight-decay parameter for the kth weight group. Substituting the right-hand side of (9.36) by the negative right-hand side of (9.40) gives

$$\frac{1}{\hat{\beta}_k} = \frac{\sum_{t=1}^{N} \pi_k(t)\left(y_t - \mu(\mathbf{x}_t;\hat{\mathbf{w}}_k) \right)^2 + 2\xi}{\sum_{t=1}^{N} \pi_k(t) + 2(\rho - 1)} \tag{9.43}$$

Finally, equations (9.37) and (9.41) lead to

$$\frac{1}{\hat{a}_k} \left(\left[\sum_{t=1}^{N} \pi_k(t) \right] + (\nu - 1) \right) = \frac{1}{\hat{a}_K} \left(\left[\sum_{t=1}^{N} \pi_K(t) \right] + (\nu - 1) \right) = (9.44)$$

the constant C being defined by the constraint $\sum_{k=1}^{K} \hat{a}_k = 1$,

$$C = \sum_{k=1}^{K} \left(\left[\sum_{t=1}^{N} \pi_K(t) \right] + (\nu - 1) \right) = \left(\sum_{t=1}^{N} \sum_{k=1}^{K} \pi_K(t) \right) + K(\nu - 1)$$
$$= N + K(\nu - 1). \tag{9.45}$$

This results in

$$\hat{a}_k = \frac{\left[\sum_{t=1}^{N} \pi_k(t) \right] + (\nu - 1)}{N + K(\nu - 1)} \tag{9.46}$$

The following abbreviation will be introduced for convenience in the notation:

$$\bar{\rho} := \rho - 1, \quad \bar{\nu} := \nu - 1 \tag{9.47}$$

As is typical for conjugate priors, prior knowledge corresponds to a set of artificial training data, which is reflected in equations (9.43) and (9.46). The action of the regularisation scheme is to shift the priors a_k towards more equal values,

$$\lim_{\nu \to \infty} \hat{a}_k = \frac{1}{K} \ \forall k \tag{9.48}$$

and to introduce a lower bound on the variance:

$$\frac{1}{\hat{\beta}_k} \geq \frac{2\xi}{\sum_{t=1}^{N} \pi_k(t) + 2\bar{\rho}} \geq \frac{2\xi}{\sum_{k=1}^{K} \sum_{t=1}^{N} \pi_k(t) + 2\bar{\rho}} = \frac{2\xi}{N + 2\bar{\rho}}. \tag{9.49}$$

It is noted that the expressions (9.43) and (9.46) are similar to those found by Ormoneit and Tresp [47] for the prediction of *unconditional* probability densities. The regularisation of the weights $\mathbf{w}$ is equivalent to the scheme discussed earlier in Section 7.5, with $\lambda_k := \frac{\alpha_k}{\beta_k}$. An application will be given later, in chapters 14 and 16.

9.6 Discussion

The Bayesian approach outlined in this chapter can be viewed as a generalisation of the standard regularisation method of simple weight decay. The
variances of the prior distributions on the weights $\mathbf{w}_k$ are directly related to
weight decay parameters, which prevent the $\mathbf{w}_k$ from taking on excessively
large values. In a similar way, the prior distributions on the remaining parameters reduce the risk of overfitting by imposing a lower bound on the kernel
widths and by equalising the output weights a_k. However, the optimal values
for the regularisation parameters (the *hyperparameters*) are not known in advance. One method for their optimisation would be to use cross-validation,
but this requires putting aside part of the data, which thus cannot be used
for training any more. Obviously, this approach is infeasible if the amount of
available data is limited.

In principle one could opimise the regularisation parameters simultaneously with the network parameters by finding the maximum a posteriori
(MAP) estimate of their joint distribution. The problem is that this MAP
estimate is not necessarily representative of the distribution as a whole. Although this is a fundamental problem of the MAP approach – one should in
fact integrate over parameters rather than optimise them – the problem is
considerably aggravated by a joint optimisation with respect to different sets
of parameters.

The following chapter will show how an improvement in this respect can
be achieved with the so-called Bayesian *evidence* scheme.

10. The Bayesian Evidence Scheme for Regularisation

This chapter generalises the Bayesian evidence scheme, introduced to the neural network community by David MacKay for the regularisation of networks under the assumption of Gaussian homoscedastic noise, to the prediction of arbitrary conditional probability densities. The idea is to optimise parameters and hyperparameters seperately, and to find the mode with respect to the hyperparameters only after the parameters have been integrated out. This integration is carried out by Gaussian approximation, which requires the calculation of the Hessian of the error function at the mode. The derivation of this matrix can be accomplished with a generalised version of the EM algorithm, as exposed in detail in the appendix.

10.1 Introduction

Consider again the simple coin flip experiment of Section 9.2. What we are ultimately interested in is the posterior conditional mean

$$\langle p|\mathbf{D} \rangle \quad := \quad \int_0^1 pP(p|\mathbf{D})dp \qquad (10.1)$$

Introducing the definition of the beta function

$$B(a,b) \quad := \quad \int_0^1 p^{a-1}(1-p)^{b-1}dp \qquad (10.2)$$

we can write equations (9.15) and (9.16) more precisely as

$$P(p) = \frac{p^{a-1}(1-p)^{b-1}}{B(a,b)}, \qquad P(p|\mathbf{D}) = \frac{p^{\tilde{a}-1}(1-p)^{\tilde{b}-1}}{B(\tilde{a},\tilde{b})} \qquad (10.3)$$

Inserting this into (10.1) gives

$$\langle p|\mathbf{D}\rangle \;=\; \int_0^1 pP(p|\mathbf{D})dp \;=\; \int_0^1 p\frac{p^{\tilde{a}-1}(1-p)^{\tilde{b}-1}}{B(\tilde{a},\tilde{b})}dp \;=\; \frac{B(\tilde{a}+1,\tilde{b})}{B(\tilde{a},\tilde{b})} \qquad (10.4)$$

This can be solved using the standard result (for example [11], p.331)

$$B(a,b) \;=\; \frac{\Gamma(a)\Gamma(b)}{\Gamma(a+b)} \qquad (10.5)$$

where $\Gamma(x)$ is the gamma function defined as

$$\Gamma(x) \;:=\; \int_0^\infty \exp\left(-t\right)t^{x-1}dt \qquad (10.6)$$

which satisfies the relation (see e.g. [11], p.331)

$$\Gamma(x+1) \;=\; x\Gamma(x) \qquad (10.7)$$

Inserting equations (10.5) and (10.7) into (10.4), we obtain

$$\langle p|\mathbf{D}\rangle \;=\; \frac{\Gamma(\tilde{a}+1)\Gamma(\tilde{a}+\tilde{b})}{\Gamma(\tilde{a}+\tilde{b}+1)\Gamma(\tilde{a})} \;=\; \frac{\tilde{a}\Gamma(\tilde{a})\Gamma(\tilde{a}+\tilde{b})}{(\tilde{a}+\tilde{b})\Gamma(\tilde{a}+\tilde{b})\Gamma(\tilde{a})} \;=\; \frac{\tilde{a}}{\tilde{a}+\tilde{b}} \qquad (10.8)$$

Now making use of definition (9.17), we get the following expression for the posterior conditional mean:

$$\langle p|\mathbf{D}\rangle \;=\; \frac{m+a}{N+a+b} \qquad (10.9)$$

It is seen that Equation (10.9) does *not* reproduce the result of equation (9.18). The observed deviation increases with the skewness of the distribution, and is related to the well-known fact that the mode is *not* invariant with respect to nonlinear parameter transformations. In fact, for strongly skewed distributions the mode is not representative of the distribution at all, which points to a major inadequacy of the *maximum a posteriori* approach. This problem is aggravated as the dimension of the parameter space increases. To illustrate this phenomenon, consider the canonical distribution in statistical physics. Although the mode is always at energy zero, it is extremely unlikely to find the system in this particular state for non-zero temperatures. The reason is that although the value of the probability density for a microstate at energy zero is orders of magnitudes larger than for a microstate at, say, the mean energy, the integral over the probability density in the neighbourhood of the zero-energy state is negligibly small. This demonstrates, as pointed out by MacKay [40], that probability maxima in high-dimensional spaces are rather irrelevant; all that counts is the attached *probability mass*.

The correct approach to learning in neural networks is therefore to integrate over the parameter space. Consider a target variable y and an input vector $\mathbf{x}$. We want to infer the conditional probability density $P(y|\mathbf{x})$ from a training set $\mathbf{D} = \{(y_t, \mathbf{x}_t)\}_{t=1}^{N}$, where the training data $(y_t, \mathbf{x}_t)$ are assumed to be independent[1]. As a prediction model we apply a neural network with parameters $\mathbf{q}$ and hyperparameters $\mathbf{r}$. The detailed nature of these two vectors will be discussed later; at the moment we will only assume that $\dim \mathbf{q} \gg \dim \mathbf{r}$. The objective of the training process is the prediction of the conditional probability of the target y conditional on the input vector $\mathbf{x}$ and the training data $\mathbf{D}$, which is given by

$$
\begin{aligned}
P(y|\mathbf{x}, \mathbf{D}) &= \int P(y, \mathbf{q}, \mathbf{r}|\mathbf{x}, \mathbf{D}) dq dr \\
&= \int P(y|\mathbf{q}, \mathbf{r}, \mathbf{x}, \mathbf{D}) P(\mathbf{q}, \mathbf{r}|\mathbf{x}, \mathbf{D}) dq dr \\
&= \int P(y|\mathbf{q}, \mathbf{r}, \mathbf{x}) P(\mathbf{q}, \mathbf{r}|\mathbf{D}) dq dr, \qquad (10.10)
\end{aligned}
$$

where the last step follows from the fact that the prediction of the network is solely determined by its (hyper)parameters,

$$
P(y|\mathbf{q}, \mathbf{r}, \mathbf{x}, \mathbf{D}) = P(y|\mathbf{x}, \mathbf{q}, \mathbf{r}) \qquad (10.11)
$$

and that for *supervised* learning the distribution of the input data is *not* modelled, that is, the distributions of $\mathbf{q}$ and $\mathbf{r}$ do not depend on $\mathbf{x}$:

$$
P(\mathbf{q}, \mathbf{r}|\mathbf{x}, \mathbf{D}) = P(\mathbf{q}, \mathbf{r}|\mathbf{D}). \qquad (10.12)
$$

The problem now is that the integral in Equation (10.10) is usually analytically intractable[2]. Therefore, a standard paradigm in neural network training is to replace the average with respect to the posterior distribution $P(\mathbf{q}, \mathbf{r}|\mathbf{D})$ by the function value at the mode $(\hat{\mathbf{q}}, \hat{\mathbf{r}})$ of $P(\mathbf{q}, \mathbf{r}|\mathbf{D})$:

$$
P(y|\mathbf{x}, \mathbf{D}) \approx P(y|\mathbf{x}, \hat{\mathbf{q}}, \hat{\mathbf{r}}). \qquad (10.13)
$$

where

$$
(\hat{\mathbf{q}}, \hat{\mathbf{r}}) = \text{argmax}_{\mathbf{q}, \mathbf{r}} P(\mathbf{q}, \mathbf{r}|\mathbf{D}) \qquad (10.14)
$$

As discussed before, this approximation may lead to suboptimal results: if the distribution is strongly skewed, the mode is not representative for the distribution as a whole (see also [40], [41]). A network whose (hyper)parameters are set to $(\hat{\mathbf{q}}, \hat{\mathbf{r}})$ then typically overfits the training data and shows only a poor generalisation performance. This problem can be alleviated by following

---

[1] Recall that this condition is satisfied for an mth order Markov process.

[2] More recent approaches try to numerically approximate this integral by sampling the parameters from the posterior distribution with hybrid Monte Carlo. This will not be covered here. The interested reader is referred to [44].

an idea introduced to the neural network community by MacKay [37]. Rather than setting $(\hat{\mathbf{q}}, \hat{\mathbf{r}})$ to the joint mode according to Equation (10.14), the mode of the hyperparameters $\mathbf{r}$ is found after integrating out the parameters $\mathbf{q}$,

$$\hat{\mathbf{r}} = \text{argmax}_{\mathbf{r}} P(\mathbf{r}|\mathbf{D}), \qquad P(\mathbf{r}|\mathbf{D}) = \int P(\mathbf{q}, \mathbf{r}|\mathbf{D}) d\mathbf{q} \qquad (10.15)$$

The idea behind this scheme is that integrating out $\mathbf{q}$ captures relevant information about the volume of the distribution, especially if dim $\mathbf{q} \gg$ dim $\mathbf{r}$. Consequently, the mode of the marginal distribution $P(\mathbf{r}|\mathbf{D})$ is likely to be more representative for the distribution as a whole than the mode of the joint distribution $P(\mathbf{q}, \mathbf{r}|\mathbf{D})$ (or, equivalently, that the skewness of $P(\mathbf{r}|\mathbf{D})$ is smaller than that of $P(\mathbf{q}, \mathbf{r}|\mathbf{D})$). The following section provides a simple, illustrative example, which was first published in [8] and [41].

10.2 A simple illustration of the evidence idea

Consider a set $\mathbf{D} = \{x_t\}_{t=1}^N$ of independently Gaussian distributed random variables

$$P(x_t|\mu, \beta) = \sqrt{\frac{\beta}{2\pi}} \exp\left(-\frac{\beta}{2}[x_t - \mu]^2\right) \qquad (10.16)$$

We want to infer the mean μ and the precision (inverse variance) β from the data $\mathbf{D}$, whose likelihood is given by

$$\begin{aligned} P(\mathbf{D}|\mu, \beta) &= \left(\frac{\beta}{2\pi}\right)^{N/2} \exp\left(-\frac{\beta}{2}\sum_{t=1}^N [x_t - \mu]^2\right) \\ &= \left(\frac{\beta}{2\pi}\right)^{N/2} \exp\left(-\frac{\beta}{2}\sum_{t=1}^N [x_t - \bar{x}]^2 - \frac{N\beta}{2}[\bar{x} - \mu]^2\right) \end{aligned} \quad (10.17)$$

where

$$\bar{x} := \frac{1}{N}\sum_{t=1}^N x_t \qquad (10.18)$$

is the empirical sample mean. Let us first find the maximum likelihood estimate for both parameters simultaneously. The log-likelihood for the data is given by

$$\ln P(\mathbf{D}|\mu, \beta) = \frac{N}{2}\ln\left(\frac{\beta}{2\pi}\right) - \frac{\beta}{2}\sum_{t=1}^N (x_t - \bar{x})^2 - \frac{N\beta}{2}[\bar{x} - \mu]^2 \qquad (10.19)$$

Taking the derivatives with respect to μ and β gives:

$$\frac{\partial}{\partial \mu} \ln P(\mathbf{D}|\mu, \beta) = N\beta(\bar{x} - \mu)$$

$$\frac{\partial}{\partial \beta} \ln P(\mathbf{D}|\mu, \beta) = \frac{N}{2\beta} - \frac{1}{2} \sum_{t=1}^{N} (x_t - \bar{x})^2 - \frac{N}{2}[\bar{x} - \mu]^2 \quad (10.20)$$

Setting these derivatives to zero, we obtain for the joint maximum likelihood estimate:

$$\hat{\mu} = \bar{x}, \qquad \frac{1}{\hat{\beta}} = \frac{1}{N} \sum_{t=1}^{N} (x_t - \bar{x})^2 \quad (10.21)$$

The mean μ is thus approximated by the empirical mean, which is an unbiased estimator for μ. However, the variance $\frac{1}{\beta}$ is approximated by $\frac{1}{N} \sum_{t=1}^{N} (x_t - \bar{x})^2$, which is *not* unbiased. The proper unbiased estimator is known to be $\frac{1}{N-1} \sum_{t=1}^{N} (x_t - \bar{x})^2$ (see, for example, [22]), since fitting the mean from the data reduces the number of degrees of freedom by one. Put in another way, the empirical mean $\bar{x}$ inevitably fits part of the noise, so the expression on the right of Equation (10.21) systematically underestimates the variance. This might not be a serious problem for this simple 2-parameter model if N is sufficiently large, but the problem is aggravated in neural networks, where considerably more parameters have to be fitted from the data.

Let us now turn to the second alternative of first integrating μ out and then finding the maximum likelihood estimate of the marginalised distribution $P(\mathbf{D}|\beta)$. Taking a uniform prior for μ, $P(\mu) = C$, we obtain:

$$P(\mathbf{D}|\beta) = \int P(\mathbf{D}, \mu|\beta) d\mu = \int P(\mathbf{D}|\mu, \beta) P(\mu) d\mu = C \int P(\mathbf{D}|\mu, \beta) d\mu \quad (10.22)$$

which after inserting Equation (10.17) yields:

$$P(\mathbf{D}|\beta) = C \int \left(\frac{\beta}{2\pi}\right)^{N/2} \exp\left(-\frac{\beta}{2} \sum_{t=1}^{N} [x_t - \bar{x}]^2 - \frac{N\beta}{2}[\bar{x} - \mu]^2\right) d\mu$$

$$= C \left(\frac{\beta}{2\pi}\right)^{N/2} \exp\left(-\frac{\beta}{2} \sum_{t=1}^{N} [x_t - \bar{x}]^2\right) \int \exp\left(-\frac{N\beta}{2}[\bar{x} - \mu]^2\right) d\mu$$

$$\propto \left(\frac{\beta}{2\pi}\right)^{(N-1)/2} \exp\left(-\frac{\beta}{2} \sum_{t=1}^{N} [x_t - \bar{x}]^2\right) \quad (10.23)$$

We obtain the maximum likelihood estimate for β by setting the derivative of the log-likelihood to zero:

$$\frac{\partial}{\partial \beta} \ln P(\mathbf{D}|\beta) = \frac{\partial}{\partial \beta} \left[\frac{N-1}{2} \ln\left(\frac{\beta}{2\pi}\right) - \frac{\beta}{2} \sum_{t=1}^{N} [x_t - \bar{x}]^2\right] = 0 \quad (10.24)$$

This leads to the correct *unbiased* estimate for the variance:

$$\frac{1}{\hat{\beta}} \quad = \quad \frac{1}{N-1} \sum_{t=1}^{N} (x_t - \bar{x})^2 \qquad (10.25)$$

10.3 Overview of the evidence scheme

The essential idea of the evidence scheme is to find the mode of the hyper-parameters $\hat{\mathbf{r}}$ after integrating out the parameters according to (10.15), and then to find the mode of the parameters $\mathbf{q}$ conditional on $\hat{\mathbf{r}}$,

$$\hat{\mathbf{q}} \quad = \quad \text{argmax}_{\mathbf{q}} \{ P(\mathbf{q}|\mathbf{D}, \hat{\mathbf{r}}) \} \qquad (10.26)$$

In practice the analytic integration in (10.15) can usually only be done by Gaussian approximation. This leads to the following iteration scheme:

1. Given a preliminary estimate for $\hat{\mathbf{r}}$, optimise the parameters $\mathbf{q}$ according to Equation (10.26), $\mathbf{q} \rightarrow \hat{\mathbf{q}}$, and approximate $P(\mathbf{q}|\mathbf{D}, \hat{\mathbf{r}})$ by a multivariate Gaussian about $\hat{\mathbf{q}}$.

2. Solve the integral in (10.15) by Gaussian approximation and find the new mode $\hat{\mathbf{r}}$. Go back to step 1.

This scheme is repeated until a self-consistent solution $(\hat{\mathbf{q}}, \hat{\mathbf{r}})$ has been found. The following subsections describe the various steps in more detail.

10.3.1 First step:
Gaussian approximation to the probability in parameter space

The objective of the first step is to approximate the distribution of the parameters $\mathbf{q}$, conditional on the hyperparameters $\mathbf{r}$ and the observed data $\mathbf{D}$, by a Gaussian about its mode $\hat{\mathbf{q}}$. Let us first apply Bayes rule

$$P(\mathbf{q}|\mathbf{D}, \mathbf{r}) \quad \propto \quad P(\mathbf{D}|\mathbf{q}, \mathbf{r}) P(\mathbf{q}|\mathbf{r}) \qquad (10.27)$$

and define

$$
\begin{aligned}
E(\mathbf{q}|\mathbf{r}) &:= -\ln P(\mathbf{D}|\mathbf{q}, \mathbf{r}) & (10.28) \\
R(\mathbf{q}|\mathbf{r}) &:= -\ln P(\mathbf{q}|\mathbf{r}) & (10.29) \\
E^*(\mathbf{q}|\mathbf{r}) &:= E(\mathbf{q}|\mathbf{r}) + R(\mathbf{q}|\mathbf{r}) & (10.30)
\end{aligned}
$$

This is in close correspondence to the definitions (9.23), (9.26), and (9.27) in Chapter 9, except that the *likelihood error function* E may also depend on $\mathbf{r}$ since we have not decided on the nature of the hyperparameters yet. As before in Chapter 9, we will refer to R as the *regularisation term*, and to E^* as the *total error function*. From (10.27)-(10.30) we see that

$$P(\mathbf{q}|\mathbf{D}, \mathbf{r}) \quad \propto \quad \exp\left(- E^*(\mathbf{q}|\mathbf{r})\right) \tag{10.31}$$

It is thus clear that finding the mode $\hat{\mathbf{q}}$ of $P(\mathbf{q}|\mathbf{D}, \hat{\mathbf{r}})$ is equivalent to finding the minimum of $E^*(\mathbf{q}|\hat{\mathbf{r}})$,

$$\hat{\mathbf{q}} \quad := \quad \mathrm{argmax}_{\mathbf{q}}\{P(\mathbf{q}|\mathbf{D}, \hat{\mathbf{r}})\} \quad = \quad \mathrm{argmin}_{\mathbf{q}}\{E^*(\mathbf{q}|\hat{\mathbf{r}})\} \tag{10.32}$$

which can be accomplished with the regularised EM algorithm described in Section 9.4. Let us further define

$$\mathbf{H} := \left[\nabla_{\mathbf{q}} \nabla_{\mathbf{q}}^{\dagger} E\right]_{\mathbf{q}=\hat{\mathbf{q}}}, \qquad \mathbf{H}^* := \left[\nabla_{\mathbf{q}} \nabla_{\mathbf{q}}^{\dagger} E^*\right]_{\mathbf{q}=\hat{\mathbf{q}}} \tag{10.33}$$

which are, respectively, the Hessians of E and E^* at the mode $\mathbf{q} = \hat{\mathbf{q}}$. Then the Gaussian approximation to (10.27) is to assume a quadratic function E^*,

$$E^*(\mathbf{q}|\hat{\mathbf{r}}) \quad = \quad E^*(\hat{\mathbf{q}}|\hat{\mathbf{r}}) + \frac{1}{2}(\mathbf{q} - \hat{\mathbf{q}})^{\dagger} \mathbf{H}^*(\mathbf{q} - \hat{\mathbf{q}}) \tag{10.34}$$

or, equivalently, approximate $P(\mathbf{q}|\hat{\mathbf{r}}, \mathbf{D}) \propto \exp\left[- E^*(\mathbf{q}|\hat{\mathbf{r}})\right]$ by

$$P(\mathbf{q}|\hat{\mathbf{r}}, \mathbf{D}) \quad \propto \quad \exp\left[- E^*(\hat{\mathbf{q}}|\hat{\mathbf{r}})\right] \exp\left(-\frac{1}{2}(\mathbf{q} - \hat{\mathbf{q}})^{\dagger} \mathbf{H}^*(\mathbf{q} - \hat{\mathbf{q}})\right) \tag{10.35}$$

which when normalised gives

$$P(\mathbf{q}|\hat{\mathbf{r}}, \mathbf{D}) \quad = \quad \sqrt{\frac{\det \mathbf{H}^*}{(2\pi)^M}} \exp\left(-\frac{1}{2}(\mathbf{q} - \hat{\mathbf{q}})^{\dagger} \mathbf{H}^*(\mathbf{q} - \hat{\mathbf{q}})\right). \tag{10.36}$$

10.3.2 Second step: Optimising the hyperparameters

The hyperparameters $\mathbf{r}$ are optimised by finding the mode of $P(\mathbf{r}|\mathbf{D})$. With Bayes' rule,

$$P(\mathbf{r}|\mathbf{D}) \quad \propto \quad P(\mathbf{D}|\mathbf{r})P(\mathbf{r}), \tag{10.37}$$

and the assumption of a (locally[3]) constant prior,

---

[3] Strictly speaking, equation (10.38) only holds locally since otherwise the probability $P(\mathbf{r})$ cannot be normalised. However, since the domain can be chosen arbitrarily large, this subtlety is irrelevant for the following derivations. In the statistics literature, one usually refers to a prior of the form (10.38), where implicitly an infinitely large domain is assumed, as an *improper* prior.

$$P(\mathbf{r}) = C_r, \tag{10.38}$$

this is equivalent to maximising $P(\mathbf{D}|\mathbf{r})$,

$$\hat{\mathbf{r}} = \text{argmax}_{\mathbf{r}}\{P(\mathbf{D}|\mathbf{r})\}. \tag{10.39}$$

Using (10.27)-(10.30) and the quadratic approximation (10.34), one gets

$$
\begin{aligned}
P(\mathbf{D}|\mathbf{r}) &= \int P(\mathbf{D}, \mathbf{q}|\mathbf{r})d\mathbf{q} = \int P(\mathbf{D}|\mathbf{q}, \mathbf{r})P(\mathbf{q}|\mathbf{r})d\mathbf{q} \\
&= \int \exp\left[-E(\mathbf{q}|\mathbf{r})\right]\exp\left[-R(\mathbf{q}|\mathbf{r})\right]d\mathbf{q} = \int \exp\left[-E^*(\mathbf{q}|\mathbf{r})\right]d\mathbf{q} \\
&= \exp\left[-E^*(\hat{\mathbf{q}}|\mathbf{r})\right]\int \exp\left(-\frac{1}{2}(\mathbf{q}-\hat{\mathbf{q}})^\dagger \mathbf{H}^*(\mathbf{q}-\hat{\mathbf{q}})\right)d\mathbf{q}
\end{aligned}
$$

$$\Longrightarrow \qquad P(\mathbf{D}|\mathbf{r}) \quad = \quad \exp\left[-E^*(\hat{\mathbf{q}}|\mathbf{r})\right]\sqrt{\frac{(2\pi)^M}{\det \mathbf{H}^*}} \tag{10.40}$$

and for the evidence

$$\ln P(\mathbf{D}|\mathbf{r}) \quad = \quad -E^*(\hat{\mathbf{q}}|\mathbf{r}) - \frac{1}{2}\ln\det \mathbf{H}^* + \frac{M}{2}\ln(2\pi) \tag{10.41}$$

The condition for the mode, $\left[\nabla_{\mathbf{r}}\ln P(\mathbf{D}|\mathbf{r})\right]_{\mathbf{r}=\hat{\mathbf{r}}} = 0$, thus leads to the following equation for the optimal hyperparameters $\hat{\mathbf{r}}$:

$$\nabla_{\mathbf{r}}E^*(\hat{\mathbf{q}}|\mathbf{r}) \quad = \quad -\frac{1}{2}\nabla_{\mathbf{r}}\ln\det \mathbf{H}^* \qquad \text{for} \quad \mathbf{r} = \hat{\mathbf{r}} \tag{10.42}$$

10.3.3 A self-consistent iteration scheme

The iteration scheme, outlined on page 152, can now be reformulated as follows:

1. Given a tentative estimate of $\hat{\mathbf{r}}$, optimise the parameters $\mathbf{q}$ according to (10.32).

2. Given a tentative estimate of $\hat{\mathbf{q}}$, re-estimate the hyperparameters according to (10.42), which implies calculating the Hessian $\mathbf{H}^*$ at $\hat{\mathbf{q}}$.

3. Goto 1.

This loop is to be iterated several times until some convergence criterion is satisfied.

10.4 Implementation of the evidence scheme

We now have to become more specific about the nature of the parameters $\mathbf{q}$ and hyperparameters $\mathbf{r}$. A straightforward approach would be to choose $\mathbf{q}$ and $\mathbf{r}$ as in Chapter 9, that is, with $\mathbf{q}$ representing all the adaptable parameters in the network, and $\mathbf{r}$ the parameters of their prior distributions. However, the derivation of the explicit update equation is rather lengthy, and a modified and simpler scheme, which follows more closely MacKay's approach in [37], can in fact be shown to achieve better results (see [26]). Rather than treating all parameters on an equal footing, we split them up into two parts. The weights $\mathbf{w}$ feeding into the $\mathcal{G}$-nodes are treated like 'normal' parameters, that is, for the remainder of this and the following chapter we set $\mathbf{q} := \mathbf{w}$. The kernel width parameters $\boldsymbol{\beta} = (\beta_1, \ldots, \beta_K)$ and the output weights $\mathbf{a} = (a_1, \ldots, a_K)$ are treated as hyperparameters, which means that they are optimised only after the weights $\mathbf{w}$ have been integrated out. The prior distribution for the weights $\mathbf{w}$ is chosen identical to Equation (9.19): a multivariate Gaussian with inverse variances (or precisions) $\boldsymbol{\alpha} = (\alpha_1, \ldots, \alpha_K)$. The total set of hyperparameters is thus given by $\mathbf{r} = (\mathbf{a}, \boldsymbol{\beta}, \boldsymbol{\alpha})$, for which we assume a uniform[4] prior distribution. To summarise:

$$\mathbf{q} := \mathbf{w} = (\mathbf{w}_1, \ldots, \mathbf{w}_K), \qquad \mathbf{r} := (\mathbf{a}, \boldsymbol{\beta}, \boldsymbol{\alpha}) \qquad (10.43)$$

and for the priors:

$$P(\mathbf{q}|\mathbf{r}) = P(\mathbf{w}|\boldsymbol{\alpha}, \boldsymbol{\beta}, \mathbf{a}) = P(\mathbf{w}|\boldsymbol{\alpha}) = \prod_{k=1}^{K} \left(\frac{\alpha_k}{2\pi}\right)^{\frac{W}{2}} \exp\left(-\frac{\alpha_k}{2}\mathbf{w}_k^\dagger \mathbf{w}_k\right) \quad (10.44)$$

$$P(\mathbf{r}) = const \qquad (10.45)$$

in which

$$W \quad := \quad \dim \mathbf{w}_k \quad \forall k \qquad (10.46)$$

Equation (10.46) results from the fact that all weight groups in the GM-RVFL network are of the same dimension (see Section 7.3, page 111), although this simplification is not essential (it does allow a convenient simplification of the notification, though). The dimensions of the parameter and hyperparameter vectors are given by

$$\dim \mathbf{q} = KW := M \qquad (10.47)$$
$$\dim \mathbf{r} = 3K - 1 := m \qquad (10.48)$$

Recall that K represents the number of weight groups in the network, and that the '-1' stems from the constraint $\sum_k a_k = 1$ on the output weights.

---

[4] It would be more precise to choose an *uninformative* prior, which for α and β is uniform on a logarithmic scale. For the derivation in this chapter this distinction is irrelevant. It needs, however, to be addressed later in connection with the *model* evidence.

10.4.1 First step:
Gaussian approximation to the probability in parameter space

From Equation (10.44) we obtain for the regularisation term (10.29) (making use of definition (10.43)):

$$R(\mathbf{w}|\mathbf{r}) = -\ln P(\mathbf{w}|\alpha) = \sum_{k=1}^{K} \left[\frac{\alpha_k}{2} \mathbf{w}_k^\dagger \mathbf{w}_k - \frac{W}{2} \ln \left(\frac{\alpha_k}{2\pi} \right) \right] \qquad (10.49)$$

which in fact only depends on α,

$$R(\mathbf{w}|\mathbf{r}) = R(\mathbf{w}|\alpha). \qquad (10.50)$$

The gradient with respect to $\mathbf{w}_k$ is identical to the expression in Equation (9.39),

$$\nabla_{\mathbf{w}_k} R = \alpha_k \mathbf{w}_k, \qquad (10.51)$$

and the derivation of the expression for the mode follows exactly Section 9.5. This reproduces the result of Equation (9.42):

$$\left(\mathbf{G} \mathbf{\Pi}_k \mathbf{G}^\dagger + \frac{\alpha_k}{\beta_k} \mathbf{I} \right) \hat{\mathbf{w}}_k = \mathbf{G} \mathbf{\Pi}_k \mathbf{y} \qquad (10.52)$$

in which $\mathbf{G}$ and $\mathbf{\Pi}_k$ are the matrices of definitions (7.31) and (7.32). Recall that, by comparison with Equation (7.64), the quotient $\lambda_k := \frac{\alpha_k}{\beta_k}$ corresponds to a weight-decay parameter for the kth weight group. We now have to derive an expression for the Hessian at the mode, $\mathbf{H} = [\nabla_{\mathbf{w}} \nabla_{\mathbf{w}}^\dagger E]_{\hat{\mathbf{w}}}$. This derivation is rather lengthy and will therefore be relegated to the appendix. The result is

$$\mathbf{H}_{\mathbf{w}_i, \mathbf{w}_k} := \nabla_{\mathbf{w}_i} \nabla_{\mathbf{w}_k}^\dagger E = \delta_{ik} \mathbf{H}_k \qquad (10.53)$$

where

$$\mathbf{H}_k := \beta_k \mathbf{G} \mathbf{\Pi}_k \mathbf{G}^\dagger - (\beta_k)^2 \sum_{t=1}^{N} \pi_k(t) \big[1 - \pi_k(t) \big] \Big(y_t - \mu(\mathbf{x}_t; \mathbf{w}_k) \Big)^2 \mathbf{g}(\mathbf{x}_t) \mathbf{g}^\dagger(\mathbf{x}_t) \qquad (10.54)$$

The vectors $\mathbf{g}(\mathbf{x}_t)$ were defined in Chapter 7, in the text over Equation (7.28) on page 111. The Hessian $\mathbf{H}^*$ of the total error function E^* follows immediately from definition (10.30),

$$\nabla_{\mathbf{w}_k} \nabla_{\mathbf{w}_k}^\dagger E^* = \nabla_{\mathbf{w}_k} \nabla_{\mathbf{w}_k}^\dagger E + \nabla_{\mathbf{w}_k} \nabla_{\mathbf{w}_k}^\dagger R. \qquad (10.55)$$

From (10.51) we obtain

$$\nabla_{\mathbf{w}_i} \nabla_{\mathbf{w}_k}^\dagger R = \delta_{ik} \alpha_k \mathbf{I} \qquad (10.56)$$

where $\mathbf{I}$ is the unit matrix. This leads to:

$$\mathbf{H}^*_{\mathbf{w}_i, \mathbf{w}_k} := \nabla_{\mathbf{w}_i} \nabla_{\mathbf{w}_k}^\dagger E^* = \delta_{ik} (\mathbf{H}_k + \alpha_k \mathbf{I}) \qquad (10.57)$$

10.4.2 Second step: Optimising the hyperparameters

The condition for the optimal hyperparameters is given by Equation (10.42), where

$$\ln \det \mathbf{H}^* \;=\; \sum_{k=1}^{K} \ln \det[\mathbf{H}_k + \alpha_k \mathbf{I}] \tag{10.58}$$

due to Equation (10.57). If we define

$$\tilde{R}(\mathbf{r}) \;:=\; R(\hat{\mathbf{w}}|\alpha) + \frac{1}{2} \sum_{k=1}^{K} \ln \det[\mathbf{H}_k + \alpha_k \mathbf{I}], \tag{10.59}$$

where $R(\hat{\mathbf{w}}|\alpha)$ was defined in equations (10.49) and (10.50), then (10.42) can be re-written as (using definition (10.30))

$$\nabla_{\mathbf{r}} E(\hat{\mathbf{q}}|\mathbf{r}) + \nabla_{\mathbf{r}} \tilde{R}(\hat{\mathbf{q}}|\mathbf{r}) \;=\; 0 \tag{10.60}$$

or

$$[\nabla_{\mathbf{a}} E]_{\hat{\mathbf{a}}} = -\Big[\nabla_{\mathbf{a}}\tilde{R}\Big]_{\hat{\mathbf{a}}}, \quad [\nabla_{\beta} E]_{\hat{\beta}} = -\Big[\nabla_{\beta}\tilde{R}\Big]_{\hat{\beta}}, \quad [\nabla_{\alpha} E]_{\hat{\alpha}} = -\Big[\nabla_{\alpha}\tilde{R}\Big]_{\hat{\alpha}}. \tag{10.61}$$

Recall that the actual network parameters are $\mathbf{w}$, β and $\mathbf{a}$. Therefore $E = -\ln P(\mathbf{D}|\mathbf{q},\mathbf{r}) = -\ln P(\mathbf{D}|\mathbf{w},\beta,\mathbf{a})$ does not explicitly depend on α, and $\nabla_{\alpha} E = 0$. Moreover, we can replace the derivatives of E by the derivatives of the EM error function U due to (7.9). Equation (10.61) can therefore be written as

$$[\nabla_{\mathbf{a}} U]_{\hat{\mathbf{a}}} = -\Big[\nabla_{\mathbf{a}}\tilde{R}\Big]_{\hat{\mathbf{a}}}, \quad [\nabla_{\beta} U]_{\hat{\beta}} = -\Big[\nabla_{\beta}\tilde{R}\Big]_{\hat{\beta}}, \quad 0 = -\Big[\nabla_{\alpha}\tilde{R}\Big]_{\hat{\alpha}}. \tag{10.62}$$

Inserting (7.18) and (7.19) into (10.62) gives

$$\sum_{t=1}^{N} \frac{\pi_k(t)}{2}\left[\big(y_t - \mu(\mathbf{x}_t;\mathbf{w})\big)^2 - \frac{1}{\beta_k}\right] \;=\; -\frac{\partial \tilde{R}}{\partial \beta_k} \tag{10.63}$$

$$-\frac{1}{a_k}\sum_{t=1}^{N}\pi_k(t) + \frac{1}{a_K}\sum_{t=1}^{N}\pi_K(t) \;=\; -\frac{\partial \tilde{R}}{\partial a_k} \tag{10.64}$$

$$0 \;=\; -\frac{\partial \tilde{R}}{\partial \alpha_k} \tag{10.65}$$

which is similar to equations (9.36) and (9.37). Therefore only the derivatives of $\tilde{R}$ remain to be calculated.

Output weights a. Equation (10.59) shows that the modified regularisation term $\tilde{R}$ only depends on α and β (the latter dependence is via $\mathbf{H}_k$, defined in (10.54)). Hence

$$\nabla_{\mathbf{a}}\tilde{R} = 0, \tag{10.66}$$

and inserting this result into (10.64) gives

$$-\frac{1}{a_k}\sum_{t=1}^{N}\pi_k(t) + \frac{1}{a_K}\sum_{t=1}^{N}\pi_K(t) = 0 \tag{10.67}$$

which is identical to (7.20). As shown on page 103, this leads to

$$\hat{a}_k = \frac{1}{N}\sum_{t=1}^{N}\pi_k(t) \tag{10.68}$$

which reproduces the result of Equation (7.23).

Regularisation hyperparameters α. From definition (10.59) we obtain

$$\frac{\partial\tilde{R}}{\partial\alpha_k} = \frac{\partial R}{\partial\alpha_k} + \frac{1}{2}\frac{\partial}{\partial\alpha_k}\ln\det[\mathbf{H}_k + \alpha_k\mathbf{I}]. \tag{10.69}$$

The first term on the right follows from (10.49):

$$\frac{\partial R}{\partial\alpha_k} = \frac{1}{2}\left(\hat{\mathbf{w}}_k^{\dagger}\hat{\mathbf{w}}_k - \frac{W}{\alpha_k}\right), \tag{10.70}$$

If we denote the W eigenvalues of $\mathbf{H}_k$ by $\{\varepsilon_k^{\nu}\}_{\nu=1}^{W}$, we get

$$\frac{\partial}{\partial\alpha_k}\ln\det(\mathbf{H}_k + \alpha_k\mathbf{I}) = \frac{\partial}{\partial\alpha_k}\sum_{\nu=1}^{W}\ln(\varepsilon_k^{\nu} + \alpha_k) = \sum_{\nu=1}^{W}\frac{1}{\varepsilon_k^{\nu} + \alpha_k} \tag{10.71}$$

With the definition of the number of *well-determined parameters* in the kth weight-group (see Section 10.5.1 for a discussion of this term),

$$\gamma_k := \sum_{\nu=1}^{W}\frac{\varepsilon_k^{\nu}}{\varepsilon_k^{\nu} + \alpha_k} \tag{10.72}$$

this can be written in the more compact form

$$\frac{\partial}{\partial\alpha_k}\ln\det(\mathbf{H}_k + \alpha_k\mathbf{I}) = \frac{W - \gamma_k}{\alpha_k}. \tag{10.73}$$

Inserting equations (10.70) and (10.73) into (10.69) leads to

$$\frac{\partial\tilde{R}}{\partial\alpha_k} = \frac{1}{2}\left(\hat{\mathbf{w}}_k^{\dagger}\hat{\mathbf{w}}_k - \frac{\gamma_k}{\alpha_k}\right) \tag{10.74}$$

Equating this expression to zero according to Equation (10.65) yields

$$\frac{1}{\hat{\alpha}_k} = \frac{1}{\gamma_k}\hat{\mathbf{w}}_k^{\dagger}\hat{\mathbf{w}}_k \tag{10.75}$$

Kernel width parameters β. Since the first term on the right of (10.59) does not depend on β, one gets

$$\frac{\partial \tilde{R}}{\partial \beta_k} = \frac{1}{2} \frac{\partial}{\partial \beta_k} \ln \det[\mathbf{H}_k + \alpha_k \mathbf{I}] = \frac{1}{2} \frac{\partial}{\partial \beta_k} \sum_{\nu=1}^{W} \ln(\varepsilon_k^\nu + \alpha_k) = \frac{1}{2} \sum_{\nu=1}^{W} \frac{1}{\varepsilon_k^\nu + \alpha_k} \frac{\partial \varepsilon_k^\nu}{\partial \beta_k}$$
(10.76)

The ε_k^ν denote the eigenvalues of $\mathbf{H}_k$, as introduced in the text over equation (10.71). They can easily be obtained from a singular value decomposition of $\mathbf{H}_k$. The problem, however, is that they are given as *values* rather than as *functions*. It is therefore impossible to obtain an analytic expression for $\frac{\partial \varepsilon_k^\nu}{\partial \beta_k}$. In principle one could obtain numerical estimates of the derivatives by repeating the singular value decomposition of $\mathbf{H}_k(\beta_k)$ for different values of β_k. But this would incur a dramatic increase in the computational costs. Therefore the following approximation will be adopted. From (10.54), the eigenvalues are of the form $\varepsilon_k^\nu = A_\nu \beta_k + \mathcal{O}(\beta_k)^2$, where A_ν is a constant. If we assume that the second term is much smaller than the first one, then ε_k is approximately homogeneous in β_k, $\varepsilon_k^\nu \propto \beta_k$, and hence

$$\frac{\partial \varepsilon_k^\nu}{\partial \beta_k} \approx \frac{\varepsilon_k^\nu}{\beta_k}.$$
(10.77)

Making this approximation and using definition (10.72) for the number of well-determined parameters γ_k leads to

$$\frac{\partial \tilde{R}}{\partial \beta_k} \approx \frac{1}{2} \sum_{\nu=1}^{W} \frac{1}{\varepsilon_k^\nu + \alpha_k} \frac{\varepsilon_k^\nu}{\beta_k} = \frac{\gamma_k}{2\beta_k},$$
(10.78)

which when inserted into (10.63) yields

$$\sum_{t=1}^{N} \frac{\pi_k(t)}{2} \left[\left(y_t - \mu(\mathbf{x}_t; \mathbf{w}_k) \right)^2 - \frac{1}{\beta_k} \right] = -\frac{\gamma_k}{2\beta_k}$$

$$\Rightarrow \quad \frac{1}{\hat{\beta}_k} = \frac{\sum_{t=1}^{N} \pi_k(t) \left(y_t - \mu(\mathbf{x}_t; \hat{\mathbf{w}}_k) \right)^2}{\sum_{t=1}^{N} \pi_k(t) - \gamma_k}.$$
(10.79)

The justification of the homogeneity approximation will be discussed in Section 10.5.2.

10.4.3 Algorithm

Summarising, we can re-write the algorithm of Section 10.3.3 in the following more explicit form:

1. **Adaptation of the weights:** Adapt the weights **w** according to (10.52), using the kernel widths $\sigma_k = (\beta_k)^{-1/2}$ and weight-decay hyperparameters α_k from the previous step.

2. **Hessian and number of well-determined parameters:** For each of the weight groups $\mathbf{w}_k$, $k \in \{1, \ldots, K\}$ (where it may be recalled that $\mathbf{w}_k$ denotes the vector of all the weights feeding into the kth $\mathcal{G}$-node), calculate the Hessian $\mathbf{H}_k$ from (10.54), and the number of well-determined parameters from (10.72). The last step requires the determination of the eigenvalues ε_k^ν of $\mathbf{H}_k$, which can be obtained with the method of singular value decomposition (using, for example, the subroutine given in [51], pp.67-70).

3. **Priors:** Obtain the new priors or output weights a_k from (10.68).

4. **Kernel widths:** Compute the new kernel widths $\sigma_k = (\beta_k)^{-1/2}$ by application of (10.79).

5. **Weight decay hyperparameters:** Update the weight decay hyperparameters α_k according to (10.75).

10.5 Discussion

10.5.1 Improvement over the maximum likelihood estimate

Recall that the ε_k^ν are the eigenvalues of the Hessian $\mathbf{H}_k$, that is, they indicate the curvature of the error function E along the eigendirections in parameter space. A directions for which $\varepsilon_k^\nu \gg \alpha_k$ will give a contribution close to one in the sum in Equation (10.72). In this case the corresponding component of the parameter vector is determined primarily by the data. If, however, the component of the parameter vector is determined mainly by the prior, then $\varepsilon_k^\nu \ll \alpha_k$, and the contribution to the sum in Equation (10.72) is small. Thus γ_k is a measure for the effective number of parameters whose values are controlled by the data rather than by the prior. MacKay [37] therefore called this term the *number of well-determined parameters*.

A comparison of Equation (10.79) with the earlier maximum likelihood estimate (7.24) shows that as a result of applying the Bayesian evidence scheme the denominator in the expression for the kernel variance is decreased by the number of well-determined parameters. This is a generalisation of the simple example of Section 10.2. In the expression for the unbiased variance estimate, Equation (10.25), the number of samples had to be reduced by the number of fitted parameters - in this case one (namely μ) - since fitting a parameter from the data 'uses up' a degree of freedom. Put in another way, when fitting a model to noisy data, it is unavoidable that some fitting of the parameters to noise will occur, and the variance estimate of Equation (10.25) takes this into account by subtracting the number of fitted parameters from the denominator. Equation (10.79) generalises this concept by subtracting from the denominator only the number of those parameters that are well determined by the data. Poorly determined parameters do not fit to noise and therefore do not need to be taken into account. For a more detailed discussion of this subject, the reader is referred to [37].

10.5.2 Justification of the approximations

Quadratic approximation in parameter space. The distribution in parameter space, $P(\mathbf{w}|\mathbf{D}, \mathbf{r})$, is approximated by a multivariate Gaussian. From equations (10.53) and (10.54) it is seen that for $K = 1$ the Hessian is independent of $\mathbf{w}$, and the quadratic approximation therefore exact. For $K > 1$, the Hessian contains a $\mathbf{w}$-dependent term; see (10.54). However, as will be discussed below, this contribution can be assumed to be small. This justifies the Gaussian approximation.

It is recalled that the derivations presented in this chapter have been applied to the GM-RVFL network. The derivation of the Hessian, which has been relegated to the appendix, assumes that the kernel centres are linear in the weights $\mathbf{w}_k$ (Equation (18.15)):

$$\mu\left(\mathbf{x}_t; \mathbf{w}_k\right) = \mathbf{w}_k^\dagger \mathbf{g}(\mathbf{x}_t). \tag{10.80}$$

This requirement is satisfied for the GM-RVFL network; see Equation (7.28). For the GM network, Equation (10.80) is only an approximation, valid if terms beyond the first order in the Taylor series expansion are small. A discussion of whether and when the Gaussian approximation is justified in this case follows closely the arguments given in [38] and [44]. Since all the applications presented in the forthcoming chapters were carried out with the GM-RVFL network, however, this topic will not be treated further here.

Homogeneity of the eigenvalues. Equation (10.79) is a generalisation of MacKay's result in [37], which reduces to the latter if $K = 1$. However, MacKay did not have to make the approximation that the eigenvalues are homogeneous in β_k, since for $K = 1$ this *does hold exactly*. This is easily seen from (10.54). If the network has only one kernel, that is, if $K = 1$, then the second, $(\beta_k)^2$-dependent term disappears and (10.77) holds exactly.

For $K > 1$, the approximation can be justified as follows. For $\beta_k \ll 1$, the second matrix on the right of (10.54) can obviously be neglected because of the factor $(\beta_k)^2$. As the kernel widths σ_k shrink during training, $\beta_k = \sigma_k^{-2}$ will become larger and the second term, scaling like β_k^2, seems to become dominating. Note, however, that the terms in the sum are weighted by the factor $\pi_k(t)[1 - \pi_k(t)]$ so that only those exemplars for which $\pi_k(t) \approx 0.5$ will give significant contributions. An increase in β_k results in a concomitant narrowing of the posterior distributions $\pi_k(t)$, leading to a steeper transition from $\pi_k(t) = 1$ to $\pi_k(t) = 0$. For large values of β_k, the number of terms that effectively contribute to the sum will therefore be small, leaving the sum of order $\mathcal{O}(1)$, whereas the first term is of order $\mathcal{O}(N)$. This suggests that the second term in (10.54) will always be considerably smaller than the first, which justifies: (1) the Gaussian approximation (as discussed above), and (2) the assumption that the eigenvalues are homogeneous in β_k.

The self-consistent iteration scheme. Recall that for optimising the hyperparameters, the parameters need to be integrated out first. This is done by Gaussian approximation, (10.40), and requires knowledge of the mode $\hat{\mathbf{w}}$. The latter, however, depends itself on the hyperparameters, which requires an iterative scheme, as outlined in Section 10.3.3. A strict mathematical proof for the convergence of this scheme has not been given yet, but in most applications – here as well as in MacKay's work – the scheme is empirically found to converge. To understand this, consider what happens if we apply the re-estimation formulae for the hyperparameters $\mathbf{r}$ with $\mathbf{w}$ close to the mode $\hat{\mathbf{w}}$. Since, as discussed above, the dependence of the Hessian $\mathbf{H}$ on $\mathbf{w}$ is rather weak, the values of γ_k will be similar as if they had been evaluated at $\hat{\mathbf{w}}$. The values of $\mathbf{w}_k^t \mathbf{w}_k$ and $\mu(\mathbf{x}_t; \mathbf{w}_k)$ in the re-estimation formulae (10.75) and (10.79) are typically slowly varying quantities with $\mathbf{w}$. So if $\mathbf{w}$ is close to $\hat{\mathbf{w}}$, the values of the hyperparameters $\mathbf{r}$ found using the algorithm will not be greatly different from those that would be found at the optimum. Note that for this reason it is advisable to preceed the iteration algorithm by a few unregularised optimisation steps. This will allow the network to quickly capture relevant features of the data, and thus will ensure that in fact the initial guess for $\mathbf{w}$ in the iterations scheme will not deviate too strongly from $\hat{\mathbf{w}}$.

The Hessian. The approximations made for calculating the Hessian are discussed at the end of Chapter 18.

10.5.3 Final remark

This chapter has presented the Bayesian evidence scheme for regularisation, where the hyperparameters, which determine the weight-decay parameters[5]

---

[5] Recall that the weight decay parameters are given by $\lambda_k = \alpha_k/\beta_k$.

for the various weight groups, are optimised on the basis of the data. Applications will be given later, in chapters 12 and 14, and will demonstrate that the generalisation performance can indeed be significantly improved over results obtained with a naive maximum likelihood approach. A consequent extension of this scheme is to integrate over the hyperparameters and obtain an expression for the model evidence. This is the subject of the following chapter.

11. The Bayesian Evidence Scheme for Model Selection

The approach of the previous chapter is extended to the derivation of the model evidence as a measure for model weighting and selection. The idea is to integrate out the hyperparameters in the likelihood term by Gaussian approximation, which requires the derivation of the Hessian at the mode. The resulting expression for the model evidence is found to be an intuitively plausible generalisation of the results obtained by MacKay for Gaussian homoscedastic noise on the target. The nature of the various Ockham factors included in the evidence is discussed. The chapter concludes with a critical evaluation of the numerical inaccuracies inherent in this scheme.

11.1 The evidence for the model

So far the predictions have been conditioned on both the given data and the employed network architecture, $\mathcal{M}$, although this has not been made explicit in the notation (strictly speaking we should have written $P(y|\mathbf{x}, \mathbf{D}, \mathcal{M})$ rather than $P(y|\mathbf{x}, \mathbf{D})$ in Equation (10.10)). We can arrive at a closer approximation to $P(y|\mathbf{x})$ by eliminating the dependence on $\mathcal{M}$, the particular model employed. Formally this is done by summing over the whole model space, which in practice implies setting up a committee of predictors,

$$P(y|\mathbf{x}, \mathbf{D}) = \sum_i P(y|\mathbf{x}, \mathbf{D}, \mathcal{M}_i) P(\mathcal{M}_i|\mathbf{x}, \mathbf{D}) \approx \sum_i \omega_i P(y|\mathbf{x}, \mathbf{D}, \mathcal{M}_i),$$

$$(11.1)$$

where the weighting factors ω_i in the sum on the right are given by

$$\omega_i = P(\mathcal{M}_i|\mathbf{x}, \mathbf{D}) = P(\mathcal{M}_i|\mathbf{D}) \propto P(\mathbf{D}|\mathcal{M}_i) P(\mathcal{M}_i).$$

If we assume that the prior probabilities for the different models, $P(\mathcal{M}_i)$, are all equal, this leads to

$$\omega_i \propto P(\mathbf{D}|\mathcal{M}_i). \qquad (11.2)$$

The logarithm of the expression on the right is called the *model evidence* for network $\mathcal{M}_i$, and can be used as a criterion for model selection. This expression is formally given by integrating out the hyperparameters,

$$P(\mathbf{D}|\mathcal{M}) \quad = \quad \int P(\mathbf{D}|\mathbf{r}, \mathcal{M}) P(\mathbf{r}|\mathcal{M}) d\mathbf{r}. \tag{11.3}$$

If we now make the quadratic approximation for $\ln P(\mathbf{D}|\mathbf{r}, \mathcal{M})$,

$$-\ln P(\mathbf{D}|\mathbf{r}, \mathcal{M}) \quad = \quad -\ln P(\mathbf{D}|\hat{\mathbf{r}}, \mathcal{M}) + \frac{1}{2}(\mathbf{r} - \hat{\mathbf{r}})^\dagger \mathbf{A}(\mathbf{r} - \hat{\mathbf{r}}), \tag{11.4}$$

in which

$$\mathbf{A} \quad := \quad -\left[\nabla_\mathbf{r}\nabla_\mathbf{r}^\dagger \ln P(\mathbf{D}|\mathbf{r}, \mathcal{M})\right]_{\mathbf{r}=\hat{\mathbf{r}}} \tag{11.5}$$

and assume a uniform prior

$$P(\mathbf{r}|\mathcal{M}) \quad = \quad \frac{1}{\Omega_\mathbf{r}}, \tag{11.6}$$

where $\Omega_\mathbf{r}$ is some constant, then Equation (11.3) becomes

$$P(\mathbf{D}|\mathcal{M}) \quad = \quad P(\mathbf{D}|\hat{\mathbf{r}}, \mathcal{M})\frac{1}{\Omega_\mathbf{r}}\sqrt{\frac{(2\pi)^m}{\det \mathbf{A}}} \tag{11.7}$$

Here and in what follows, m denotes the dimension of the hyperparameter vector, as defined in (10.48).

For the model evidence $\ln P(\mathbf{D}|\mathcal{M})$ one gets with this expression and the evidence for the hyperparameters $\ln P(\mathbf{D}|\mathbf{r}, \mathcal{M})$, Equation (10.41) (recall that the dependence on $\mathcal{M}$ has not been made explicit in the notation of Chapter 10):

$$\ln P(\mathbf{D}|\mathcal{M}) = \ln P(\mathbf{D}|\hat{\mathbf{r}}, \mathcal{M}) + \frac{m}{2}\ln(2\pi) - \frac{1}{2}\ln\det\mathbf{A} - \ln\Omega_\mathbf{r} \tag{11.8}$$

$$= -E^*(\hat{\mathbf{q}}|\hat{\mathbf{r}}) - \frac{1}{2}\ln\det\mathbf{H}^* - \frac{1}{2}\ln\det\mathbf{A} + \frac{M+m}{2}\ln(2\pi) - \ln\Omega_\mathbf{r}$$

From (10.41) and (11.5) we obtain for $\mathbf{A}$:

$$\mathbf{A} \quad = \quad \left[\nabla_\mathbf{r}\nabla_\mathbf{r}^\dagger E^*(\hat{\mathbf{q}}|\mathbf{r}) + \frac{1}{2}\nabla_\mathbf{r}\nabla_\mathbf{r}^\dagger \ln\det\mathbf{H}^*\right]_{\mathbf{r}=\hat{\mathbf{r}}} \tag{11.9}$$

Now recall that $E^* = E + R$ and apply definition (10.59) to re-write (11.9) in the form

$$\mathbf{A} \quad = \quad \left[\nabla_\mathbf{r}\nabla_\mathbf{r}^\dagger E(\hat{\mathbf{q}}|\mathbf{r}) + \nabla_\mathbf{r}\nabla_\mathbf{r}^\dagger \tilde{R}(\hat{\mathbf{q}}|\mathbf{r})\right]_{\mathbf{r}=\hat{\mathbf{r}}} \tag{11.10}$$

The Hessian $\mathbf{A}$ is composed of two additive sub-parts. The second part, $\nabla_\mathbf{r}\nabla_\mathbf{r}^\dagger \tilde{R}(\hat{\mathbf{q}}|\mathbf{r})$, is straightforward to derive. Since $\tilde{R}$ does not depend on a_k, as seen from (10.66), all derivatives with respect to these variables disappear:

$$\frac{\partial^2 \tilde{R}}{\partial a_i \partial a_k} = \frac{\partial^2 \tilde{R}}{\partial a_i \partial \beta_k} = \frac{\partial^2 \tilde{R}}{\partial a_i \partial \alpha_k} = 0. \tag{11.11}$$

For the derivations of the remaining terms, two approximations will be introduced. Recall that the number of well-determined parameters in the kth weight group, defined in (10.72), is given by

$$\gamma_k = \sum_{\nu=1}^{W} \frac{\varepsilon_k^\nu}{\varepsilon_k^\nu + \alpha_k}. \tag{11.12}$$

It will now be assumed that the derivatives of this term with respect to α_k and β_k can be neglected,

$$\frac{\partial \gamma_k}{\partial \alpha_k} \approx 0, \qquad \frac{\partial \gamma_k}{\partial \beta_k} \approx 0. \tag{11.13}$$

The justification is

$$\frac{\partial \gamma_k}{\partial \alpha_i} = \frac{\partial}{\partial \alpha_i} \sum_{\nu=1}^{W} \frac{\varepsilon_k^\nu}{\varepsilon_k^\nu + \alpha_k} = -\delta_{ik} \sum_{\nu=1}^{W} \frac{\varepsilon_k^\nu}{(\varepsilon_k^\nu + \alpha_k)^2} = -\frac{\delta_{ik}}{\alpha_k} \sum_{\nu=1}^{W} \frac{\alpha_k \varepsilon_k^\nu}{(\varepsilon_k^\nu + \alpha_k)^2} \approx 0 \tag{11.14}$$

This follows a similar approximation made by MacKay [37] (see also [7], chapter 10]). The reason is that the argument in the last sum, $\frac{\alpha_k \varepsilon_k^\nu}{(\varepsilon_k^\nu + \alpha_k)^2}$, is only significantly different from zero for $\varepsilon_k^\nu \approx \alpha_k$, and the number of eigenvalues that satisfy this condition is typically small. In the same way, making use of approximation (10.77), one obtains for the derivative with respect to β_k:

$$\frac{\partial \gamma_k}{\partial \beta_i} = \frac{\partial}{\partial \beta_i} \sum_{\nu=1}^{W} \frac{\varepsilon_k^\nu}{\varepsilon_k^\nu + \alpha_k} = \delta_{ik} \sum_{\nu=1}^{W} \left[\frac{1}{\varepsilon_k^\nu + \alpha_k} \frac{\partial \varepsilon_k^\nu}{\partial \beta_k} - \frac{\varepsilon_k^\nu}{(\varepsilon_k^\nu + \alpha_k)^2} \frac{\partial \varepsilon_k^\nu}{\partial \beta_k} \right] \tag{11.15}$$

$$= \delta_{ik} \sum_{\nu=1}^{W} \frac{\alpha_k}{(\varepsilon_k^\nu + \alpha_k)^2} \frac{\partial \varepsilon_k^\nu}{\partial \beta_k} \approx \frac{\delta_{ik}}{\beta_k} \sum_{\nu=1}^{W} \frac{\alpha_k \varepsilon_k^\nu}{(\varepsilon_k^\nu + \alpha_k)^2} \approx 0. \tag{11.16}$$

With these approximations, we obtain for the remaining second derivatives, from (10.74):

$$\frac{\partial^2 \tilde{R}}{\partial \alpha_i \partial \alpha_k} = -\frac{1}{2} \frac{\partial}{\partial \alpha_i} \frac{\gamma_k}{\alpha_k} = -\frac{1}{2\alpha_k} \frac{\partial \gamma_k}{\partial \alpha_i} + \frac{\gamma_k}{2} \frac{\delta_{ik}}{(\alpha_k)^2} \approx \frac{\delta_{ik} \gamma_k}{2(\alpha_k)^2} \tag{11.17}$$

and from (10.78):

$$\frac{\partial^2 \tilde{R}}{\partial \beta_i \partial \beta_k} = \frac{\partial}{\partial \beta_i} \frac{\gamma_k}{2\beta_k} = \frac{1}{2\beta_k} \frac{\partial \gamma_k}{\partial \beta_i} - \frac{\delta_{ik} \gamma_k}{2(\beta_k)^2} \approx -\frac{\delta_{ik} \gamma_k}{2(\beta_k)^2} \tag{11.18}$$

$$\frac{\partial^2 \tilde{R}}{\partial \alpha_i \partial \beta_k} = \frac{\partial}{\partial \alpha_i} \frac{\gamma_k}{2\beta_k} = \frac{1}{2\beta_k} \frac{\partial \gamma_k}{\partial \alpha_i} \approx 0 \tag{11.19}$$

We now have to turn to the first term, $\nabla_{\mathbf{r}}\nabla_{\mathbf{r}}^{\dagger}E(\hat{\mathbf{q}}|\mathbf{r})$. Since E does not explicitly depend on $\boldsymbol{\alpha}$ – the neural network parameters are $\mathbf{w}$, $\boldsymbol{\beta}$, and $\mathbf{a}$, whereas the α_k determine the prior distribution in $\mathbf{w}$-space – we have, with the notation $\mathbf{H}_{r_i,r_k} := \frac{\partial^2 E}{\partial r_i \partial r_k}$:

$$\mathbf{H}_{\alpha_i,\alpha_k} = \mathbf{H}_{\alpha_i,\beta_k} = \mathbf{H}_{\alpha_i,a_k} = 0 \tag{11.20}$$

The derivation of the remaining matrix elements is more demanding and will be relegated to the appendix. The result is, from (18.49), (18.50), and (18.51) (using the notation $\mathbf{H}_{r_i,r_k} := \frac{\partial^2 E}{\partial r_i \partial r_k}$ again):

$$\mathbf{H}_{\beta_i,\beta_k} = \frac{\delta_{ik}}{2(\beta_k)^2}(Na_k - B_k) \tag{11.21}$$

$$\mathbf{H}_{a_i,a_k} = \delta_{ik}\sum_{t=1}^{N}\left(\frac{\pi_K(t)}{a_K} - \frac{\pi_k(t)}{a_k}\right)^2 \tag{11.22}$$

$$\mathbf{H}_{a_i,\beta_k} = 0 \tag{11.23}$$

where

$$B_k := \frac{(\beta_k)^2}{2}\sum_{t=1}^{N}\pi_k(t)[1 - \pi_k(t)]\left[\left(y_t - \mu(\mathbf{x}_t;\mathbf{w}_k)\right)^2 - \frac{1}{\beta_k}\right]^2 \tag{11.24}$$

Inserting equations (11.11) and (11.17)-(11.23) into (11.10) yields

$$A_{a_i,a_k} = \delta_{ik}\sum_{t=1}^{N}\left(\frac{\pi_K(t)}{a_K} - \frac{\pi_k(t)}{a_k}\right)^2 \tag{11.25}$$

$$A_{\beta_i,\beta_k} = \frac{\delta_{ik}}{2(\beta_k)^2}(Na_k - B_k - \gamma_k) \tag{11.26}$$

$$A_{\alpha_i,\alpha_k} = \frac{\gamma_k}{2}\frac{\delta_{ik}}{(\alpha_k)^2} \tag{11.27}$$

$$A_{a_i,\beta_k} = A_{\alpha_i,\beta_k} = A_{\alpha_i,a_k} = 0 \tag{11.28}$$

which when inserted into (11.8) gives an expression for the model evidence. There is, however, a further subtlety which we need to address first.

11.2 An uninformative prior

It was mentioned earlier that we choose a uniform prior for the hyperparameters $\mathbf{r}$, that is, $P(\mathbf{r}) = const$. The motivation was that we have no idea about the proper values for the hyperparameters in advance and should therefore choose the least committed distribution, so that basically the data will decide on their most appropriate values. Hence what we are actually looking for is an *uninformative* prior, which is not necessarily identical to a *uniform* prior. Take, for example, the kernel widths. There are three equivalent ways of specifying their values, namely in terms of their standard deviations σ_k, their variances σ_k^2 or, as done most of the time in the present book, their precisions $\beta_k = \sigma_k^{-2}$ (this is a typical property of so-called *scale* parameters). If we choose a constant prior for β_k, then obviously the prior is not uniform in σ_k or σ_k^2. In the same way, a constant prior for σ_k^2 implies that the prior is not constant for σ_k or β_k, and the same holds for σ_k. Since all three forms of description are identical, the quality of uninformativeness should be satisfied in every case. This is accomplished if and only if the prior is uniform on a logarithmic scale, that is, if $P(\ln \beta_k) = const$. Obviously, a transformation $\ln \beta_k \rightarrow \ln \sigma_k$ or $\ln \beta_k \rightarrow \ln \sigma_k^2$ now only changes the value of the constant, but leaves the functional form of the prior unaffected. The same holds for α_k, which is the precision (inverse variance) of the prior distribution in weight space. Hence an uninformative[1] prior for the hyperparameters has the following form:

$$
\begin{aligned}
P(\mathbf{a}, \ln\beta, \ln\alpha) &= P(a_1, \ldots, a_{K-1}, \ln\beta_1, \ldots, \ln\beta_K, \ln\alpha_1, \ldots, \ln\alpha_K) \\
&= \prod_{k=1}^{K-1} P(a_k) \prod_{k=1}^{K} P(\ln\beta_k) \prod_{k=1}^{K} P(\ln\alpha_k) \\
&= \left(\frac{1}{\Omega_a}\right)^{K-1} \left(\frac{1}{\ln\Omega_\beta}\right)^{K} \left(\frac{1}{\ln\Omega_\alpha}\right)^{K} := \frac{1}{\Omega_C} \quad (11.29)
\end{aligned}
$$

where Ω_a, Ω_β and Ω_α are positive constants[2]. For the following derivations, note that we have, for any differentiable function F:

$$
\frac{\partial F}{\partial \ln\beta_i} = \beta_i \frac{\partial F}{\partial \beta_i}, \tag{11.30}
$$

$$
\frac{\partial^2 F}{\partial(\ln\beta_i)\partial(\ln\beta_k)} = \beta_i \frac{\partial}{\partial\beta_i}\left[\beta_k \frac{\partial F}{\partial\beta_k}\right] = \delta_{ik}\beta_i\frac{\partial F}{\partial\beta_i} + \beta_i\beta_k\frac{\partial^2 F}{\partial\beta_i\partial\beta_k}. \tag{11.31}
$$

Equation (11.30) implies that for finding the mode it does not matter whether we take the derivative with respect to β_i or $\ln\beta_i$,

$$
\frac{\partial F}{\partial(\ln\beta_i)} = 0 \quad \Leftrightarrow \quad \frac{\partial F}{\partial\beta_i} = 0. \tag{11.32}
$$

---

[1] The given derivation is only intuitive reasoning; a proper mathematical treatment of this subject can be found in [9], Chapter 1. In fact, the prior of equation (11.29) is not uninformative in the output weights $\mathbf{a}$, the implications of which are discussed on page 176.

[2] Strictly speaking, this equation only holds locally within intervals of length Ω_a, $\ln\Omega_\beta$, and $\ln\Omega_\alpha$. However, these details have no influence on the following analysis and will therefore not be further considered here.

From Equation (11.31) we find that the following identity holds at the mode:

$$\left[\frac{\partial^2 F}{\partial(\ln\beta_i)\partial(\ln\beta_k)}\right]_{\hat{\beta}} = \beta_i\beta_k\left[\frac{\partial^2 F}{\partial\beta_i\partial\beta_k}\right]_{\hat{\beta}} \tag{11.33}$$

Moreover, note that for any invertible function f it does not make any difference whether we specify x or $f(x)$ in the conditional part of a conditional probability. Thus $P(x|\beta) = P(x|f(\beta))$ and, in particular,

$$P(x|\beta) = P(x|\ln\beta). \tag{11.34}$$

The only effect of replacing the uniform prior of Equation (11.6) by the proper *uninformative* prior (11.29) is thus the replacement of the Hessian

$$\mathbf{A} := -\left[\nabla_{\mathbf{a},\beta,\alpha}\nabla_{\mathbf{a},\beta,\alpha}^{\dagger}\ln P(\mathbf{D}|\mathbf{a},\beta,\alpha,\mathcal{M})\right]_{\mathbf{a}=\hat{\mathbf{a}},\beta=\hat{\beta},\alpha=\hat{\alpha}} \tag{11.35}$$

by the modified Hessian

$$\tilde{\mathbf{A}} := -\left[\nabla_{\mathbf{a},\ln\beta,\ln\alpha}\nabla_{\mathbf{a},\ln\beta,\ln\alpha}^{\dagger}\ln P(\mathbf{D}|\mathbf{a},\beta,\alpha,\mathcal{M})\right]_{\mathbf{a}=\hat{\mathbf{a}},\beta=\hat{\beta},\alpha=\hat{\alpha}} \tag{11.36}$$

which according to (11.31) is related to the matrix $\mathbf{A}$ of (11.35) via

$$\begin{aligned}
\tilde{A}_{a_i,a_k} &= A_{a_i,a_k} & \tilde{A}_{\alpha_i,\alpha_k} &= \alpha_i\alpha_k A_{\alpha_i,\alpha_k} \\
\tilde{A}_{\beta_i,\beta_k} &= \beta_i\beta_k A_{\beta_i,\beta_k} & \tilde{A}_{a_i,\alpha_k} &= \alpha_k A_{a_i,\alpha_k} \\
\tilde{A}_{a_i,\beta_k} &= \beta_k A_{a_i,\beta_k} & \tilde{A}_{\alpha_i,\beta_k} &= \alpha_i\beta_k A_{\alpha_i,\beta_k}
\end{aligned} \tag{11.37}$$

Applying these transformation rules to equations (11.25)-(11.28), we obtain

$$\tilde{A}_{a_i,a_k} = \delta_{ik}\sum_{t=1}^{N}\left(\frac{\pi_K(t)}{a_K} - \frac{\pi_k(t)}{a_k}\right)^2 \tag{11.38}$$

$$\tilde{A}_{\beta_i,\beta_k} = \frac{\delta_{ik}}{2}(Na_k - B_k - \gamma_k) \tag{11.39}$$

$$\tilde{A}_{\alpha_i,\alpha_k} = \frac{\delta_{ik}\gamma_k}{2} \tag{11.40}$$

$$\tilde{A}_{a_i,\beta_k} = \tilde{A}_{\alpha_i,\beta_k} = \tilde{A}_{\alpha_i,a_k} = 0 \tag{11.41}$$

As a consequence of approximation (11.13), this matrix is diagonal. Its determinant can thus easily be computed, and we obtain from (11.38)-(11.41):

$$\ln\det\tilde{\mathbf{A}} = \tag{11.42}$$

$$\sum_{k=1}^{K-1}\ln\sum_{t=1}^{N}\left(\frac{\pi_K(t)}{a_K} - \frac{\pi_k(t)}{a_k}\right)^2 + \sum_{k=1}^{K}\ln\left(\frac{Na_k - B_k - \gamma_k}{2}\right) + \sum_{k=1}^{K}\ln\left(\frac{\gamma_k}{2}\right)$$

Inserting this term (instead of $\ln\det\mathbf{A}$) into (11.8) and making use of (10.58) gives for the *model evidence:*

$$\ln P(\mathbf{D}|\mathcal{M}) = -\hat{E}^* - \frac{1}{2}\sum_{k=1}^{K}\ln\det[\mathbf{H}_k+\alpha_k\mathbf{I}]-\frac{1}{2}\ln\det\tilde{\mathbf{A}}+\frac{M+m}{2}\ln(2\pi)-\ln\Omega_C$$

$$(11.43)$$

Note the change of the constant, $\Omega_\mathbf{r} \rightarrow \Omega_C$, which results from replacing the uniform prior of Equation (11.6) by the *uninformative* prior (11.29). The hat over E^* indicates that the function value is taken at the mode, and the same notation will also be used for E and R:

$$\hat{E}^* := E^*(\hat{\mathbf{q}}|\hat{\mathbf{r}}), \quad \hat{E} := E(\hat{\mathbf{q}}|\hat{\mathbf{r}}), \quad \hat{R} := R(\hat{\mathbf{q}}|\hat{\mathbf{r}}) \qquad (11.44)$$

Making the expression for $\ln\det\tilde{\mathbf{A}}$ explicit and substituting $M = KW$, $m = 3K - 1$ (see equations (10.47) and (10.48)) leads to

$$\ln P(\mathbf{D}|\mathcal{M}) = \qquad\qquad\qquad\qquad\qquad\qquad\qquad\qquad (11.45)$$

$$-\hat{E} - \hat{R} - \frac{1}{2}\sum_{k=1}^{K}\ln\det[\mathbf{H}_k + \alpha_k\mathbf{I}] + \frac{K(W+3)-1}{2}\ln 2\pi$$

$$-\frac{1}{2}\sum_{k=1}^{K-1}\ln\sum_{t=1}^{N}\left(\frac{\pi_K(t)}{a_K} - \frac{\pi_k(t)}{a_k}\right)^2 - \frac{1}{2}\sum_{k=1}^{K}\ln\left(\frac{Na_k - B_k - \gamma_k}{2}\right)$$

$$-\frac{1}{2}\sum_{k=1}^{K}\ln\left(\frac{\gamma_k}{2}\right) - \ln\Omega_C$$

The terms $\mathbf{H}_k$, B_k, and C_Ω are defined in (10.54), (11.24), and (11.29), respectively.

11.3 Comparison with MacKay's work

Recall that for $K = 1$ the chosen probability model reduces to a simple Gaussian with $\mathbf{x}$-dependent centre (the *interpolant*) and input-independent variance. This case was studied by MacKay [37], [38], and it will now be shown that his results are included as a special case in the more general expressions derived in this chapter. Obviously, for $K = 1$ the set $\{a_1, \ldots, a_K\}$ has only one element, a_1, which is *not* a free parameter, $a_1 \equiv 1$. This implies $\pi_1(t) = 1$ $\forall t$. The update rules for the hyperparameters thus become, from (10.75) and (10.79):

$$\frac{1}{\hat{\alpha}_1} = \frac{1}{\gamma}\hat{\mathbf{w}}_1^\dagger\hat{\mathbf{w}}_1 \qquad\qquad (11.46)$$

$$\frac{1}{\hat{\beta}_1} = \frac{\sum_{t=1}^{N}\left(y_t - \mu(\mathbf{x}_t;\hat{\mathbf{w}}_1)\right)^2}{N - \gamma}. \qquad\qquad (11.47)$$

This is identical to the hyperparameter update equations (2.4) and (2.5) in [38]. Now $B_1 = 0$, from (11.24). Thus (11.45) becomes:

$$\ln P(\mathbf{D}|\mathcal{M}) =$$
$$-\hat{E} - \hat{R} - \frac{1}{2}\ln\det[\mathbf{H}_1 + \alpha_1\mathbf{I}] + \frac{W+2}{2}\ln 2\pi$$
$$-\frac{1}{2}\ln\left(\frac{N-\gamma_1}{2}\right) - \frac{1}{2}\ln\left(\frac{\gamma_1}{2}\right) - \ln\Omega_C \tag{11.48}$$

In order to compare this expression with the one derived by MacKay [37], note the slight difference in the notation. MacKay introduced an error function E_D and a regularisation term E_W in the following way[3]:

$$P(\mathbf{D}|\mathbf{w}, \beta, \mathcal{M}) = \left(\frac{\beta}{2\pi}\right)^{\frac{N}{2}} \exp\left[-\beta E_D(\mathbf{w})\right], \tag{11.49}$$

$$P(\mathbf{w}|\alpha, \mathcal{M}) = \left(\frac{\alpha}{2\pi}\right)^{\frac{W}{2}} \exp\left[-\alpha E_W(\mathbf{w})\right], \tag{11.50}$$

where $E_D = \frac{1}{2}\sum_{t=1}^{N}(y_t - \mu(\mathbf{x}; \mathbf{w}))^2$ and $E_W = \frac{1}{2}\mathbf{w}^\dagger\mathbf{w}$. A comparison with the definitions of the error function E and the regularisation term R introduced in the previous chapter, (10.28) and (10.49), gives the following relation:

$$E = \beta E_D - \frac{N}{2}\ln\left(\frac{\beta}{2\pi}\right) \tag{11.51}$$

$$R = \alpha E_W - \frac{W}{2}\ln\left(\frac{\alpha}{2\pi}\right) \tag{11.52}$$

Inserting these results into (11.48) leads to the same expression for the evidence as in [37][4]. (Recall that $\mathbf{H}^* = \mathbf{H} + \alpha\mathbf{I}$. Constant terms have been collected in $Const$, and the indices for α_1 and γ_1 have been dropped.):

$$\ln P(\mathbf{D}|\mathcal{M}) =$$
$$-\beta\hat{E}_D - \alpha\hat{E}_W + \frac{1}{2}\ln\left(\frac{\det(\mathbf{H}^*)^{-1}}{\alpha^{-W}}\right)$$
$$-\frac{1}{2}\ln\left(\frac{N-\gamma}{2}\right) - \frac{1}{2}\ln\left(\frac{\gamma}{2}\right) + \frac{N}{2}\ln\beta + Const \tag{11.53}$$

11.4 Interpretation of the model evidence

An interpretation of the terms contributing to the evidence, equation (11.53), can be found in [37] and [7], Chapter 10. The term βE_D represents the misfit

---

[3] Note that W (the number of weights) is denoted by m in MacKay's work.
[4] See also [7], chapter 10.

of the interpolant to the data, and obviously rewards models that give a good fit. However, overcomplex models are penalised by the third term on the right of equation (11.53),

$$\text{Ock}_{\mathbf{w}} \quad := \quad \frac{1}{2} \ln \left(\frac{\det(\mathbf{H}^*)^{-1}}{\alpha^{-W}} \right), \tag{11.54}$$

which MacKay [37] refers to as the *Ockham factor*. Recall that α is the inverse variance of the prior distribution $P(\mathbf{w}|\alpha)$ in weight space, $\sigma_{\mathbf{w}}^2 = \alpha^{-1}$, and that the inverse of the Hessian, $\mathbf{H}^{*-1}$, is equal to the covariance matrix $\mathbf{Cov}_{\mathbf{w}}$ of the posterior distribution in weight space, $P(\mathbf{w}|\mathbf{D}, \alpha, \beta)$. The Ockham factor thus measures the ratio of the posterior to the prior accessible volume in weight space:

$$\text{Ock}_{\mathbf{w}} = \ln \left(\frac{\sqrt{\det(\mathbf{Cov}_{\mathbf{w}})}}{\sigma_{\mathbf{w}}^W} \right) = \frac{\text{posterior accessible volume}}{\text{prior accessible volume}}. \tag{11.55}$$

Overcomplex models whose parameters are finely tuned to the data and whose posterior distributions become very narrow give rise to a small ratio and are consequently penalised by a strong negative contribution of the Ockham factor to the evidence. If we now turn to the more general expression for the evidence for arbitrary conditional distributions, equation (11.45), we can find $K + 2$ Ockham factors: for the K weight groups $(\mathbf{w}_1, \ldots, \mathbf{w}_K)$, for the set of kernel width parameters $\{\beta_k\}$, and for the priors or output weights $\{a_k\}$.

11.4.1 Ockham factors for the weight groups

The Ockham factors for the weight groups are given by the terms

$$\text{Ock}_{\mathbf{w}_k} := -\frac{1}{2} \ln \det[\mathbf{H}_k + \alpha_k \mathbf{I}] = \frac{1}{2} \ln \det[\mathbf{H}^*_{\mathbf{w}_k, \mathbf{w}_k}]^{-1}. \tag{11.56}$$

These terms are similar to the Ockham factor of (11.54), except that the denominator is missing. To understand this deviation, note that the denominator in (11.54) is proportional to the normalisation factor in (11.50), which, as seen from (11.52), is included in the regularisation terms R. The matrix $\mathbf{H}^*_{\mathbf{w}_k, \mathbf{w}_k}$ denotes the submatrix of the Hessian with respect to the weights $\mathbf{w}_k$, and its inverse is consequently the covariance matrix for the distribution over the sub-space spanned by the parameters $\mathbf{w}_k$. The Ockham factor defined by (11.56) is thus seen to be a measure of the posterior accessible volume in the corresponding sub-space. This follows the idea outlined above and discussed at length in [37] and [7]: a small posterior accessible volume in parameter space suggests a fine-tuning of the network parameters and thus points to overfitting. This is penalised by a small Ockham factor, which decreases the evidence.

11.4.2 Ockham factors for the kernel widths

The Ockham factor for the noise level β in (11.53) is given by the term

$$\text{Ock}_\beta \quad = \quad -\frac{1}{2}\ln\left(\frac{N-\gamma}{2}\right). \tag{11.57}$$

The argument of the logarithm contains the difference between the number of data points and the number of well-determined parameters in the network. This is the *effective* number of training data available for the determination of β. Obviously, with increasing value for $(N-\gamma)$, the determination of the noise level becomes more accurate. This leads to a decrease of the posterior variance with respect to this parameter and, consequently, a decrease of the Ockham factor (as seen from the above equation, (11.57)). At first it might seem surprising that large data sets are penalised. Note, however, that the likelihood term also scales like N, as seen from (11.49), so an effective penalty only arises when an increase in the number of training data is *not* accompanied by a sufficient decrease of the modelling error.

Let us now turn to the generalisation of this term, given by the sixth component in the expression for the generalised evidence, equation (11.45),

$$\text{Ock}_\beta \quad := \quad -\frac{1}{2}\sum_{k=1}^{K}\ln\left(\frac{Na_k - B_k - \gamma_k}{2}\right) \tag{11.58}$$

which by comparison with (11.39) is seen to measure the posterior accessible volume in the parameter subspace spanned by β. It is recalled that B_k, defined in (11.24), is given by

$$B_k \quad := \quad \frac{(\beta_k)^2}{2}\sum_{t=1}^{N}\pi_k(t)\left[1-\pi_k(t)\right]\left[\left(y_t-\mu(\mathbf{x}_t;\mathbf{w}_k)\right)^2-\frac{1}{\beta_k}\right]^2 \tag{11.59}$$

In order to arrive at a better understanding of this expression, let us first assume that we have a *hard partitioning* of the target data y_t such that $\pi_k(t) = \delta_{ik}$. This implies that each of the K kernels or network branches focuses exclusively on a subset of $n_k = Na_k$ data points, $\mathbf{D}_k = \{(\mathbf{x}_{t_i}, y_{t_i})_{i=1}^{n_k}\}$, while ignoring all the others, $(\mathbf{x}_t, y_t) \notin \mathbf{D}_k$. Since $B_k = 0$ in this case (as seen from (11.59)), equation (11.58) reduces to

$$\text{Ock}_\beta \quad := \quad -\frac{1}{2}\sum_{k=1}^{K}\ln\left(\frac{n_k - \gamma_k}{2}\right). \tag{11.60}$$

Each weight group thus gives rise to an Ockham factor identical to (11.57), and the total Ockham factor is just the sum of the individual contributions.

In general, however, the partitioning of the data is *soft*, which gives rise to the additional contribution B_k. Since this term is non-negative, it decreases the argument of the log in (11.58) and therefore increases the Ockham factor Ock_β. A closer inspection of (11.59) reveals when this is going to have a significant influence: The second factor, $[(y_t - \mu(\mathbf{x}_t; \mathbf{w}_k))^2 - 1/\beta_k]^2$, is large when the squared deviation of the target y_t from the interpolant $\mu(\mathbf{x}_t; \mathbf{w}_k)$ deviates significantly from the predicted variance $\frac{1}{\beta_k}$. The first factor, $\pi_k(t)[1 - \pi_k(t)]$, is large when the posterior $\pi_k(t)$ is significantly different from 0 and 1, that is, when the assignment of the tth exemplar is fuzzy and non-distinctive. Consequently, Ock_β is increased by exemplars that are not clearly classified as having been generated by the kth component, and whose squared deviation from the interpolant $\mu(\mathbf{x}_t; \mathbf{w}_k)$ deviates significantly from the variance $\frac{1}{\beta_k}$. Obviously, this scenario increases the uncertainty about β_k, so the increase of Ock_β is consistent with the interpretation given above at the beginning of Section 11.4.

11.4.3 Ockham factor for the priors

The Ockham factor for the priors or output weights is given by

$$\text{Ock}_{\mathbf{a}} := -\frac{1}{2} \sum_{k=1}^{K-1} \ln \sum_{t=1}^{N} \left(\frac{\pi_K(t)}{a_K} - \frac{\pi_k(t)}{a_k} \right)^2 \tag{11.61}$$

which measures the posterior accessible volume in a-space, as seen from (11.38). For a better understanding of this term, consider the special case of two kernels, $K = 2$,

$$\text{Ock}_{\mathbf{a}} = -\frac{1}{2} \ln \sum_{t=1}^{N} \left(\frac{\pi_2(t)}{a_2} - \frac{\pi_1(t)}{a_1} \right)^2 = -\frac{1}{2} \ln \sum_{t=1}^{N} \left(\frac{1 - \pi_1(t)}{1 - a_1} - \frac{\pi_1(t)}{a_1} \right)^2$$

$$= -\frac{1}{2} \ln \sum_{t=1}^{N} \left(\frac{a_1 - \pi_1(t)}{a_1(1 - a_1)} \right)^2. \tag{11.62}$$

The argument of the logarithm is maximised and, consequently, the Ockham factor minimised when

$$\pi_1(t) = \begin{cases} 1 & \text{if} \quad \pi_1(t) > a_1 \\ 0 & \text{if} \quad \pi_1(t) < a_1 \end{cases}$$

that is, when the partitioning of the data is *hard*. This corresponds to a complex model whose parameters are well determined by the data such that different network branches focus on distinct state-space domains. According to the above philosophy, this is penalised by a small Ockham factor. Conversely, the Ockham factor is maximised for $\pi_1(t) = \pi_2(t)$ $\forall t$ and $a_1 = a_2$. In this case the network parameters are poorly determined by the data and the model is very vague: the functions implemented in the kernels overlap and no partitioning of the state-space is found. Following the evidence philosophy again, this low degree of model complexity is rewarded by a large Ockham factor.

There is, however, an obvious difficulty with the **a**-dependent Ockham factor of equation (11.61). First, its computation becomes numerically unstable when some of the priors a_k are close to zero. This can be remedied by a pruning scheme that disposes of network branches whose output weights have fallen below a certain threshold θ, $a_k := 0$ if $a_k < \theta$. The second more fundamental problem arises in the last scenario where all the network branches implement nearly identical functions. For $a_k \approx a_K$ and $\pi_k(t) \approx \pi_K(t)$ $\forall t$, the Ockham factor $Ock_\mathbf{a}$ diverges to infinity and thus gives large positive contributions to the evidence (11.45), which will dominate over all other terms. The reason for this problem is the fact that the prior for **a** is *not* uninformative (the prior of Equation (11.29) is in fact only uninformative for β and α). Note that if we choose a uniform rather than an uninformative prior on the α_k, then the corresponding Ockham factor obtains the additional term $\ln \alpha_k$, resulting from (11.27), which in the limit of vanishing weight-decay, $\alpha_k \rightarrow 0$, also becomes singular. We can remedy this problem by using an *uninformative* prior on the a_k, as outlined in [26]. However, this will not be discussed here, since there is a more fundamental bottleneck in the evidence approach.

11.5 Discussion

In practical applications, the determination of the Bayesian model evidence is hampered by the requirement to calculate the determinant of the matrices $(\mathbf{H}_k + \alpha_k \mathbf{I})$. Recall that the determinant is given by the product of the eigenvalues of the matrix. Its expression is therefore dominated by the smallest eigenvalues, which are most susceptible to numerical inaccuracies and roundoff errors. For this reason MacKay, in his successful time series prediction application [39], resorted to conventional cross-validation for model selection, rather than applying the evidence scheme. Similar problems with the model evidence have been reported in [61]. The increased complexity of the Gaussian mixture model with the requirement to calculate K determinants $\det[\mathbf{H}_k + \alpha_k \mathbf{I}]$ is likely to compound the numerical problems. For this reason the application of the model evidence scheme is, in practice, restricted to small, low-dimensional problems.

Note that these problems do *not* apply to the Bayesian evidence scheme for *regularisation*, which was discussed in the previous chapter. Recall from

Equation (10.71) that the optimisation of the hyperparameters α_k is based on the calculation of $\frac{\partial}{\partial \alpha_k} \ln \det(\mathbf{H}_k + \alpha_k \mathbf{I})$. However, this expression does *not* require the explicit calculation of the determinant, since $\ln \det(\mathbf{H}_k + \alpha_k \mathbf{I}) = \operatorname{tr} \ln(\mathbf{H}_k + \alpha_k \mathbf{I})$ and therefore $\frac{\partial}{\partial \alpha_k} \ln \det(\mathbf{H}_k + \alpha_k \mathbf{I}) = \operatorname{tr}(\mathbf{H}_k + \alpha_k \mathbf{I})^{-1}$. The evidence-based regularisation scheme thus depends on the *trace* of the inverse Hessian, which is the *sum* of the eigenvalues of $(\mathbf{H}_k + \alpha_k \mathbf{I})^{-1}$ and therefore dominated by the largest eigenvalues. Small eigenvalues, which are susceptible to numerical inaccuracies and cause problems when calculating the determinant, effectively drop out.

12. Demonstration of the Bayesian Evidence Scheme for Regularisation

The Bayesian evidence approach to regularisation, derived in the previous chapter, is applied to the stochastic time series generated from the logistic-kappa map. The scheme is found to prevent overfitting and lead to a stabilisation of the training process with respect to changes in the length of training time. For a small training set, it is also found to include an automatic pruning scheme: the network complexity is reduced until all remaining parameters are well-determined by the data.

12.1 Method and objective

The objective of the study summarised in the present chapter was the application of the Bayesian evidence scheme for regularisation, derived in Chapter 10, to the prediction of the stochastic time series generated from the logistic-kappa map (4.16). The GM-RVFL network employed had the same architecture as in the study of Section 8.2, that is $m = 1, H = 10, K = 10$, with a bias unit connected to all nodes in the two hidden layers (but without direct connections between the input and the $\mathcal{G}$-layer). As before, the random weights (that is, the weights between the input and the $\mathcal{S}$-layer) were drawn independently from $N(0, \sigma_{rand})$. Training was carried out with the EM-algorithm (Chapter 7), and different regularisation schemes were applied, as described below.

12.1.1 Initialisation

The output weights were set to the inverse number of nodes in the $\mathcal{G}$-layer, $a_k = 1/K$, and the kernel widths were chosen sufficiently large: $\sigma_k = 1$. The adaptable weights $\mathbf{w}$ between the $\mathcal{S}$-layer and the $\mathcal{G}$-layer were initialised with small random values drawn from $N(0, 0.1)$. The fixed weights feeding into the $\mathcal{S}$-layer were independently drawn from $N(0, \sigma_{rand})$, where two different values for the distribution width σ_{rand} were employed: $\ln \sigma_{rand} = 0$, and $\ln \sigma_{rand} = -1$.

12.1.2 Different training and regularisation schemes

Unregularised training. A standard *unregularised* training simulation with the EM algorithm iterates the following adaptation steps:

– **Priors:** Adapt the priors or output weights, a_k, according to (7.23).

– **Kernel widths:** Apply (7.24) for the adaptation of the kernel widths, σ_k.

– **Remaining weights:** Adapt the remaining weights, **w**, according to (7.36).

Quasi-unregularised training. The problem of the unregularised training scheme, outlined above, is that the matrix on the left of Equation (7.36) may be singular, in which case there is no unique solution for the weights **w**. As discussed earlier in Chapter 7, this can be remedied by adding a small constant to the diagonal elements of the matrix, and to adapt the weights **w** according to (9.42) rather than (7.36). In the simulations, the value of α_k in (9.42) was set to $\alpha_k = 10^{-6}$ (the same value was used for all weight groups).

Weight decay. Equation (9.42) for adapting the weights **w** implies a simple weight decay scheme (see Section 7.5), which is identical to a Gaussian prior distribution on **w** (as discussed in Chapter 9). While the objective of the quasi-unregularised scheme is merely to prevent any ambiguities in the adaptation of **w**, a simple regularisation method can be effected by choosing larger values for α_k . In the simulations for this chapter, a value of $\alpha_k = 0.1\ \forall k$ was chosen.

Eigenvalue cutoff. An alternative regularisation scheme is given by discarding the contributions of small eigenvalues, following Equation (7.63). Details of this approach were discussed earlier, in sections 7.4 and 7.5.

Bayesian evidence scheme. The algorithm for the Bayesian evidence scheme was given in Section 10.4.3[1] Recall that for the reasons discussed

---

[1] For the simulations reported here, this algorithm was slightly modified as follows. The order of steps 3 and 5 was inverted, and the hyperparameters α_k were updated by finding the root of (10.75). Note that this equation is nonlinear in α_k since the number of well-determined parameters γ_k on the right-hand side of (10.75) depends on α_k via (10.72). In practice, a solution can easily be obtained by invocation of a root-finding algorithm, like Brent's method ([51], Chapter 9). In this way the speed of the standard algorithm can be slightly improved.

in Section 10.5.2, it is advisable not to apply the evidence scheme at an early stage of the training process. For the first $n = 3$ epochs, the parameters were therefore adapted with the quasi-unregularised method.

12.1.3 Pruning

As discussed in Section 8.1, too small a value for a_k leads to numerical instabilities. Therefore a pruning scheme should be applied, which switches a network branch off when a_k has fallen below a certain threshold:

$$a_k < \epsilon \quad \Rightarrow \quad a_k := 0. \tag{12.1}$$

In the simulations, a value of $\epsilon = 10^{-6}$ was used.

Regularisation scheme	$E_{gen}(1)$	$E_{gen}(2)$
Eigenvalue cutoff	-1.029	-0.971
Weight decay, $\alpha_k = 10^{-6}$ (quasi unregularised)	-0.992	-1.027
Weight decay, $\alpha_k = 0.1$	-1.028	-1.026
Bayesian evidence scheme	-1.039	-1.046

Table 12.1. Different regularisation schemes, large data set. The table shows the estimated generalisation 'error' E_{gen} for a GM-RVFL network ($m = 1, H = 10, K = 10$) trained on $N_{train} = 1000$ data points generated from the logistic-kappa map. The two right columns show the results obtained from two different selections of the random weights (in either case from $N(0,1)$, i.e. $\sigma_{rand} = 1$). Smaller values of E_{gen} correspond to a better generalisation performance.

12.2 Large Data Set

In a first experiment, the above scheme was applied to a 'large' training set of size $N_{train} = 1000$, generated from (4.16). Figure 12.1 shows the evolution of the training and cross-validation 'error'[2] for two different, independent selections of the random weights from $N(0,1)$. The figure on the top left shows the results of applying a simple weight decay term with a very small weight-decay hyperparameter[3], $\alpha_k = 10^{-6}$. In this case the network is *quasi unregularised*,

---

[2] The cross-validation 'error' was estimated on an independent test set of the same size as the training set.

[3] This terminology is slightly imprecise, because the actual weight decay 'constant' is given by the ratio α_k/β_k

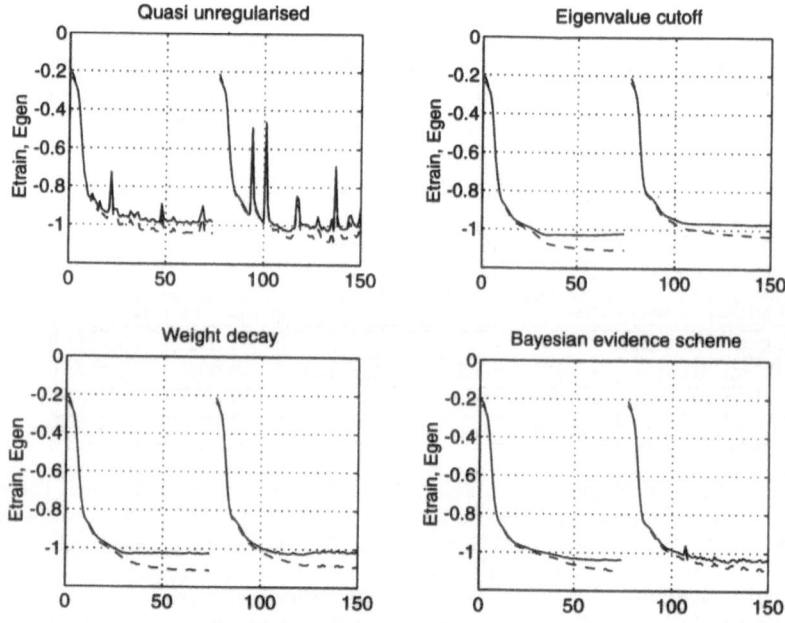

Fig. 12.1. Evolution of E, large data set. The graphs show the evolution of the training 'error' E_{train} (dashed line) and the cross-validation 'error' E_{cross} (solid line) for a GM-RVFL network ($m = 1, H = 10, K = 10$) applied to the logistic-kappa map problem. The abscissa shows the number of EM-steps. The network was trained on a *'large'* data set of $N_{train} = 1000$ data points, and the cross-validation 'error' was measured on an independent set of the same size. Each figure shows the results of *two* simulations, using different values for the random weights between the input and the $\mathcal{S}$-layer (all drawn from a standard Normal distribution, $N(0, 1)$). Four different regularisation schemes were applied. *Top left:* Simple weight decay with $\alpha_k = 10^{-6}\,\forall k$ (quasi unregularised). *Bottom left:* Simple weight decay with $\alpha_k = 0.1\,\forall k$. *Top right:* Eigenvalue cutoff. *Bottom right:* Bayesian evidence scheme. Note that $\alpha_k = 10^{-6}$ is strongly under-regularised, and that the best results are achieved with the Bayesian evidence scheme.

and strong oscillations occur, resulting from the ill-conditioning of the Hessian. This is remedied by any of the three regularisation methods: (i) weight decay with a larger weight-decay hyperparameter, $\alpha_k = 0.1$ (bottom left), (ii) eigenvalue cutoff (top right), and the Bayesian evidence scheme (bottom right). However, the latter case also results in a slight reduction of the generalisation 'error', estimated as follows. From the graphs of Fig.12.1, the minimum of the cross-validation 'error' was found, and the network in this configuration applied to a third independent test set (of equal size). The results are listed in Tab.12.1 and suggest that the Bayesian evidence scheme indeed leads to a slight improvement of the generalisation performance, with values for E_{gen} as good as for the committee prediction of Section 8.2 (compare with Tab. 8.2). However, the achieved improvement is only marginal. This is mainly due to the large size of the training set, which does not make overfitting so much of a problem in the first place. The simulations were therefore repeated on a training set of reduced size.

Regularisation scheme	$\ln \sigma_{rand} = -1$		$\ln \sigma_{rand} = 0$	
	$E_{gen}(1)$	$E_{gen}(2)$	$E_{gen}(1)$	$E_{gen}(2)$
Weight decay, $\alpha_k = 10^{-6}$	-0.85	-0.86	-0.81	-0.86
Weight decay, $\alpha_k = 0.1$	-0.85	-0.32	-0.86	-0.83
Bayesian evidence scheme	-0.90	-0.88	-0.82	-0.88

Table 12.2. Different regularisation schemes, small data set. The table shows the estimated generalisation 'error' E_{gen} for a GM-RVFL network ($m = 1, H = 10, K = 10$) trained on $N_{train} = 200$ data points generated from the logistic-kappa map. Smaller values of E_{gen} correspond to a better generalisation performance. The random weights were drawn independently from a zero-mean Gaussian distribution with (i) $\ln \sigma_{rand} = -1$ and (ii) $\ln \sigma_{rand} = 0$. Each simulation was carried out twice, for two different sets of the random weights.

12.3 Small Data Set

Figure 12.2 shows the evolution of the training and cross-validation 'error' for a sparse training set of only $N_{train} = 200$ data points. For each regularisation scheme the simulation was repeated four times, with four different sets of the random weights, drawn from the two distributions $N(0, 0.368)$ and $N(0, 1)$ (corresonding to $\ln \sigma_{rand} = -1$ and $\ln \sigma_{rand} = 0$). Otherwise, the details of the simulations were as described in the previous section. As before, too small a weight decay constant of $\alpha_k = 10^{-6}$ leads to oscillations resulting from the ill-conditioning of the Hessian (upper row of Fig.12.2). This is rectified by increasing α_k (middle row). However, the regularisation scheme still

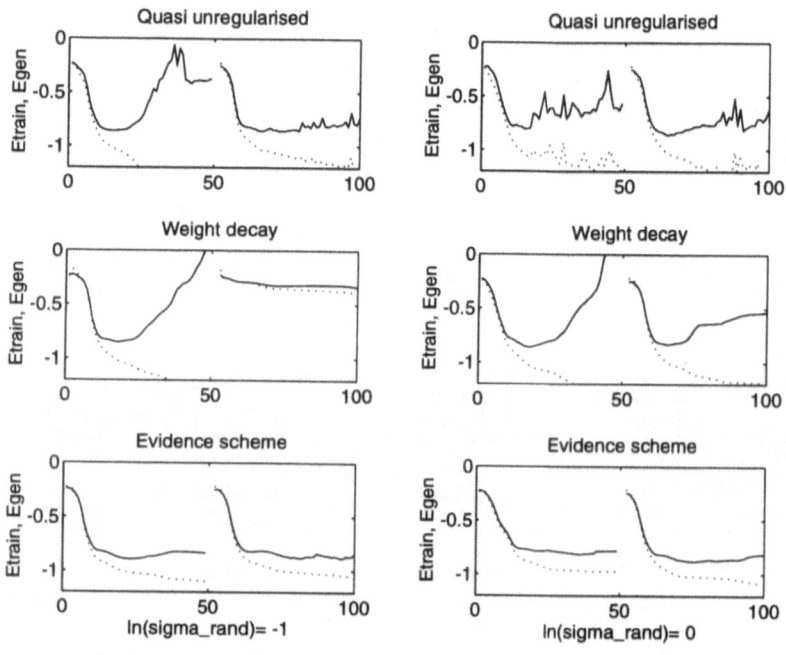

Fig. 12.2. Evolution of E, small data set. The graphs show the evolution of the training 'error' E_{train} (dotted line) and the cross-validation 'error' E_{cross} (solid line) for a GM-RVFL network ($m = 1, H = 10, K = 10$) applied to the logistic-kappa map problem (abscissa: number of EM-steps). The network was trained on a 'small' data set of $N_{train} = 200$ data points. (The cross-validation 'error' was measured on an independent set of size $N_{cross} = 1000$ data points.) The random weights between the input and the S-layer were drawn from a zero-mean Gaussian distribution with standard deviation $\sigma_{rand} = 0.368$ (ln $\sigma_{rand} = -1$) for the graphs on the left, and $\sigma_{rand} = 1.0$ (ln $\sigma_{rand} = 0$) for the graphs on the right. Each simulation was repeated twice, with two different sets of the random weights. *Top:* Simple weight decay with $\alpha_k = 10^{-6} \, \forall k$ (quasi unregularised). The oscillations are caused by the ill-conditioning of the Hessian. *Middle:* Simple weight decay with $\alpha_k = 0.1 \, \forall k$. The larger value of α_K solves the problem of ill-conditioning, but still leaves the regularisation scheme sub-optimal. *Bottom:* Bayesian evidence. Note that this scheme leads to a considerable reduction of the overfitting problem, and stabilises the training process with respect to changes in the number of adaptation steps.

remains sub-optimal, as clearly seen from the graphs in the middle of the figure. In three cases a sharp increase of the cross-validation 'error' points to the model being drastically under-regularised, whereas in one simulation the 'error' hardly decreases at all (thus suggesting that for this particular configuration of the random weights the model is over-regularised). Compare this with the graphs at the bottom of the figure, which were obtained with the Bayesian evidence scheme. Now the cross-validation 'error' increases only slightly, if at all, thus suggesting that the regulariser is well-matched to the data. In order to assess the generalisation performance of the different schemes, the procedure of the previous section was followed. The network configuration with minimal cross-validation error was identified, and then applied to a third independent test set (the same as before, i.e. $N_{gen} = 1000$). The results are listed in Tab.12.2 and suggest that, overall, Bayesian regularisation does result in a decrease of the generalisation 'error' (note that the 'errors' are larger than those of Tab.12.1 as a consequence of the smaller training set size), but this improvement is only marginal and not significant. However, the identification of the optimal network configuration when training with the under-regularised scheme requires a cross-validation set, which reduces the amount of data available for training. In contrast, the Bayesian evidence scheme does not depend on a cross-validation set, since the risk of over-fitting is considerably reduced, and variations in the length of training time are less critical. Consequently, a larger set of data is available for training, which in general can be expected to lead to enhanced generalisation results.

12.4 Number of well-determined parameters and pruning

12.4.1 Automatic self-pruning

Consider the number of adaptable parameters in the GM-RVFL network. For each of the K nodes in the $\mathcal{G}$-layer, H weights feed in from the $\mathcal{S}$-layer and 1 weight from the bias unit. Moreover, there are K adaptable kernel widths σ_k and $(K-1)$ adaptable priors a_k. The total number of adpatable network parameters is thus, with $H = K = 10$:

$$\text{Number of parameters} = K(H+1) + K + (K-1) = K(H+3) - 1 = 129$$

This leads to the following ratio of training exemplars per parameter:

$$N = 200 \quad \Rightarrow \quad \frac{N}{\text{Number of parameters}} = 1.6$$

$$N = 1000 \quad \Rightarrow \quad \frac{N}{\text{Number of parameters}} = 7.7$$

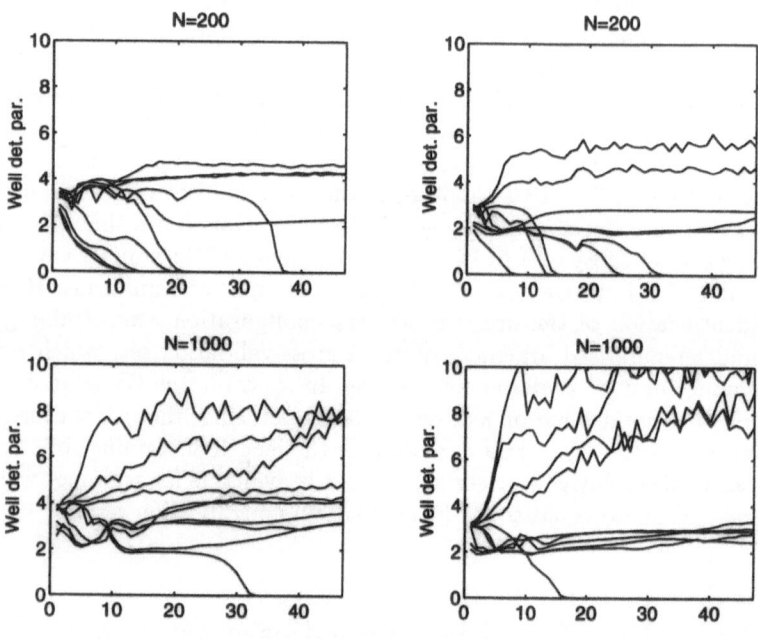

Fig. 12.3. Number of well-determined parameters. The figures show the evolution of the number of well-determined parameters in the K weight groups, $\gamma_k, k \in \{1, \ldots, K\}$, for a GM-RVFL network of architecture $m = 1, H = 10, K = 10$. The abscissa represents the number of EM-steps. The simulations were performed twice, for two different sets of the random weights (drawn from $N(0, 1)$ in each case). *Upper row:* Small training set of size $N_{train} = 200$. The number of well-determined parameters, γ_k, is always considerably smaller than the total number of parameters $(H + 1)$. For several weight groups, γ_k even decays to zero. *Bottom row:* Large training set with $N_{train} = 1000$. The number of well-determined parameters, γ_k, is on average considerably larger than for the small training set and approaches, for several weight groups, even the maximum possible value of $H + 1$. Note that only for one weight group γ_k decays to zero.

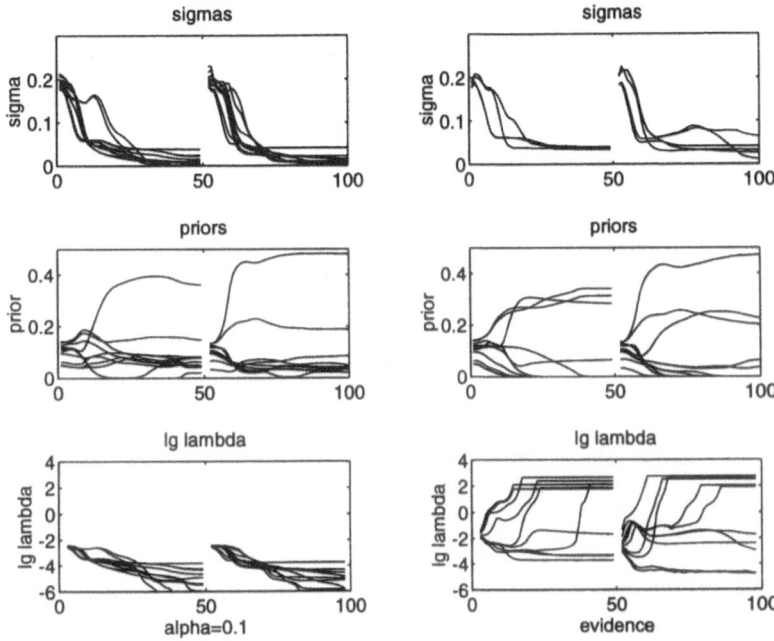

Fig. 12.4.   Evolution of the priors, kernel widths and weight decay constants for a small training set. A GM-RVFL network with architecture $m = 1, H = 10, K = 10$ was applied to the logistic-kappa map problem and trained on a *'small'* training set of size $N_{train} = 200$. The figures show the evolution of different entities during training, repeated for two different selections of the random weights (drawn from $N(0,1)$). The abscissa shows the number of EM steps. *Top:* kernel widths σ_k. *Middle:* prior probabilities or output weights a_k. *Bottom:* weight-decay 'constants' $\lambda_k = \frac{\alpha_k}{\beta_k} = \alpha_k \sigma_k^2$. The graphs on the *left* were obtained with a simple weight-decay regulariser ($\alpha_k = 0.1$), whereas the graphs on the *right* were obtained by applying the Bayesian evidence scheme. Note that in the second case several priors decay, which is correlated with a concomitant divergence of the corresponding weight-decay 'constants' λ_k. Hence the network is effectively pruned as a consequence of the regularisation scheme. (The contribution of a kernel was 'switched off' when $a_k < \epsilon := 10^{-6}$. This is indicated by the kinks in the graphs of the bottom-right figure. The top-right figure shows only the σ_ks of those kernels that had not been pruned away.)

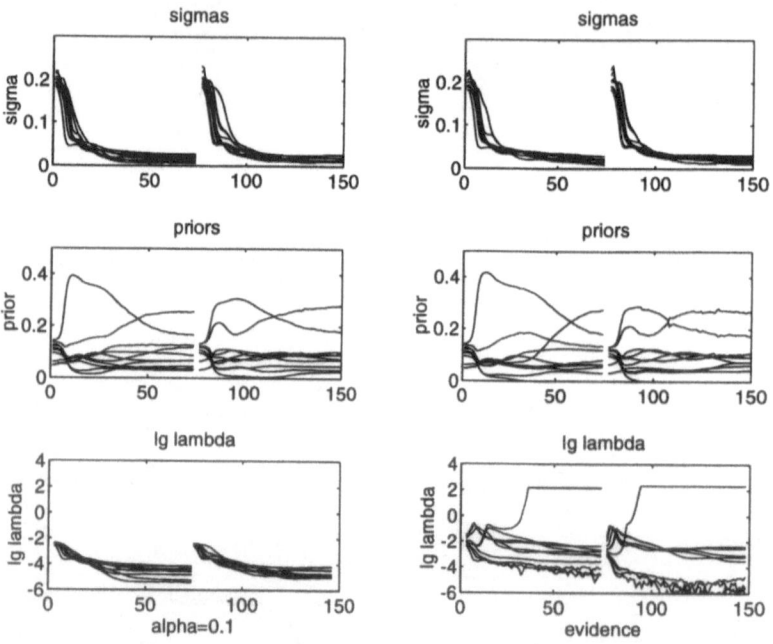

Fig. 12.5. Evolution of the priors, kernel widths and weight decay constants for a large training set. The figures are equivalent to those in Fig.12.4, except that the size of the training set was $N_{train} = 1000$ rather than $N_{train} = 200$. Note that the Bayesian regularisation scheme only leads to the pruning of one kernel.

For the small data set, the obtained ratio of only 1.6 exemplars per adaptable parameter is very small, and not all parameters in the network can be well-determined by the data. This is, interestingly, recognised and rectified by the Bayesian regularisation scheme. Figure 12.3 shows the evolution of the number of well-determined parameters, γ_k, for the K weight groups. When the data set is small, several of these values decay to zero. This is correlated with a concomitant divergence of the corresponding weight-decay 'constant' λ_k and a decay of the respective output weight[4] a_k, as seen from Fig.12.4. The Bayesian evidence approach thus effectively includes a pruning scheme that reduces an over-complex network structure to a smaller one in which all the parameters are well-determined by the data. In this way the remaining kernels can fit the data in an 'appropriate way', as suggested by Fig.12.4,

---

[4] Note that with a diverging weight-decay 'constant', λ_k, all the weights in the respective weight group decay to zero and leave the mapping implemented in the kth network branch completely misplaced. Consequently, the posterior probability for the kth component in the mixture will always be small, leading to a decay of the prior a_k when updated with the EM algorithm.

top right, where none of the remaining kernel widths becomes 'too small'. Compare this with Fig.12.4, top left, which shows the evolution of the σ_ks for the network regularised with the simple weight decay term. In this case all the network branches 'stay alive' (as seen from Fig.12.4, middle left), but several σ_ks become very small. This points to serious overfitting, caused by some kernel centres $\mu(\mathbf{x}, \mathbf{w}_k)$ fitting only a few outliers more or less exactly. When the training set size is increased to $N = 1000$, the ratio of the number of parameters per data point increases from 1.6 to 7.7. This larger ratio is reflected in the number of well-determined parameters γ_k, shown in the bottom part of Fig.12.3, which on the average are significantly larger than in the previous case of the sparse data (compare with Fig.12.3, top part). Moreover, γ_k only decays for one weight group, leading to the pruning of only one kernel (Fig.12.5, middle right, bottom right). Keeping this network complexity does not degrade the generalisation performance, as supported from Fig.12.5, top, which shows the evolution of the kernel widths σ_k during the training process and suggests, since all the σ_ks remain 'sufficiently large' (as opposed to the graphs depicted in Fig.12.4, left), that overfitting is effectively avoided.

12.4.2 Mathematical elucidation of the pruning scheme

From Equation (10.75) it is not immediately plausible how a divergence $\alpha_k \to \infty$ can occur with a concomitant decay $\gamma_k \to 0$. Therefore, a simple one-dimensional illustration will be given here. Consider a hypothetical network that can be described completely by a scalar weight w. The maximum-likelihood error function E is assumed to be quadratic in w,

$$E(w) \quad = \quad \frac{\varepsilon}{2}(w - w_0)^2 \qquad (12.2)$$

and the regularisation term R is given by

$$R(w) \quad = \quad \frac{\alpha}{2}w^2. \qquad (12.3)$$

The total error function is given by the sum of these two expressions:

$$E^*(w) \quad = \quad E(w) + R(w) \quad = \quad \frac{\varepsilon}{2}(w - w_0)^2 + \frac{\alpha}{2}w^2. \qquad (12.4)$$

The mode $\hat{w}$ is easily obtained by application of Newton's rule:

$$\hat{w} \quad = \quad w - \left(\frac{\partial^2 E^*}{\partial w^2}\right)^{-1}\left(\frac{\partial E^*}{\partial w}\right) = -\left(\frac{\partial^2 E^*}{\partial w^2}\right)^{-1}\left(\frac{\partial E^*}{\partial w}\right)_{w=0}$$

$$= \quad (\varepsilon + \alpha)^{-1}\varepsilon w_0 = \frac{\varepsilon w_0}{\varepsilon + \alpha}. \qquad (12.5)$$

The update equation for the hyperparameter α, Equation (10.75), reads for this one-dimensional problem:

$$\frac{1}{\alpha} = \frac{\hat{w}^2}{\gamma}, \tag{12.6}$$

where the number of well-determined parameters follows from (10.72):

$$\gamma = \frac{\varepsilon}{\varepsilon + \alpha}. \tag{12.7}$$

From (12.5), (12.6), (12.7) we obtain:

$$\alpha = \frac{\gamma}{\hat{w}^2} = \frac{1}{\hat{w}^2}\frac{\varepsilon}{\varepsilon + \alpha} = \left[\frac{\varepsilon w_0}{\varepsilon + \alpha}\right]^{-2}\frac{\varepsilon}{\varepsilon + \alpha} = \frac{\varepsilon + \alpha}{\varepsilon}\frac{1}{(w_0)^2}. \tag{12.8}$$

Solving for α leads to the expression

$$\alpha = \frac{\varepsilon}{\varepsilon(w_0)^2 - 1}. \tag{12.9}$$

which has a pole at $\varepsilon = (w_0)^{-2}$. Consequently, α diverges as ε approaches $(w_0)^{-2}$ from above. For $\varepsilon < (w_0)^{-2}$, there is no solution to α at all since in this case the expression in (12.9) would become negative. This can be interpreted in a slightly different way. With the definition

$$F(\alpha) := \frac{1}{(w_0)^2}\frac{\varepsilon + \alpha}{\varepsilon} \tag{12.10}$$

Equation (12.8) can be written as

$$\alpha = F(\alpha). \tag{12.11}$$

Now recall that in higher dimensions α can only be obtained in an iterative process,

$$\alpha_{n+1} = F(\alpha_n), \tag{12.12}$$

in which the subscript n indicates the iteration number. Equation (12.12) defines an iterative discrete map. From the theory of dynamical systems (see, for example, [15], [33]) it is known that a fixed point $\hat{\alpha}$ of such a map is stable only if the modulus of the derivative of F satisfies the condition

$$|F'(\hat{\alpha})| \leq 1 \tag{12.13}$$

Taking the derivative of the function in (12.10) gives

$$F'(\alpha) = \frac{1}{\varepsilon(w_0)^2}. \tag{12.14}$$

Consequently, if

$$\varepsilon < \frac{1}{(w_0)^2} \tag{12.15}$$

then $|\Gamma'(\alpha)| > 1$ and the iterative map given by (12.12) cannot have a stable fixed point. This implies that α diverges and, as a consequence of (12.7), the number of well-determined parameters decays to zero. So altogether we have

$$\varepsilon < \frac{1}{(w_0)^2} \quad \Rightarrow \quad \begin{array}{l} \alpha \to \infty \\ \gamma \to 0 \end{array}. \tag{12.16}$$

Now recall from (12.2) and (12.4) that the parameter ε defines the weight of the likelihood term relative to the regularisation term (12.3). The former scales as N, the number of training data:

$$\varepsilon \quad \propto \quad N. \tag{12.17}$$

It is thus seen that a decrease of the training set size N leads to a decrease in ε, which increases the likelihood for the scenario described by (12.16). This is exactly what was observed in the experiment and led to the performance-improving self-pruning effect.

12.5 Summary and Conclusion

The results of the simulations suggest that the adoption of the Bayesian evidence scheme for regularisation considerably reduces the tendency of overfitting and stabilises the training process with respect to changes in the number of parameter adaptation steps. This is partly achieved by an automatic pruning of oversized network complexity, which reduces the model to a smaller one that only contains *well-determined* parameters. As a consequence, overfitting resulting from too long a training process, as typically seen in graphs like Fig.12.2, top and middle, can be greatly reduced (Fig.12.2, bottom). This avoids the need for a cross-validation set and thus allows the use of all the available data for training.

13. Network Committees and Weighting Schemes

It is well-known that a combination of many different predictors can improve predictions. This combination is usually effected by majority in classification or by simple averaging in regression, but one can also use a weighted combination of the networks. The first section of this chapter summarises the main ideas of a recent study by Krogh and Vedelsby on network committees for simple interpolation tasks. The generalisation performance of the committee is seen to depend on the variation of the output of ensemble members, which the authors refer to as the ambiguity. The second section generalises the results to the prediction of conditional probability densities. A similar dependence on an ambiguity term is found, which suggests that diversity in the committee is a crucial requirement for improving the generalisation performance. The third section discusses aspects of the weighting scheme. The optimisation of the weights on the basis of the training data is shown to lead to a sub-optimal behaviour, and the evidence scheme is suggested as an approach that arises naturally within a Bayesian framework. However, numerical deficiencies of the latter are pointed out, and the chapter concludes with the introduction of an alternative scheme.

13.1 Network committees for interpolation

Consider an ensemble of N_{Com} conventional networks $\{\mathcal{M}_i\}_{i=1}^{N_{Com}}$ for predicting the conditional mean

$$f(\mathbf{x}) := \langle y|\mathbf{x} \rangle = \int y P_0(y|\mathbf{x}) dy \qquad (13.1)$$

of an unknown distribution $P_0(y, \mathbf{x}) = P_0(y|\mathbf{x}) P_0(\mathbf{x})$. The output of the ith network is denoted by $\mu_i(\mathbf{x})$, and the committee prediction is given by

$$\overline{\mu(\mathbf{x})} := \sum_{i=1}^{N_{Com}} c_i \mu_i(\mathbf{x}), \qquad (13.2)$$

where the weighting constants c_i are subjected to the constraint

$$c_i \geq 0 \ \forall i, \quad \sum_{i=1}^{N_{Com}} c_i = 1. \tag{13.3}$$

The mean square prediction error of the ith network in the committee is given by

$$
\begin{aligned}
\left\langle \left(\mu_i(\mathbf{x}) - y \right)^2 \right\rangle_{y,\mathbf{x}} &= \int \left(\mu_i(\mathbf{x}) - y \right)^2 P_0(y, \mathbf{x}) dy d\mathbf{x} \\
&= \int \left(\mu_i(\mathbf{x}) - f(\mathbf{x}) + f(\mathbf{x}) - y \right)^2 P_0(y|\mathbf{x}) P_0(\mathbf{x}) dy d\mathbf{x} \\
&= \int \left(\mu_i(\mathbf{x}) - f(\mathbf{x}) \right)^2 P_0(\mathbf{x}) d\mathbf{x} + \\
& \quad \int \left(f(\mathbf{x}) - y \right)^2 P_0(y|\mathbf{x}) P_0(\mathbf{x}) dy d\mathbf{x}.
\end{aligned}
$$

Note that the mixed term disappears as a result of (13.1). The second term in the last expression is the mean variance of the intrinsic noise, which does not depend on the network parameters. In dealing with last, it is therefore reasonable to neglect consideration of this term and define a generalisation error for the single network as

$$\mathcal{E}_i := \left\langle \left(\mu_i(\mathbf{x}) - f(\mathbf{x}) \right)^2 \right\rangle_{\mathbf{x}} = \int \left(\mu_i(\mathbf{x}) - f(\mathbf{x}) \right)^2 P_0(\mathbf{x}) d\mathbf{x}. \tag{13.4}$$

In the same way, a generalisation error for the committee is defined as

$$\mathcal{E}_{Com} := \left\langle \left(\overline{\mu(\mathbf{x})} - f(\mathbf{x}) \right)^2 \right\rangle_{\mathbf{x}} = \int \left(\overline{\mu(\mathbf{x})} - f(\mathbf{x}) \right)^2 P_0(\mathbf{x}) d\mathbf{x}. \tag{13.5}$$

This is to be compared with the average error of the individual predictors,

$$\overline{\mathcal{E}} := \left\langle \overline{\left(\mu_i(\mathbf{x}) - f(\mathbf{x}) \right)^2} \right\rangle_{\mathbf{x}} = \left\langle \sum_{i=1}^{N_{Com}} c_i \left(\mu_i(\mathbf{x}) - f(\mathbf{x}) \right)^2 \right\rangle_{\mathbf{x}} \tag{13.6}$$

which can be rewritten as

$$
\begin{aligned}
\overline{\mathcal{E}} &:= \left\langle \overline{\left(\mu_i(\mathbf{x}) - \overline{\mu(\mathbf{x})} + \overline{\mu(\mathbf{x})} - f(\mathbf{x}) \right)^2} \right\rangle_{\mathbf{x}} \\
&= \left\langle \overline{\left(\mu_i(\mathbf{x}) - \overline{\mu(\mathbf{x})} \right)^2} \right\rangle_{\mathbf{x}} + \left\langle \left(\overline{\mu(\mathbf{x})} - f(\mathbf{x}) \right)^2 \right\rangle_{\mathbf{x}} \tag{13.7}
\end{aligned}
$$

The second term is the generalisation error of the network committee, defined in (13.5). The first term, referred to as the *ambiguity term* by Krogh and Vedelsby [36], is the average variance of the interpolants implemented in the individual networks and, consequently, always non-negative:

$$A := \left\langle \left(\mu_i(\mathbf{x}) - \overline{\mu(\mathbf{x})} \right)^2 \right\rangle_{\mathbf{x}} = \left\langle \overline{\mu^2(\mathbf{x})} - \overline{\mu(\mathbf{x})}^2 \right\rangle_{\mathbf{x}} \geq 0. \qquad (13.8)$$

Putting (13.5), (13.7) and (13.8) together leads to

$$\mathcal{E}_{Com} = \overline{\mathcal{E}} - A \leq \overline{\mathcal{E}}, \qquad (13.9)$$

which shows that the generalisation error of the network committee is never larger than the average error of the individual predictors. In order to achieve a considerable reduction in the generalisation error, the ambiguity term must be large. Obviously, if the committee consists of identical networks that all make the same predictions, the ambiguity term disappears and nothing is gained. Only if the predictors are different is there something to be gained from using an ensemble. In fact, since the ambiguity term (13.8) increases with increasing variance of the interpolants implemented in the individual committee members, it is seen that disagreement between the predictors is crucial. The more the predictors differ, the lower the ensemble error will be, provided the individual errors remain constant.

The problem, therefore, is how to generate a committee with a large degree of diversity. One important factor is obviously the flexibility of the prediction model employed. Models with a low degree of flexibility will implement very similar functions, which will give rise to a small ambiguity term. On the other hand, a complex model with a high degree of flexibility makes overfitting of the individual models more likely, so that $\overline{\mathcal{E}}$ is likely to increase. The problem is therefore reminiscent of the bias-variance dilemma. Krogh and Vedelsby [36] write:

[Equation (13.9)] expresses the tradeoff between bias and variance in the ensemble, but in a different way than the common bias-variance relation in which the averages are over possible training sets instead of ensemble averages. If the ensemble is strongly biased the ambiguity will be small, because the networks implement very similar functions and thus agree on inputs even outside the training set. Therefore the generalization error will be essentially equal to the weighted average of the generalization errors of the individual networks. If, on the other hand, there is a large variance, the ambiguity is high and in this case the generalization error will be smaller than the average generalization error.

As a further means of increasing the ambiguity, Krogh et al. [35], [36], [57] study the training of networks on different subsets of the available data. When part of the data is thus set aside, the generalisation errors of the individual members of the ensemble will increase, but the conjecture is that the ambiguity might increase more and thus might lead to a decrease in the overall generalisation error. The approach taken in the present study is always to train the networks on the whole training set, but to take advantage of the inherent stochasticity of the RVFL approach, by which a larger degree of ambiguity per se is achieved.

13.2 Network committees for modelling conditional probability densities

The results of the previous section can easily be generalised to the modelling of arbitrary conditional probability densities by making use of Jensen's inequality,

$$c_i \geq 0, \quad \sum_i c_i = 1, \quad \Rightarrow \quad \ln\left(\sum_i c_i z_i\right) \geq \sum_i c_i \ln(z_i) \qquad (13.10)$$

which follows from the convexity of the logarithm. With the definition

$$\bar{z} \quad := \quad \sum_i c_i z_i \qquad (13.11)$$

this can be rewritten in the following more compact form:

$$\sum_i c_i \ln\left(\frac{\bar{z}}{z_i}\right) \quad \geq \quad 0. \qquad (13.12)$$

Similar to (13.2), the committee prediction for the conditional probability density is given by

$$\overline{P}(y|\mathbf{x}) = \sum_{i=1}^{N_{Com}} c_i P_i(y|\mathbf{x}), \qquad (13.13)$$

in which the abbreviated notation $P_i(y|\mathbf{x}) := P(y|\mathbf{x}, \mathbf{q}_i, \mathcal{M}_i)$ has been introduced. The generalisation 'error' of a single model has been defined in (3.7):

$$\mathcal{E}_i \quad = \quad -\left\langle \ln P_i(y|\mathbf{x}) \right\rangle = -\int P_0(y,\mathbf{x}) \ln P_i(y|\mathbf{x}) dy d\mathbf{x}. \qquad (13.14)$$

In a similar way, the generalisation 'error' of the network committee reads

$$\mathcal{E}_{Com} \quad = \quad -\left\langle \ln \overline{P}(y|\mathbf{x}) \right\rangle = -\int P_0(y,\mathbf{x}) \ln \overline{P}(y|\mathbf{x}) dy d\mathbf{x}. \qquad (13.15)$$

A comparison with the average generalisation error of the individual predictors,

$$\overline{\mathcal{E}} \quad = \quad \sum_{i=1}^{N_{Com}} c_i \mathcal{E}_i, \tag{13.16}$$

yields

$$\mathcal{E}_{Com} - \overline{\mathcal{E}} \quad = \quad -\Big\langle \ln \overline{P}(y|\mathbf{x}) \Big\rangle + \sum_{i=1}^{N_{Com}} c_i \Big\langle \ln P_i(y|\mathbf{x}) \Big\rangle$$

$$= \quad -\left\langle \sum_{i=1}^{N_{Com}} c_i \Big(\ln \overline{P}(y|\mathbf{x}) - \ln P_i(y|\mathbf{x}) \Big) \right\rangle$$

and

$$\mathcal{E}_{Com} \quad = \quad \overline{\mathcal{E}} - \left\langle \sum_{i=1}^{N_{Com}} c_i \ln \left(\frac{\overline{P}(y|\mathbf{x})}{P_i(y|\mathbf{x})} \right) \right\rangle. \tag{13.17}$$

The second term is non-negative as a consequence of Jensen's inequality (13.12) and can be interpreted as a generalisation of the ambiguity term (13.8):

$$A \quad := \quad \left\langle \sum_{i=1}^{N_{Com}} c_i \ln \left(\frac{\overline{P}(y|\mathbf{x})}{P_i(y|\mathbf{x})} \right) \right\rangle \quad \geq \quad 0. \tag{13.18}$$

Inserting (13.18) into (13.17) gives

$$\mathcal{E}_{Com} \quad = \quad \overline{\mathcal{E}} - A \quad \leq \quad \overline{\mathcal{E}}. \tag{13.19}$$

Consequently, the generalisation 'error' of the network committee is less than or equal to the average single-network prediction. In order to achieve a *considerable* reduction in the generalisation error, the ambiguity term needs to be large. Obviously, if the predictions of the networks in the committee are all very similar, $P_i(y|\mathbf{x}) \approx \overline{P}(y|\mathbf{x}) \ \forall i$, then the argument in the logarithm in (13.18) is close to 1, and the ambiguity term approaches 0. Nothing is gained when there is no disagreement between the models! Conversely, when the network predictions are very different, then the ambiguity term can become large and thus lead to a considerable decrease of the generalisation 'error'. It can therefore be concluded, as in Section 13.1, that a sufficient degree of diversity in the ensemble is a crucial requirement for a significant improvement of the generalisation performance.

Comparison with the best member of the committee
Relation (13.19) compares the committee performance with the average single-model performance. However, when the ambiguity term A becomes sufficiently large, the committee can also outperform the *best* single-model predictor. This happens, for instance, if for each committee member $\mathcal{M}_i$ there is a $(\mathbf{x}, y)$ such that $P_i(y|\mathbf{x}) \to 0$. In this case $\mathcal{E}_i \to \infty \ \forall i$ and consequently $\min_i \{\mathcal{E}_i\} \to \infty$. However, if for each $(y, \mathbf{x})$ there exists at least one model $\mathcal{M}_i$ such that $P_i(y|\mathbf{x})$ is 'sufficiently large', then $\overline{P}(y|\mathbf{x})$ will also be 'sufficiently large', and hence $E_{Com} \ll \min_i \{E_{single}(i)\}$.

13.3 Weighting Schemes for Predictors

13.3.1 Introduction

Let us now turn to the weighting factors c_i. The simplest approach is to choose them all equal in size, $c_i := \frac{1}{N_{Com}}$ $\forall i$. However, we would expect to be able to reduce the generalisation 'error' still further if their setting could be optimised in some way. The ideal choice, of course, would be to adapt them such that the overall generalisation error (13.15) is minimised. Unfortunately this is not possible since the true distribution $P_0(y|\mathbf{x})$ is not known. A straightforward approximation therefore seems to be the minimisation of the overall training 'error'

$$E_{train}^{Com} := -\frac{1}{N}\sum_{t=1}^{N}\ln\left(\sum_{i=1}^{N_{Com}}c_i P_i(y_t|\mathbf{x}_t)\right),\qquad(13.20)$$

which is equivalent to the method reviewed in [7], Section 9.6. The disadvantage of this approach is that the training data $\mathbf{D}$ are used in two places: in optimising the parameters of the prediction model $P_i(y|\mathbf{x})$ and in evaluating its prediction performance. Inserting the expression for the ith prediction model (2.19)

$$P_i(y|\mathbf{x}) = \sum_k a_k^i \mathcal{G}_{\beta_k^i}\left[y - \mu_k^i(\mathbf{x})\right]\qquad(13.21)$$

into (13.13) gives

$$\overline{P}(y|\mathbf{x}) = \sum_{i=1}^{N_{Com}} c_i \sum_{k=1}^{K} a_k^i \mathcal{G}_{\beta_k^i}\left[y - \mu_k^i(\mathbf{x})\right] = \sum_{i=1}^{N_{Com}}\sum_{k=1}^{K} c_i a_k^i \mathcal{G}_{\beta_k^i}\left[y - \mu_k^i(\mathbf{x})\right].$$

$$(13.22)$$

Since the parameters c_i and a_k^i are both optimised on the same data set, they can be replaced by a single parameter $b_{ik} := c_i a_k^i$, leading to

$$\overline{P}(y|\mathbf{x}) = \sum_{i=1}^{N_{Com}}\sum_{k=1}^{K} b_{ik} \mathcal{G}_{\beta_k^i}\left[y - \mu_k^i(\mathbf{x})\right].\qquad(13.23)$$

This expression is identical to a mixture model of the form (2.19) with $K N_{Com}$ kernels. The only effect of employing a network committee is therefore a considerable increase in the number of adaptable parameters, which has the opposite effect to that desired: rather than improving the generalisation performance, overfitting is only made more likely. Consequently, the weighting parameters c_i should not be optimised on the basis of the training data.

13.3.2 A Bayesian approach

The proper choice for the weighting factors c_i follows from the Bayesian approach discussed in Chapter 11. Recall from page 165 that the prediction $P(y|\mathbf{x})$ is formally obtained by summing over the whole model space

$$P(y|\mathbf{x}, \mathbf{D}) = \sum_i P(y, \mathcal{M}_i|\mathbf{x}, \mathbf{D}) = \sum_i P(y|\mathbf{x}, \mathbf{D}, \mathcal{M}_i)P(\mathcal{M}_i|\mathbf{x}, \mathbf{D})$$

$$= \sum_i P(y|\mathbf{x}, \mathbf{D}, \mathcal{M}_i)P(\mathcal{M}_i|\mathbf{D}), \tag{13.24}$$

where the last step results from the fact that the distribution of the input data $\mathbf{x}$ is not modelled by the network. In practice the sum in (13.24) over the whole model space is approximated by a sum over a network committee of limited size:

$$P(y|\mathbf{x}, \mathbf{D}) = \sum_{i=1}^{N_{Com}} P(\mathcal{M}_i|\mathbf{D})P(y|\mathbf{x}, \mathbf{D}, \mathcal{M}_i). \tag{13.25}$$

Applying Bayes' rule and assuming a constant prior on the models, $P(\mathcal{M}) = $ const, gives

$$P(\mathcal{M}_i|\mathbf{D}) = \frac{P(\mathbf{D}|\mathcal{M}_i)P(\mathcal{M}_i)}{P(\mathbf{D})} = \frac{P(\mathbf{D}|\mathcal{M}_i)P(\mathcal{M}_i)}{\sum_j P(\mathbf{D}|\mathcal{M}_j)P(\mathcal{M}_j)} = \frac{P(\mathbf{D}|\mathcal{M}_i)}{\sum_j P(\mathbf{D}|\mathcal{M}_j)}. \tag{13.26}$$

Denote the evidence for the model $\mathcal{M}_i$ by

$$\mathrm{Ev}_i := \ln P(\mathbf{D}|\mathcal{M}_i). \tag{13.27}$$

The sum in (13.25) can then be written as

$$P(y|\mathbf{x}, \mathbf{D}) = \sum_{i=1}^{N_{Com}} c_i P(y|\mathbf{x}, \mathbf{D}, \mathcal{M}_i), \tag{13.28}$$

in which the weighting factors c_i are given by

$$c_i = \frac{\exp(\mathrm{Ev}_i)}{\sum_j \exp(\mathrm{Ev}_j)}. \tag{13.29}$$

Their explicit calculation is therefore possible by making use of the expression for the evidence (11.45), which was derived in Chapter 11.

13.3.3 Numerical problems with the model evidence

In practice, the accurate evaluation of the model evidence can prove to be very difficult. As discussed before in Section 11.5, the evidence depends on the determinant of the Hessian, which is dominated by the small eigenvalues and therefore very susceptible to numerical roundoff errors. For this reason, the weighting scheme (13.29) has hardly been used in practical applications. For example, MacKay writes in the report on the energy prediction competition [39]:

> As for the evaluation of the evidence (for model comparison), it was found that the evidence for small models was often well correlated with cross-validation performance. However, it was disappointing to find that the numerics of evidence evaluation were hampered in larger models where there were many ill-determined parameters.

And for the actual model selection task he resorted to cross-validation:

> Since I had numerical problems with evaluating the posterior probabilities [of the models], I used for each task the following ad hoc procedure [...]. A 'committee' of the best k models was created (ranked by performance on the validation set), and the mean of their predictions was evaluated. The size k of the committee was chosen so as to minimise the validation error of the mean predictions.

Thodberg [61] suggested another heuristic selection scheme, which is based on the evidence but takes numerical roundoff errors into consideration in that the evidence is only used for selecting a committee rather than weighting its members:

> As an approximation to this, we define the committee as the models with evidence larger than $\log \mathrm{Ev}_{max} - \Delta \log \mathrm{Ev}$, where Ev_{max} is the largest evidence obtained, and all members enter with the same weight.

In this book, another scheme is introduced. First, define the evidence *per training exemplar*:

$$\mathrm{ev} \quad := \quad \frac{\mathrm{Ev}}{N_{train}}, \tag{13.30}$$

where N_{train} denotes the size of the training set. Equation (13.29) is then modified by introduction of a positive parameter $T > 0$,

$$c_i \quad = \quad \frac{\exp(\mathrm{ev}_i/T)}{\sum_j \exp(\mathrm{ev}_j/T)}. \tag{13.31}$$

The additional parameter T is akin to a temperature and allows some extra flexibility in the weighting scheme. For example, for $T \to 0$, the committee reduces to the prediction with only a single model, namely the one with the maximum evidence. If $T \rightarrow \infty$, all predictors in the committee contribute with equal weights. The original weighting scheme (13.29) is included in (13.31) as the special case $T = \frac{1}{N_{train}}$. However, in the simulations discussed in the forthcoming chapters, the latter scheme was found to assign significant weights c_i only to a small number of models. Considering the numerical problems in the evaluation of the evidence, this behaviour is obviously sub-optimal. Larger values of T lead to larger effective committee sizes, which can be assumed to compensate to some extent for the numerical noise inherent in the weighting procedure. In fact, the simulation results reported in the following chapter support this conjecture, and confirm that a considerable improvement in the generalisation performance is achieved with temperatures $T \gg \frac{1}{N_{train}}$.

13.3.4 A weighting scheme based on the cross-validation performance

When a sufficient amount of training data is available, it might be advantageous to abandon the evidence scheme altogether and base the weighting scheme on the cross-validation 'error' (see Equation (5.3))

$$E_{cross}(i) \quad = \quad -\frac{1}{N_{cross}} \sum_{t=1}^{N_{cross}} \ln P_i(y_t | \mathbf{x}_t). \qquad (13.32)$$

The disadvantage is a reduction in the number of available training exemplars, but this might be counteracted by the avoidance of the numerical problems discussed above. The justification is given by the fact that, for models well-matched to the data, the evidence has empirically been found to be (anti-) correlated with the cross-validation error [37], [38]. This suggests introduction of the following weighting scheme as an alternative to (13.31):

$$c_i \quad = \quad \frac{\exp(-E_{cross}(i)/T)}{\sum_j \exp(-E_{cross}(j)/T)}. \qquad (13.33)$$

Again, for $T \rightarrow 0$ the committee reduces to the prediction with a single model (the one with the minimum cross-validation 'error'), whereas $T \rightarrow \infty$ assigns equal weights to all networks. The temperature itself can be chosen so as to optimise the performance of the committee on the cross-validation set. The application of this scheme will be discussed in chapters 14 and 16.

14. Demonstration: Committees of Networks Trained with Different Regularisation Schemes

An ensemble of GM-RVFL networks is applied to the stochastic time series generated from the logistic-kappa map, and the dependence of the generalisation performance on the regularisation method and the weighting scheme is studied. For a single-model predictor, application of the Bayesian evidence scheme is found to lead to superior results. However, when using network committees, under-regularisation can be advantageous, since it leads to a larger model diversity, as a result of which a more substantial decrease of the generalisation 'error' can be achieved.

14.1 Method and objective

It was demonstrated earlier, in Chapter 12, that the generalisation performance of a single model can be enhanced by application of the *Bayesian regularisation* scheme. In Chapter 13, it was argued that the generalisation performance can be improved by combining several prediction models in a *committee*. In this section, a report on the results of a study that compares and combines these two approaches is given. As a benchmark problem, the stochastic process of Section 4.2 was chosen again. A small training set of $N_{train} = 200$ data points and two larger sets of $N_{cross} = N_{gen} = 1000$ data points for cross-validation[1] and testing were created from (4.16). As a prediction model, a GM-RVFL network with $m = 1$, $H = 10$, and $K = 10$ was chosen. Parameter adaptation was performed with the EM algorithm (Chapter 7), and the training simulations were carried out over six fixed numbers of epochs, $n = 20, 30, 40, 50, 60, 70$. Four different regularisation schemes were applied:

---

[1] This partitioning of the available data into a small training set and a large cross-validation set is not realistic for practical applications. The small training set size was chosen for testing the effects of overfitting. The large cross-validation set was used for getting a reliable estimate of the weighting scheme (13.33), with which the alternative weighting scheme (13.31) and a uniform weighting scheme are to be compared.

1. Eigenvalue cutoff

2. Weight decay

3. Bayesian evidence scheme

4. Simple Bayesian regularisation

For the simple Bayesian regularisation scheme, described in Chapter 9, the prior of Section 9.3 was applied, with the following choice of the hyperparameters: $\alpha_k = 0.1 \; \forall k$, $\xi = 0.1$, $\rho = 1$, and $\nu = 1$. This corresponds to a Gaussian prior on the weights $\mathbf{w}_k$, a gamma prior on the kernel precisions β_k, and a uniform prior on the output weights a_k. The first three regularisation schemes were as described in Section 12.1.2, except that two different values of α_k in the simple weight decay scheme were tried out: $\alpha_k = 0.01$, and $\alpha_k = 0.1$ (identical for all weight groups). For each method, an ensemble of $N_{Com} = 80$ networks was created in the following way. The random weights in the GM-RVFL were drawn from four different Gaussian distributions $N(0, \sigma_{rand})$ with $\ln \sigma_{rand} \in \{-1, 0, 1, 2\}$ (i.e. $\sigma_{rand} \in \{0.37, 1.0, 2.72, 7.39\}$). For each of the four distribution widths σ_{rand}, 20 weight configurations were drawn, based on different random number generator seeds. Each set of networks with identical σ_{rand} was further subdivided into four subsets, which differed with respect to the initialisation of the adaptable weights $\mathbf{w}$, drawn from (i) $N(0, 0.1)$, (ii) $N(0, 0.25)$, (iii) $N(0, 0.5)$, or (iv) $N(0, 1.0)$. The remaining parameters, β_k and a_k, were initialised as described in Section 5.2.1.

The subject of the first part of this study, summarised in Section 14.2, was the dependence of the single-network performance on the regularisation scheme. In the second part, discussed in Section 14.3, the generalisation performance of network committees was studied. Two different weighting schemes were applied, based on the cross-validation 'error' (13.33) and, for networks regularised with the Bayesian evidence scheme, on the Bayesian model evidence (13.31). The expression for the latter is given in Equation (11.45). This term was slightly modified, though, for reasons discussed in Section 11.4.3: in order to avoid numerical instability, the Ockham factor for the output weights, (11.61), was omitted.

The results of the simulations can be found in Tab.14.1-14.5, parts of which are displayed in Fig.14.1-14.4.

14.2 Single-model prediction

Figure 14.1 shows a scatter plot of the generalisation 'error' E_{gen} (see 5.5) against the training 'error' E_{train} (see (5.2)). For the first two regularisa-

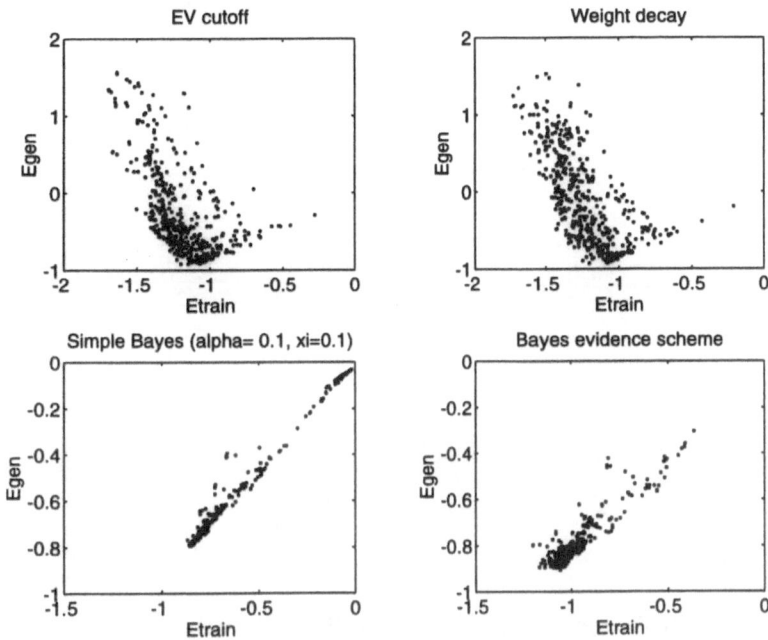

Fig. 14.1. Correlation between the training 'error' E_{train} and the generalisation 'error' E_{gen} for different regularisation schemes. 80 GM-RVFL networks were trained over 6 different training times (see text), leading to 480 pairs (E_{train}, E_{gen}) altogether. *Top left:* Eigenvalue cutoff. The performance results on the training set and the test set are strongly anti-corelated, with small training 'errors' E_{train} giving rise to large generalisation errors 'E_{gen}', and, conversely, large training 'errors' E_{train} leading to small generalisation 'errors' E_{gen}. Note that this points to a considerable under-regularisation. *Top right:* Weight decay, $\alpha_k = 0.01\ \forall k$. As before, the performance results on the training set and the test set are strongly anti-correlated, pointing again to a drastic under-regularisation. *Bottom left:* Simple Bayesian regularisation with a Gaussian prior on the weights **w**, a gamma prior on the kernel precisions β_k, and a uniform prior on the putput weights a_k: $\alpha_k = 0.1\ \forall k,\ \xi = 0.1,\ \rho = \nu = 1\ \bar{\rho} = \bar{\nu} = 0$. Training and generalisation 'error', E_{train} and E_{gen}, are now strongly correlated, but the best generalisation performance is considerably worse than in the previous two cases. This suggests that the model is over-regularised. *Bottom right:* Bayesian evidence scheme. As for the simple Bayesian regularisation scheme, E_{train} and E_{gen} are correlated, with an improvement of the performance on the training set normally accompanied by an improvement of the generalisation performance. A comparison with the graph on the bottom left shows that the whole distribution 'cloud' is shifted towards smaller generalisation 'errors' E_{gen}, suggesting that the problem of over-regularisation has been considerably reduced.

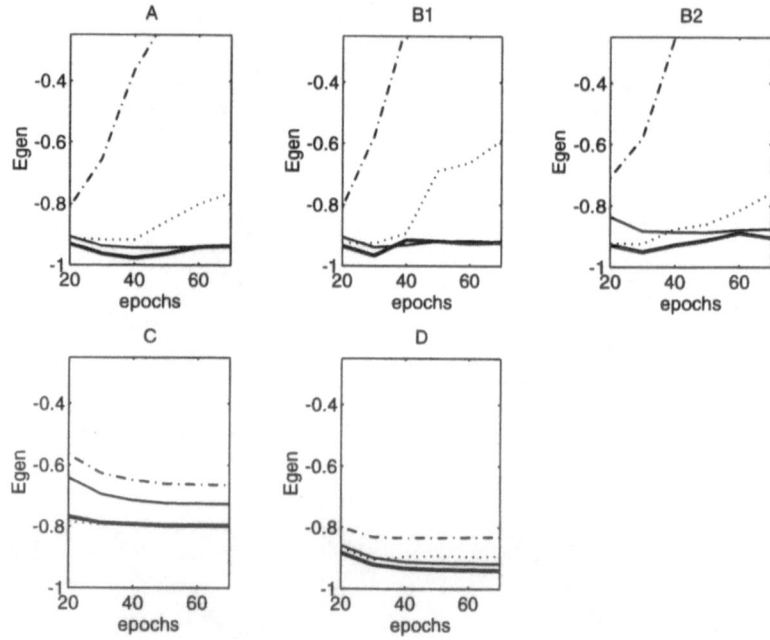

Fig. 14.2. Evolution of different generalisation 'errors' during training.
A committee of 80 GM-RVFL networks was trained on the time series generated
from the logistic-kappa map (4.16), as described in the text. The graphs show the
evolution of the generalisation 'error' E_{gen} for different regularisation schemes during training, where the abscissa represents the number of training epochs. *Dashed-
dotted line:* mean single network performance, $\overline{E_{gen}}$, averaged over the whole committee. *Dotted line:* best single network performance, $\min_i\{E_{gen}^i\}$. *Narrow solid
line:* committee performance at $T = \infty$, $E_{gen}^{Com}(T = \infty)$. *Bold solid line:* committee performance at $T = 0.1$, $E_{gen}^{Com}(T = 0.1)$. Five different regularisation schemes
were applied, as described in the text. **A:** eigenvalue cutoff. **B1:** weight decay with
$\alpha_k = 0.01 \; \forall k$. **B2:** weight decay with $\alpha_k = 0.1 \; \forall k$. **C:** simple Bayesian regularisation, $\alpha_k = 0.1 \; \forall k$, $\xi = 0.1$, $\rho = \nu = 1$. **D:** Bayesian evidence scheme.

tion schemes, (i) *eigenvalue cutoff* (top left) and (ii) *weight decay* (top right),
the scatter plot shows a clearly discernible anti-correlation between E_{train}
and E_{gen}. This suggests that the model is under-regularised, in accordance
with earlier findings discussed in Chapter 12 and displayed in Fig.12.2: as
training progresses, the training 'error' reaches very small values, but this is
accompanied by a dramatic increase in the generalisation 'error'. This overfitting problem can be staved off by Bayesian regularisation, leading to the
two graphs at the bottom of Fig.14.1, which show a clear *positive* correlation
between E_{train} and E_{gen}. However, the best generalisation 'error' obtained
with the simple Bayesian scheme (Fig.14.1, bottom left) is found to be considerably larger than the best generalisation 'error' in the previous two cases

(Fig.14.1 top). This suggests that with the simple Bayesian approach the model is *over-regularised*, most probably due to the rather poor *ad hoc* choice of the hyperparameters[2]. A more systematic grid search in the space of hyper-parameters can be expected to lead to better results. However, the Bayesian evidence scheme is superior in this respect, since it adjusts the hyperparameters automatically during training. In fact, a comparison between the two bottom graphs in Fig.14.1 shows that for the Bayesian evidence scheme, the whole distribution of (E_{train}, E_{gen})-pairs is moved towards significantly smaller generalisation 'errors'. A study of the results given in the first two rows of Tab.14.1-14.5 (which show the average and the best single-network generalisation 'error') and an inspection of the dotted and dashed-dotted curves in Fig.14.2 suggest:

- The generalisation 'error' obtained with the evidence approach is robust with respect to changes in the number of adaptation steps, n. This does not hold for the under-regularised models, where, due to over-fitting, E_{gen} increases sharply with n.

- The generalisation 'errors' obtained with the Bayesian evidence method are significantly smaller than the values obtained with the simple Bayesian scheme.

- The best generalisation 'error' obtained with the under-regularised models is slightly better than the best result achieved with the Bayesian evidence scheme. However, the identification of this optimal network configuration requires a cross-validation scheme, and thus the holding-out of a further independent subset of the data. Since such a cross-validation set is not required with the Bayesian approach, extra data are available for network training. A fair comparison should take this into account, and can then be expected to lead to a further improvement of the generalisation performance of the Bayesian evidence scheme.

14.3 Committee prediction

14.3.1 Best and average single-model performance

Figure 14.2 shows the evolution of the *best* and the *average* generalisation 'error' of the single model. From the graphs in the upper row it is seen that

---

[2] Values that had achieved good results in the simulations of Chapter 16 were simply used again.

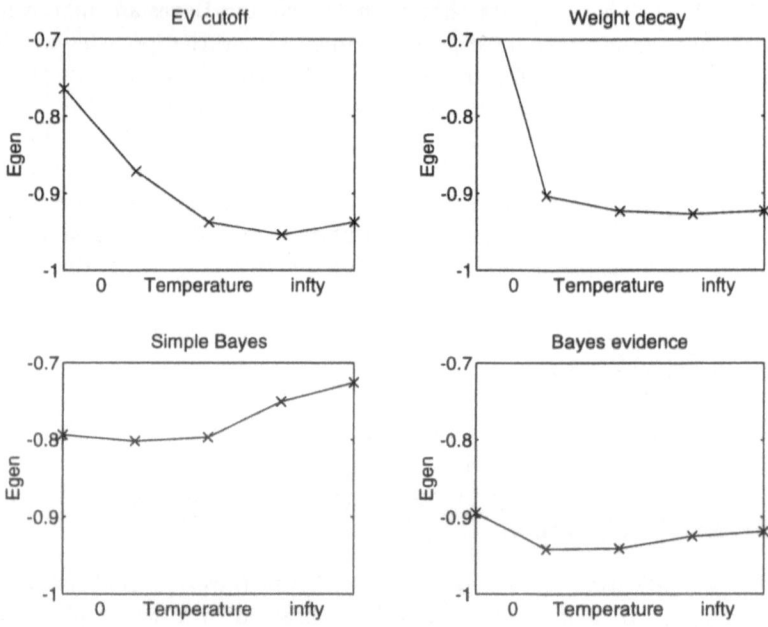

Fig. 14.3. Dependence of the generalisation performance on the temperature (1). An ensemble of 80 GM-RVFL networks were trained with the EM algorithm and different regularisation schemes over 70 epochs, as described in the text. For the ensemble prediction, the weighting scheme of Equation (13.33) was applied. The graphs show the dependence of the generalisation 'error' on the temperature for four different regularisation methods: eigenvalue cutoff (top left), weight decay with $\alpha_k = 0.01$ (top right), simple Bayesian regularisation (bottom left), and the Bayesian evidence scheme (bottom right). The simulations were carried out for five different temperatures, $T \in \{0, 0.05, 0.1, 1.0, \infty\}$, indicated by the crosses. Note that the scale of the abscissa is not linear. For the under-regularised models, shown in the top row, the best generalisation performance is achieved for large temperatures, and a dramatic deterioration can be observed as T approaches 0. In contrast, for the Bayesian regularised models, the optimal temperature is at a much smaller value, and the deterioration for $T \to 0$ is much less dramatic.

these 'errors' increase drastically during training when the model is under-regularised. This overfitting can be prevented with the Bayesian approach, as seen from the graphs in the bottom row. The Bayesian evidence scheme (bottom right) is seen to achieve a considerably better generalisation performance than the simple Bayesian regularisation technique (bottom left), since the latter, as discussed above, is over-regularised.

A different picture emerges when the performance of the whole network *committee* is studied. The narrow solid line shows the generalisation 'error' of a committee with equal weights for all members, equivalent to a temperature

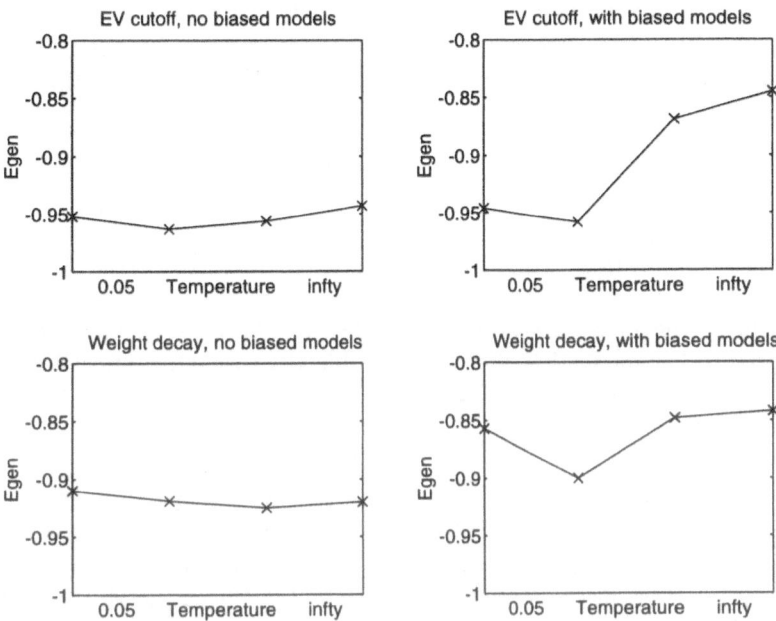

Fig. 14.4. Dependence of the generalisation performance on the temperature (2). An ensemble of 80 GM-RVFL networks were trained over 50 epochs, as described in the text. The networks were regularised with the method of *eigenvalue cutoff*, top row, and *weight decay* ($\alpha_k = 0.01$), bottom row. The ensemble prediction followed the weighting scheme (13.33). Two experiments were carried out. In the first one, shown in the left column, the distribution widths for the random weights were chosen from the set $\ln \sigma_{rand} \in \{-1, 0, 1, 2\}$ so that, generically, all networks were unbiased. The dependence of the generalisation performance of the ensemble on the temperature is found to be weak. In the second study, shown on the right, the ensemble contained models with $\ln \sigma_{rand} = -2$. For this value, the networks are, generically, biased, and their inclusion into the ensemble leads to a much more critical dependence of the generalisation performance on the temperature.

of $T = \infty$ in the weighting scheme (13.33). The bold solid line represents E_{gen} of a committee with a weighting scheme based on the cross-validation error, (13.33), at a temperature of $T = 0.1$. The following observations can be made:

14.3.2 Improvement over the average single-model performance

The generalisation 'error' of the committee, E_{gen}^{Com}, is always found to be significantly smaller than the average single-model generalisation 'error',

$\overline{E_{gen}}$. This observation confirms the theoretical result of Section 13.2, Equation (13.19), which states that the generalisation 'error' of the committee is equal to the average generalisation 'error' of the single network minus a positive so-called *ambiguity* or *diversity* term. The latter, defined in Equation (13.18), measures the variation of the outputs over the ensemble and is responsible for the decrease in the generalisation 'error' by using network committees. Interestingly, the graphs in Fig. 14.2 suggest that for the under-regularised models, shown in the top row, a considerably larger reduction in the generalisation 'error' by using network committees is achieved than for those regularised with one of the Bayesian schemes (bottom row). This is akin to the bias-variance dilemma [18], and intuitively understandable: Under-regularised models are more flexible than their regularised counterparts, and therefore give rise to a larger *diversity* term.

14.3.3 Improvement over the best single-model performance

As discussed in Section 13.2, the generalisation 'error' of the committee can also become smaller than that of the *best individual* network if the diversity term becomes sufficiently large. This is indeed observed in the training simulations for the *under-regularised* networks, shown in the top graphs of Fig.14.2. It is interesting to compare the results obtained from the two simulations with *weight decay*, B1 and B2. When augmenting the weight-decay hyperparameter from $\alpha_k = 0.01$ (B1) to $\alpha_k = 0.1$ (B2), the regularisation effect is increased. Consequently, we would expect the degree of overfitting by the individual network to be decreased and, as a consequence of reduced model flexibility, the size of the diversity term - and hence the improvement achieved with the committee - also to be reduced. This conjecture is in fact borne out by the simulations: For $\alpha_k = 0.01$, the network committee at $T = \infty$ outperforms the best individual network throughout, whereas for $\alpha_k = 0.1$ this only happens when the training time is longer than $n \approx 40$ epochs. Note that for an over-regularised model, shown in graph C of Fig.14.2, the best single-network predictor is never worse than the committee. This is intuitively plausible, since in this case the diversity term is expected to be small.

14.3.4 Robustness of the committee performance

In Chapter 12 it was found that one of the main advantages of the Bayesian evidence scheme is its robustness with respect to variations in the length of training time (i.e. the number of epochs n). The same is seen to hold for the application of a network committee. While the single-model predictor overfits drastically in graphs A, B1, and B2 of Fig.14.2, the network committee shows a nearly constant generalisation performance irrespective of the number of epochs. This suggests that the application of network committees can be as effective as the Bayesian evidence scheme.

14.3.5 Dependence on the temperature

Recall from Equation (13.19) that two terms contribute to the generalisation 'error' of the network ensemble, E_{gen}^{Com}: the average generalisation 'error' of the individual network (averaged over the ensemble), and the ambiguity or diversity term, defined in (13.18). Now consider a temperature-dependent weighting scheme, as defined in equations (13.31) and (13.33). For low temperatures, $T \to 0$, only the 'best' networks have weights significantly different from zero. Consequently, the average generalisation 'error' of the single model, $\overline{E_{gen}}$, will be close to that of the best individual model, but the diversity term, too, will be small, and nothing is gained by using a committee. As T increases, $\overline{E_{gen}}$ will increase due to the assignment of larger weights to poorer models, but the ambiguity term will also increase, and this latter effect can counteract the deterioration caused by the first and lead to a net decrease of E_{gen}^{Com}. Consequently, the best generalisation performance is expected to be obtained for some nonzero temperature, $T > 0$.

The objective of this study, therefore, was to investigate the dependence of E_{gen}^{Com} on T more thoroughly. The networks were trained over different numbers of epochs n, and the generalisation error E_{gen}^{Com} was calculated for five different temperatures $T = 0, 0.05, 0.1, 1.0, \infty$. (Recall that for $T = 0$ the committee prediction reduces to the prediction with only *one* network, namely that showing the best cross-validation performance, whereas $T = \infty$ gives uniform weights.) The results can be found in Tab.14.1-14.5, and are visualised in Fig.14.2 (for $T \in \{0.1, \infty\}$ and different n) and Fig.14.3 (for $n = 70$ and different values of T).

Figure 14.3 shows, for $n = 70$, the dependence of E_{gen}^{Com} on T for the different regularisation schemes employed. It is seen that for the under-regularised schemes *eigenvalue cutoff* and *weight decay*, shown in the top row of the figure, the best generalisation performance is obtained for large values of T. In contrast, for $T \to 0$, the generalisation error is found to increase drastically. This behaviour can be understood from Fig.14.2, which shows that a training time of $n = 70$ epochs leads to considerable over-fitting of the individual networks. A significant enhancement of the generalisation performance can therefore only be achieved with a large diversity term. Obviously, this requires contributions from many different models and, therefore, calls for a high temperature.

These results do not carry over to the networks regularised with the two Bayesian schemes, as seen from the bottom row of Fig.14.3. Here, the generalisation error of the committee, E_{gen}^{Com}, takes on its minimum value at rather a low temperature of $T = 0.05$. As T increases, E_{gen}^{Com} becomes larger, in case of the over-regularised networks regularised with the *simple Bayesian* scheme considerably. This behaviour can be understood from the fact that, as a result of regularisation, the flexibility of the networks and, therefore, the size of the diversity term is greatly reduced. Consequently, the contribution from the average single-network performance, $\overline{E_{gen}}$, in the expression for E_{gen}^{Com}, Equation (13.19), tends to dominate. Since $\overline{E_{gen}}$ is minimised for $T \to 0$, the optimal temperature for minimal E_{gen}^{Com} is decreased.

14.3.6 Dependence on the temperature when including biased models

In the simulations reported above, the distribution widths for the random weights were chosen from the set $\ln \sigma_{rand} \in \{-1, 0, 1, 2\}$. From Section 8.2 it is known that, for these values of σ_{rand}, the nonlinear complexity is sufficiently large for the GM-RVFL network to be, generically, capable of modelling the underlying data-generating processes. The impact on the generalisation performance of the single-network predictor is caused by overfitting and too high a degree of nonlinear complexity. This is depicted in Fig.8.4 and can be expected to give rise to a large diversity term (as discussed above).

A different picture emerges when the distribution widths are decreased to $\ln \sigma_{rand} = -2$. From Section 8.2 it is known that for this small a value of σ_{rand} the GM-RVFL network is, generically, not sufficiently complex to approximate the data-generating processes. A typical state-space plot is shown in Fig.8.2 and suggests that the model is strongly *biased* towards linear functions. The inclusion of biased models was tested in the following simulation: For each of the values $\ln \sigma_{rand} \in \{-2, -1, 0, 1, 2\}$, 16 GM-RVFL networks were created and trained as before (over $n = 50$ epochs). The overall size of the ensemble, $N_{Com} = 80$, thus remained unchanged. However, a fifth of the networks were now, generically, incapable of modelling the non-linearities in the time series. The results of the simulations, carried out with the under-regularised schemes *eigenvalue cutoff* and *weight decay*, are given in Tab.14.6 and are visualised in Fig.14.4.

The graphs on the left of Fig.14.4 show the dependence of E_{gen}^{Com} on T for an ensemble of *unbiased* models. It is seen that this dependence is rather weak (note that as opposed to Fig.14.3 the performance for $T = 0$ is not shown here). Especially, $T \to \infty$, i.e. the limit of uniform weights, can be chosen without incurring a serious impact on the generalisation performance. The explanation for this behaviour was already given above. Non-selective inclusion of all networks into the committee leads to a deterioration in $\overline{E_{gen}}$

due to a 'contamination' with poor predictors, but this is counteracted by the increasing diversity term, which dominates in (13.19).

A different behaviour, however, is expected when *biased* models are included. Since *under-complex models do not give rise to a large diversity*, the ambiguity term A in (13.19) will be less dominant. Again, a large temperature will lead to a deterioration in $\overline{E_{gen}}$, but now this will not be sufficiently compensated for by an increasing diversity term. Consequently, a fine-tuned trade-off between the two terms in Equation (13.19) will be more crucial, leading to a more critical dependence on the temperature. This conjecture was in fact borne out by the simulations, as seen from the graphs on the right of Fig.14.4.

14.3.7 Optimal temperature

The boldface figures in Tab.14.1-14.5 show the lowest generalisation 'error' for each number of epochs n. It is seen that for the under-regularised models, regularised with *eigenvalue cutoff* (Tab.14.1) and *weight decay* (especially with $\alpha_k = 0.01$, Tab.14.2), these lowest values are shifted towards larger temperatures as n increases. The intuitive explanation for this tendency is that for small training times n, there are still several good models available, so the generalisation performance of the committee can be improved by selecting the best ones. This requires a small temperature T. For large training times, however, all committee members overfit drastically. In this case any selection scheme becomes inefficient, and the improvement in the generalisation performance of the committee results from a large diversity term, which requires larger values of T. Note that a temperature of $T = 0.1$ always leads to results with the best or close to the best performance. Interestingly, this empirical observation will be made again in Chapter 16 for a problem of a completely different nature.

14.3.8 Model selection and evidence

Figure 14.5 shows a scatter plot of the Bayesian model evidence, obtained from (11.45), and the generalisation 'error' E_{gen}. Crosses symbolise values obtained for a training time of $n = 70$ epochs, dots refer to shorter training times, $n \in \{20, 30, 40, 50, 60\}$. It is seen that a satisfactory degree of (anti-) correlation between the evidence and E_{gen} exists. This also holds for the shorter training times of $n \leq 60$, where the training process has not yet converged, and Equation (11.45) for the calculation of the evidence, strictly speaking, cannot be applied.

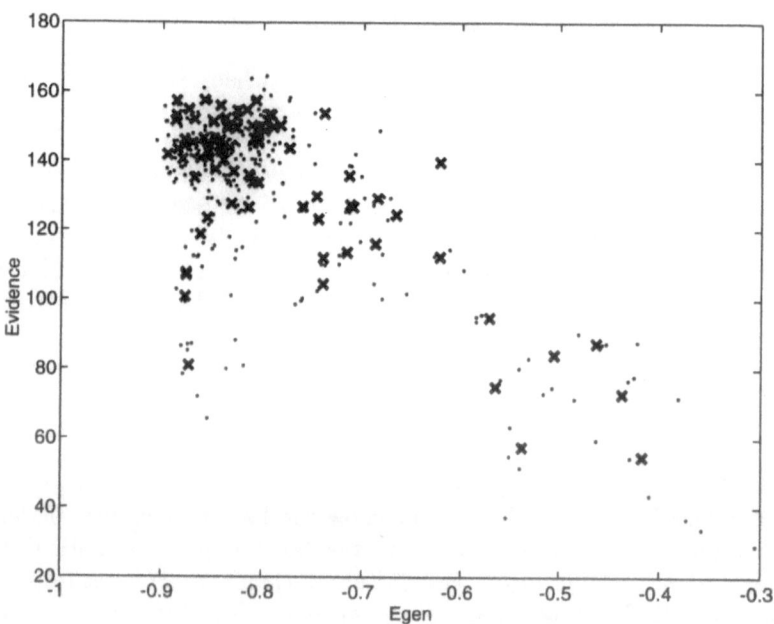

Fig. 14.5. Correlation between the evidence and the generalisation 'error'. The figure shows the scatter plot of the evidence and the generalisation 'error' E_{gen}, where the latter was estimated on an independent test set of $N = 1000$ data points. 80 GM-RVFL networks were trained over different training times. Crosses show the (E_{gen}, evidence)-pairs after $n = 70$ epochs, dots represent shorter training times ($n = 20, \ldots, 60$).

Table 14.7 compares the generalisation 'error' obtained from a weighting scheme based on the cross-validation 'error', (13.33), with that obtained by weighting the committee members according to their evidence, (13.31). The results show that the latter scheme leads to a generalisation performance which is only slightly worse than that obtained with the cross-validation-based method, and noticeably better than achieved with uniform weights (corresponding to $T = \infty$).

The simulations therefore suggest that, at least for the present problem, the Bayesian model evidence provides a reliable entity for model weighting and model selection. The optimal temperature, however, is found to be considerably larger than the theoretical one. It is recalled that in the weighting scheme of Equation (13.31), the evidence *per training exemplar*, (13.30), is used. The theoretically expected temperature is therefore given by $T = 1/N_{train} = 1/200 = 0.005$, which is considerably smaller than the empirically found optimal value of $T \approx 0.1$.

14.3.9 Advantage of under-regularisation and over-fitting

Table 14.8 shows, for each of the five regularisation methods employed, the minimal generalisation 'error', $\text{argmin}_{T,n}\{E_{gen}^{Com}(T,n)\}$. It is seen that committees of *under-regularised* models achieve slightly better results than a committee comprising of models regularised with the Bayesian evidence scheme. This trend persists, though to a lesser extent, even when no cross-validation set is available for weighting the predictors. The latter finding, however, pertains to a committee whose members are generically *unbiased*. Recall from Section 14.3.6 and Fig. 14.4 that the inclusion of *biased* models in a committee of *under-regularised* models leads to a deterioration in the generalisation performance at $T = \infty$, that is, in the absence of a cross-validation set for model weighting. The Bayesian model evidence, on the other hand, still allows weighting of the committee members, which due to the correlation observed in Fig.14.5 was found to achieve results similar to that of a cross-validation-based weighting scheme (Tab.14.7). Consequently, the Bayesian evidence scheme should be superior to the alternative approaches when the committee contains biased models and no cross-validation set is available. This is confirmed in Tab.14.9.

14.4 Conclusions

The main results of this chapter can be summarised as follows:

- The training process for a single-model predictor is stabilised with respect to changes in the length of training time by applying the *Bayesian evidence scheme*. The alternative methods were found to be either over-regularised or to lead to serious overfitting.

- For a committee of *unbiased* models and an *under-regularised* training scheme, the same stabilisation can be achieved by using committees at a sufficiently high temperature: overfitting of the individual models is counteracted by an increasing diversity term.

- The enhancement of the generalisation performance by applying network committees is a result of the *ambiguity term* (13.18). Consequently, the largest effect is obtained for *under-regularised* models, which give rise to a large diversity in the ensemble. In contrast, hardly any improvement can be expected for committees of *over-regularised* models, since for them the diversity will usually be small.

- For a committee of *unbiased* models, an *under-regularised* training scheme achieves a generalisation performance slightly better than obtained with the Bayesian evidence scheme. This even holds for the absence of a cross-validation set (corresponding to $T = \infty$). However, if the committee contains *biased* models, the Bayesian evidence scheme is superior.

Predictor	T	n=20	n=30	n=40	n=50	n=60	n=70
Average single	–	-0.810	-0.653	-0.364	-0.188	-0.076	-0.029
Best single	0	-0.912	-0.916	-0.917	-0.857	-0.800	-0.764
Committee	0.05	**-0.933**	-0.962	-0.967	-0.952	-0.877	-0.871
Committee	0.1	-0.930	**-0.963**	**-0.977**	**-0.963**	-0.941	-0.937
Committee	1.0	-0.911	-0.943	-0.956	-0.956	**-0.955**	**-0.954**
Committee	∞	-0.906	-0.937	-0.944	-0.943	-0.940	-0.937

Table 14.1. Eigenvalue cutoff. The table shows the generalisation 'error' E_{gen}, measured on a test set of $N = 1000$ data points, for a GM-RVFL network regularised with the eigenvalue-cutoff scheme (7.63). The different columns show the dependence on the training time, expressed in number of epochs, $n = 20, 30, \ldots, 70$. First row: mean single-network performance, averaged over the whole committee of $N_{Com} = 80$ networks. Second row: best single-network performance. Remaining rows: committee performance at different temperatures. The weighting scheme was based on the cross-validation set. For each number of epochs n, the lowest value of E_{gen} is printed in bold.

Predictor	T	n=20	n=30	n=40	n=50	n=60	n=70
Average single	–	-0.808	-0.577	-0.203	0.047	0.109	0.167
Best single	0	-0.925	-0.926	-0.892	-0.690	-0.663	-0.591
Committee	0.05	**-0.937**	**-0.971**	-0.906	-0.910	-0.910	-0.903
Committee	0.1	-0.935	-0.966	-0.915	-0.919	**-0.926**	-0.923
Committee	1.0	-0.913	-0.944	**-0.932**	**-0.925**	-0.925	**-0.927**
Committee	∞	-0.905	-0.939	**-0.932**	-0.920	-0.918	-0.923

Table 14.2. Weight decay, alpha=0.01. The table shows the generalisation 'error' E_{gen} for a GM-RVFL network regularised with weight decay, $\alpha_k = 0.01$. For further details, see caption of Tab.14.1.

Predictor	T	n=20	n=30	n=40	n=50	n=60	n=70
Average single	–	-0.708	-0.580	-0.253	-0.079	0.022	0.073
Best single	0	-0.923	-0.925	-0.876	-0.860	-0.815	-0.758
Committee	0.05	**-0.930**	**-0.959**	**-0.930**	-0.866	-0.869	-0.896
Committee	0.1	-0.929	-0.951	-0.929	**-0.913**	**-0.889**	**-0.905**
Committee	1.0	-0.874	-0.903	-0.892	-0.880	-0.875	-0.876
Committee	∞	-0.836	-0.883	-0.886	-0.887	-0.879	-0.876

Table 14.3. Weight decay, alpha=0.1. Results of a study similar to that of Tab.14.3, but with $\alpha_k = 0.1$ rather than $\alpha_k = 0.01$. For further details, see caption of Tab.14.1.

Predictor	T	n=20	n=30	n=40	n=50	n=60	n=70
Average single	–	-0.567	-0.626	-0.649	-0.661	-0.663	-0.665
Best single	0	-0.784	-0.794	-0.794	-0.794	-0.794	-0.794
Committee	0.05	**-0.788**	**-0.794**	**-0.796**	**-0.802**	**-0.802**	**-0.802**
Committee	0.1	-0.767	-0.788	-0.793	-0.797	-0.797	-0.797
Committee	1.0	-0.678	-0.726	-0.742	-0.750	-0.750	-0.751
Committee	∞	-0.641	-0.694	-0.714	-0.724	-0.725	-0.726

Table 14.4. Simple Bayesian regularisation. Repetition of the study summarised in the caption of Fig.14.1, using a simple Bayesian regularisation scheme with a Gaussian prior on the weights $\mathbf{w}_k$, a gamma prior on the β_k, and a constant prior on the output weights a_k: $\alpha = 0.1$, $\xi = 0.1$, $\rho = \nu = 1$.

Predictor	T	n=20	n=30	n=40	n=50	n=60	n=70
Average single	–	-0.800	-0.832	-0.835	-0.834	-0.833	-0.832
Best single	0	**-0.867**	-0.906	-0.896	-0.893	-0.896	-0.895
Committee	0.05	-0.886	**-0.924**	**-0.935**	**-0.939**	**-0.940**	**-0.942**
Committee	0.1	-0.883	-0.921	-0.934	-0.938	-0.939	-0.941
Committee	1.0	-0.865	-0.905	-0.919	-0.922	-0.923	-0.925
Committee	∞	-0.859	-0.899	-0.913	-0.917	-0.918	-0.919

Table 14.5. Bayesian evidence scheme. The table shows the generalisation 'error' E_{gen} for a GM-RVFL network regularised with the Bayesian evidence scheme. For further details, see caption of Tab.14.1.

Range of $\ln \sigma_{rand}$	Regularisation	$T = 0.05$	$T = 0.1$	$T = 1.0$	$T = \infty$
$-2, \ldots, 2$	Eigenvalue cutoff	-0.946	**-0.958**	-0.869	-0.845
$-1, \ldots, 2$	Eigenvalue cutoff	-0.952	**-0.963**	-0.956	-0.943
$-2, \ldots, 2$	Weight decay	-0.857	**-0.900**	-0.848	-0.842
$-1, \ldots, 2$	Weight decay	-0.910	-0.919	**-0.925**	-0.920

Table 14.6. Dependence of the committee performance on the temperature. The table shows the generalisation 'error' E_{gen} for a committee of 80 GM-RVFL networks at different temperatures, where the weighting scheme is based on the cross-validation 'error' E_{cross}. All networks were trained for $n = 50$ epochs, and were regularised either with the method of eigenvalue cutoff or weight decay ($\alpha_k = 0.01$). The column on the left indicates the range of distribution widths for the random weights in the GM-RVFL network. For $\ln \sigma_{rand} \geq -1$, all networks are generically of sufficient complexity to model the underlying processes of the time series, whereas for $\ln \sigma_{rand} = -2$ they are not. It is seen that in the latter case a stronger dependence of E_{gen} on T results.

Weighting scheme	T	n=30	n=50	n=70
Cross-validation	0.05	-0.924	-0.939	-0.942
Bayesian evidence	0.05	-0.916	-0.927	-0.932
Cross-validation	0.1	-0.921	-0.938	-0.941
Bayesian evidence	0.1	-0.916	-0.930	-0.934
Uniform weights	∞	-0.899	-0.917	-0.919
Best single model	0	-0.906	-0.893	-0.895

Table 14.7. Comparison between the evidence-based and the cross-validation-based weighting scheme. The table shows the generalisation 'error' E_{gen} for a committee of 80 GM-RVFL networks regularised with the Bayesian evidence scheme. Details of the simulations can be found in the text. Two weighting schemes, based on (i) the cross-validation 'error' E_{cross} (13.33) and (ii) the Bayesian evidence (13.31), are compared with each other for different temperatures T and different training times (measured in epochs, n). For comparison, the generalisation 'error' of a committee with uniform weights(equivalent to $T = \infty$) and the performance of the best single model (equivalent to $T = 0$) are also listed. It is seen that the evidence-based weighting scheme performs only slightly worse than the cross-validation-based method, and noticeably better than the best single model or the committee with uniform weights.

Regularisation scheme	Best Performance	
	with cross-validation	without cross-validation
Eigenvalue cutoff	-0.977	-0.944
Weight decay, $\alpha_k = 0.01$	-0.971	-0.939
Weight decay, $\alpha_k = 0.1$	-0.959	-0.887
Bayesian evidence scheme	-0.942	-0.934
Simple Bayesian regularisation	-0.802	-0.726

Table 14.8. Comparison between the best generalisation errors for a committee of unbiased estimators. The table shows, for each of the five regularisation methods, the best committee generalisation 'error' found, $\text{argmin}_{T,n}\{E_{gen}^{Com}(T, n)\}$. The values listed in the middle column were obtained with a weighting scheme based on the cross-validation 'error', (13.33). The right column shows the results without cross-validation, where either equal weights were assigned to all committee members or, in case of the evidence scheme, the weighting scheme was based on the Bayesian model evidence (13.31). Note that the under-regularised models in the first two rows achieve a better performance than the Bayesian evidence approach, though the difference is only small when no cross-validation set is used. The network committee contained 80 GM-RVFL networks with $\ln \sigma_{rand} \in \{-1, 0, 1, 2\}$, that is, the individual models were generically unbiased.

| Regularisation | Best Performance | |
scheme	with cross-validation	without cross-validation
Eigenvalue cutoff	-0.962	-0.845
Weight decay, $\alpha_k = 0.01$	-0.941	-0.842
Weight decay, $\alpha_k = 0.1$	-0.940	-0.787
Bayesian evidence scheme	-0.929	-0.908

Table 14.9. Comparison between the best generalisation 'errors' for a committee that includes biased estimators. The table is equivalent to Tab.14.8, except that for 20% of the GM-RVFLs the random weights were drawn from a Gaussian distribution with $\ln \sigma_{rand} = -2$. As discussed in Section 8.2, these models are generically incapable of approximating the non-linearities of the target function, thus the committee contains biased models. It is seen that when employing a weighting scheme based on the cross-validation 'error', the committee of under-regularised models still achieves a slightly better performance than the Bayesian evidence scheme (middle column), as in Tab.14.8. However, when no cross-validation set is available, the Bayesian evidence approach performs significantly better (right column).

15. Automatic Relevance Determination (ARD)

A scheme for the systematic adaptation of the random-parameter distribution widths is introduced. Weights exiting the same input node are combined into a weight group, and the distribution widths of the weight groups are adjusted during training by a method similar to Manhattan updating. A practical algorithm is derived, and an empirical demonstration shows that irrelevant inputs are detected and effectively switched off. The whole scheme was inspired by and is akin to Neal's and MacKay's automatic relevance determination. It will therefore be referred to by the same name.

15.1 Introduction

In Chapter 8, it was demonstrated that the choice of an appropriate value for σ_{rand}, the standard deviation of the random-weight distribution, is crucial to a good generalisation performance of the model. It was suggested that several simulations with different values for σ_{rand} be carried out, and then a weighting scheme of the form (13.31) or (13.33) applied to the resulting committee. However, it cannot be assumed in general that an optimal model can be found by a simultaneous scaling of all random weights. This is especially the case if there are several inputs of different nature and different relevance level. The objective of this chapter is therefore to improve the original scheme in two respects. Firstly, rather than drawing all the weights feeding into the S-layer from the same distribution, it is advantageous to introduce several *weight groups* such that weights exiting different input nodes are drawn from different distributions. This will allow different scales and 'relevance levels' (see below) of different inputs to be taken into account. Secondly, an adaptation scheme will be proposed, which will allow a more systematic search in the space of the respective standard deviations than just by a random walk. Since the following ideas have been inspired mainly by the work of Neal [44] and MacKay [39], who introduced a Bayesian approach for automatically determining the relevance of a given input, their terminology will be borrowed and the method to be described referred to as *Automatic Relevance Determination* (ARD).

Assume we have m input variables. Let all the (random) weights exiting the same input node constitute a weight group. Introduce for each of the m weight groups a hyperparameter[1] $\rho_g, g \in \{1, ..., m\}$, such that $\sigma_{rand}^g = \exp(\rho_g)$ defines the standard deviation[2] of the Gaussian distribution from which the random weights in the gth weight group are drawn. Alternatively, one can introduce *unscaled* weights $\mathbf{u}$ to be drawn from a standard normal distribution (having unit variance), and define the actual weights in the ith weight group to be given by $\sigma_{rand}^g \mathbf{u}_i$. In what follows, the latter concept will be adopted. Obviously, this scheme allows different relevance levels of the inputs to be taken into account; for instance, when the data-generating function does not depend on a particular input x_g, one would expect the best generalisation performance to be achieved with a very small value of $\sigma_g \approx 0$ such that any variation in x_g will effectively leave the total input to the network unchanged.

A major problem with the proposed concept of several weight groups, however, is the increased dimensionality of the hyperparameter space. Whereas the optimisation of a single σ_{rand} can easily be performed by a random search, as carried out in the previous simulations, such a scheme becomes practically infeasible when the dimension of the input vector, m, becomes larger than about 2 or 3. What is required is a more systematic adaptation, e.g. so as to minimise the training error

$$E = \sum_{t=1}^{N} \varepsilon_t, \qquad \varepsilon_t = -\ln P(y_t | \mathbf{x}_t) \tag{15.1}$$

with respect to $\boldsymbol{\rho} = (\rho_1, \ldots, \rho_m)$. However, since the ρ_gs feed in *before* the sigmoidal nonlinearities, this minimum cannot be found in a single step. The only feasible approach would be to follow the gradient $\nabla_{\boldsymbol{\rho}} E$, but this would lead again to a slow successive iteration scheme, in obvious contradiction to the RVFL concept: recall that the very reason for choosing the weights between the input and the S-layer at random was to avoid such an iterative scheme. In order to allow the inclusion of gradient information in the adaptation process while avoiding an iterative steepest descent scheme that would run afoul of the RVFL concept, the hybrid approach proposed below is introduced here.

---

[1] The term *hyperparameter* will be borrowed since the σ_{rand}^g are parameters of a *prior* probability distribution, and are closely related to the hyperparameters α_g in MacKays's work [39].

[2] By defining σ as the exponential of ρ its positivity is always ensured. The other reason for introducing ρ is that σ is a *scale* parameter. Since a non-informative prior for a scale parameter is uniform on a logarithmic scale (as discussed in Section 11.2), ρ is the natural parameter for any adaptation scheme.

First, let us impose a restriction on the parameter space by defining ρ to be a discrete, rather than a continuous, variable. For example, ρ_g can be defined as an integer, or a subset of integers. In general, define

$$\rho_g \quad \in \quad \Upsilon := \{\rho_{min}, \ \rho_{min} + \Delta\rho, \ \ldots, \ \rho_{max} - \Delta\rho, \ \rho_{max}\}. \qquad (15.2)$$

Given a vector of hyperparameters ρ, a network is trained in the usual way until a termination criterion (convergence or early stopping) is satisfied. Only then is the gradient $\nabla_\rho E$ calculated, and the ρ_gs are updated according to

$$\rho_g^{new} \quad = \quad \rho_g^{old} - \text{sign}\left(\frac{\partial E}{\partial \rho_g}\right)\Delta\rho \qquad (15.3)$$

such that $\rho_g^{new} \in \Upsilon$ is satisfied, that is, the updating is skipped if the latter condition does not hold. Mathematically, this scheme is similar to Manhattan updating (see, e.g., [52]). The difference, however, is that the ρ_gs are *not* adapted during the training process for a given network, but only *after* this training process has finished. This implies that, with respect to ρ, current networks do not themselves learn directly, but only from the 'errors' of their 'predecessors'.

15.2 Two alternative ARD schemes

Since the weights exiting the gth input node belong to the same weight group, they have the factor $\exp(\rho_g)$ in common. This can be interpreted in a slightly different way: the weights are left unchanged, and the input x_g is multiplied by the factor $\exp(\rho_g)$. Obviously, for $\rho_g \ll 0$, an input is effectively switched off. There is, however, a drawback in the above scheme, namely, the different scaling properties between (i) the weights from the input to the $\mathcal{S}$-layer, and (ii) the direct weights between the input layer and the $\mathcal{G}$-layer. To illustrate this, consider the case where the training data are subjected to a linear transformation such that the target, y, is multiplied by a factor λ. It can be arranged for the mapping implemented in the network to be unchanged by scaling the μ_ks, i.e. the centres of the $\mathcal{G}$-units, by the same factor, which is achieved by scaling all the weights feeding into the $\mathcal{G}$-units by λ while leaving the remaining weights unchanged. Consequently, weights between the input and the $\mathcal{S}$-layer scale differently from direct weights between the input and the $\mathcal{G}$-layer and therefore require different scaling factors[3]. This suggests the introduction of a change in the ARD scheme such that direct weights are

---

[3] The nature of the inconsistency of scheme ARD2 (vide infra) becomes clearer when the update rule for the ρ_gs is analysed. As will be shown shortly in (15.15), the gradient of E with respect to ρ depends on all the weights exiting the input units, that is both the weights feeding into the $\mathcal{S}$-layer and those feeding into the

taken out of the weight groups, which implies that the gth weight group now comprises only those weights that exit the gth input node *and* feed into the S-layer. Though being consistent with scaling laws, this alternative scheme is more restricted than the previous one in that it no longer allows the linear contributions of irrelevant inputs to be switched off (it is only their *nonlinear* contribution that can be switched off).

To summarise, there are two alternative ARD schemes, henceforth referred to as ARD1 and ARD2, both of which have certain advantages and disadvantages.

ARD1: Direct weights between the input layer and the G-layer are *not* included in the weight groups. *Advantage:* Consistent with scaling laws. *Disadvantage:* Linear contributions of irrelevant inputs cannot be switched off.

ARD2: Direct weights between the input layer and the G-layer are included in the weight groups. *Advantage:* Irrelevant inputs can be switched off completely. *Disadvantage:* Inconsistent with scaling properties.

The derivation of the gradients for the two schemes is given in the following section. A comparison of their performance can be found in Chapter 16.3.

15.3 Mathematical implementation

The derivative of E (see 15.1) with respect to ρ_g can be obtained by use of the chain rule,

$$-\frac{\partial E}{\partial \rho_g} = -\sum_{t=1}^{N} \frac{\partial \varepsilon_t}{\partial \rho_g} = -\sum_{t=1}^{N} \sum_{k=1}^{K} \frac{\partial \varepsilon_t}{\partial \mu_k} \frac{\partial \mu_k(\mathbf{x}_t)}{\partial \rho_g}. \tag{15.4}$$

The first factor in the argument of the sum on the right,

$$\delta_k^0(t) \quad := \quad -\frac{\partial \varepsilon_t}{\partial \mu_k}, \tag{15.5}$$

G-layer. However, as illustrated above and discussed in a more general way in [7], pp.340-342, these weights scale differently when the training data are subjected to a linear transformation. Consequently, the sign of the gradient in (15.15), and hence the network's 'assumption' about the significance of the different inputs, can change as the result of such a linear transformation. This is a striking inconsistency, since linear transformations of the data should lead to equivalent networks which differ only by the linear transformation of the weights.

is the error at a $\mathcal{G}$-node, which is given by (3.54) and (3.55) (note, however, that in practice these expressions should be replaced by (3.69) and (3.70) for reasons discussed in Section 3.2.5). Inserting (15.5) into (15.4) gives

$$-\frac{\partial E}{\partial \rho_g} \;=\; \sum_{t=1}^{N}\sum_{k=1}^{K}\delta_k^0(t)\frac{\partial \mu_k(\mathbf{x}_t)}{\partial \rho_g}. \tag{15.6}$$

Denote the weights between the gth input node and the ith unit in the $\mathcal{S}$-layer by u_{ig}, the weight between the ith $\mathcal{S}$-node and the kth $\mathcal{G}$-unit by w_{ki}, and the direct weights between the gth input unit and the kth $\mathcal{G}$-node by v_{kg} (note that only the weights w_{kj} and v_{kg} are adapted by the supervised training scheme). Further, denote the input to the ith $\mathcal{S}$-node by h_i,

$$h_i(\mathbf{x}, \boldsymbol{\rho}) \;=\; \sum_{g=1}^{m} u_{ig}\exp(\rho_g)x_g, \tag{15.7}$$

the output of the ith node in the $\mathcal{S}$-layer by z_i,

$$z_i \;=\; z\Big(h_i(\mathbf{x}, \boldsymbol{\rho})\Big) \tag{15.8}$$

and let μ_k be the centre of the kth $\mathcal{G}$-unit,

$$\mu_k(\mathbf{x}) \;=\; \Big\{\sum_{g=1}^{m} v_{kg}x_g\Big\}_1 + \Big\{\sum_{g=1}^{m} v_{kg}\exp(\rho_g)x_g\Big\}_2 + \sum_{i=1}^{H} w_{ki}z\Big(h_i(\mathbf{x}, \boldsymbol{\rho})\Big). \tag{15.9}$$

The expression in the first curly brackets, $\{\ldots\}_1$, only occurs in scheme ARD1, whereas the expression in the second curly brackets, $\{\ldots\}_2$, is non-zero only in scheme ARD2. For the derivative of μ_k with respect to ρ_g, one obtains by applying the chain rule (using the notation $z' = \frac{dz}{dh}$):

$$\frac{\partial \mu_k}{\partial \rho_g} \;=\; \Big\{v_{kg}\exp(\rho_g)x_g\Big\}_2 + \sum_{i=1}^{H} w_{ki}z_i'\frac{\partial h_i}{\partial \rho_g} \tag{15.10}$$

From (15.7) we obtain

$$\frac{\partial h_i}{\partial \rho_g} \;=\; \frac{\partial}{\partial \rho_g}\sum_{g'=1}^{m} u_{ig'}\exp(\rho_{g'})x_{g'} \;=\; u_{ig}\exp(\rho_g)x_g. \tag{15.11}$$

Inserting (15.11) into (15.10) yields

$$\frac{\partial \mu_k}{\partial \rho_g} \;=\; \Big\{v_{kg}\exp(\rho_g)x_g\Big\}_2 + \sum_{i=1}^{H} w_{ki}z_i'u_{ig}\exp(\rho_g)x_g \tag{15.12}$$

which is inserted into (15.6) to give

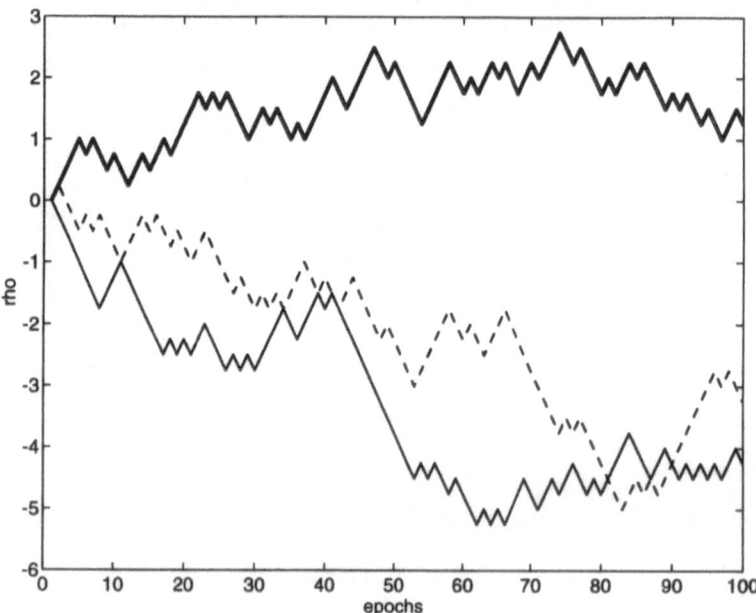

Fig. 15.1. Automatic relevance determination. A GM-RVFL network with $m = 3$ inputs was trained on the stochastic time series generated from the logistic-kappa map (4.16), and an ARD scheme with $\Delta\rho = 0.25$ was applied. The graphs show the evolution of the $\rho_g = \ln \sigma_g^{rand}$, where the bold line corresponds to the input x_t, the narrow dashed line to the input x_{t-1}, and the narrow solid line to the input x_{t-2}. It is seen that, for the two redundant inputs, x_{t-1} and x_{t-2}, the logarithmic distribution widths are decreased considerably, so their influence is systematically reduced by the ARD scheme.

$$-\frac{\partial E}{\partial \rho_g} = \sum_{t=1}^{N} \sum_{k=1}^{K} \delta_k^0(t) \left[\left\{ v_{kg} \exp(\rho_g) x_{g,t} \right\}_2 + \sum_{i=1}^{H} w_{ki} z'_{i,t} u_{ig} \exp(\rho_g) x_{g,t} \right]$$

$$(15.13)$$

This expression can be simplified by defining the error of an S-node in the standard backpropagation sense:

$$\delta_i^1(t) \quad := \quad \sum_{k=1}^{K} \delta_k^0(t) w_{ki} z'_{i,t} \qquad (15.14)$$

(note that this definition includes the derivative of the transfer function of the S-node). Inserting (15.14) into (15.13) results in

$$-\frac{\partial E}{\partial \rho_g} = \sum_{t=1}^{N} \left[\sum_{k=1}^{K} \delta_k^0(t) \Big\{ v_{kg} \exp(\rho_g) x_{g,t} \Big\}_2 + \sum_{i=1}^{H} \delta_i^1(t) u_{ig} \exp(\rho_g) x_{g,t} \right].$$

$$(15.15)$$

This update rule is of the familiar delta form and can be summarised verbally as: for scheme ARD2, set up a loop over all the wires exiting a given input node, get the error at each target node, and sum up the products of this error multiplied by the respective input and by the weight of the connecting wire. The result is the gradient of ε_t with respect to the hyperparameter ρ_g. Summing up the contribution of all the training exemplars gives the gradient of E with respect to ρ_g. Scheme ARD1 is similar, except that the loop is restricted to those wires that feed into the S-layer.

15.4 Empirical demonstration

In order to test the ARD scheme, the simulations of Chapter 14 were repeated with two further inputs, x_{t-1} and x_{t-2}, presented to the input layer of the GM-RVFL network. Note that these additional inputs are redundant since the given time series is first-order Markov, that is, the distribution of x_{t+1} only depends on x_t:

$$P(x_{t+1}|x_t, x_{t-1}, x_{t-2}, \ldots) = P(x_{t+1}|x_t) \qquad (15.16)$$

The network architecture was similar to that of Chapter 14, $H = 10$, $K = 10$, except that now $m = 3$ rather than $m = 1$. Following the ARD paradigm, wires leaving the same input node constituted a weight group, with their weights drawn from the same distribution, $N(0, \exp(\rho_g))$. The ρ_gs were adapted according to (15.3), starting from $\rho_g = 0$ $\forall g$. Note that since no direct weights between the input and the G-layer are employed, there is no difference between the schemes ARD1 and ARD2. After each adaptation of the ρ_gs, the adaptable weights $\mathbf{w}$ were initialised from $N(0, 0.25)$, a_k and β_k were initialised as described in Section 5.2.1, and the network was trained over $n = 30$ epochs with the EM-algorithm[4]. Figure 15.1 shows the evolution of the ρ_gs during network training. It is seen that, for the two redundant inputs, x_{t-1} and x_{t-2}, the logarithmic distribution widths ρ_g are decreased considerably, so their influence is systematically reduced by the ARD scheme.

---

[4] The method of simple weight decay, with $\alpha_k = 0.01$ for all weight groups, was applied for regularisation; see Section 12.1.2 for details.

16. A Real-World Application: The Boston Housing Data

An ensemble of GM-RVFL networks is applied to the prediction of housing prices in the Boston metropolitan area on the basis of various socio-economic explanatory variables. The ARD scheme is tested and found to succeed in identifying and effectively switching off two redundant dummy inputs added to the data. The employment of a network committee leads to significantly better results than achieved with an individual network. A simple Bayesian regularisation scheme is applied, but found to decrease only the generalisation 'error' of the single-model predictor. For a committee, the best generalisation performance is achieved when employing over-complex, under-regularised models that, individually, overfit the training data.

Preliminary remark

This study was carried out *before* the Bayesian evidence scheme for mixture models, presented in Chapter 10, had been derived. In what follows, the term *Bayesian regularisation* always refers to the simple Bayesian scheme of Chapter 9, with a conjugate prior on the network parameters of the form introduced in Section 9.3. The results presented in this chapter were published, in part, in [31] and [28].

Notation

In the present chapter, the symbol ρ is used with two different meanings. Without index – ρ – it refers to the hyperparameter for the prior distribution of the kernel precisions β_k (equation (9.20)). With index – ρ_g – it represents the logarithm of the distribution width for the gth weight group, introduced in Chapter 15. Also recall the notation introduced in Equation (9.47) for the hyperparameters of the prior: $\bar{\rho} = \rho - 1$, $\bar{\nu} = \nu - 1$.

16.1 A real-world regression problem: The Boston house-price data

In a final study, the GM-RVFL network and the ARD scheme derived in the previous chapter were applied to the Boston house-price data of Harrison and Rubinfeld [21]. This is a real-world regression problem that has been studied by several authors before and can be used as a benchmark set[1]. For each of the 506 census tracts within the Boston metropolitan area the data give 13 socio-economic explanatory variables, as listed in Tab.16.1. The target is to predict the median housing price for that tract. The data set is interesting for two reasons. First, the conditional probability distribution deviates strongly from a Gaussian. Neal writes in his thesis [44]:

> *The data set is messy in several respects. Some of the attributes are not actually measured on a per-tract basis, but only for larger regions. The median prices for the highest-priced tracts appear to be censored. Considering these potential problems, it seems unreasonable to expect that the distribution of the target variable (median price), given the input variables, will be nicely Gaussian. Instead, one would expect the error distribution to be heavily tailed.*

Second, the data are rather sparse, which calls for sophisticated measures to be taken against overfitting. Neal [44] addressed the first problem by modelling the conditional probability density of the target with a t-distribution, which is heavy-tailed and thus more appropriate than a Gaussian if several far outliers are present in the data. For the second problem, he adopted a Bayesian approach and drew the network parameters from the posterior distribution $P(\mathbf{w}|\mathbf{D})$, using Gibbs sampling and a hybrid Monte Carlo scheme. In the present study, a GM-RVFL network was applied to modelling the conditional probability distribution $P(y|\mathbf{x})$. This approach can be expected to achieve an improvement in the accuracy of the predictions *if* the conditional distribution is skewed or multimodal. Concerning the second problem, overfitting, it will be tested in how far the generalisation performance can be enhanced by applying network committees rather than a single network.

CRIME	per capita crime rate by town
ZN	proportion of residential land zoned for lots over 25,000 sq.ft.
INDUS	proportion of non-retail business acres per town
RIVER	river variable (1 if tract bounds river; 0 otherwise)
NOX	nitric oxides concentration (parts per 10 million)
RM	average number of rooms per dwelling
AGE	proportion of owner-occupied units built prior to 1940
DIS	weighted distances to five Boston employment centres
RAD	index of accessibility to radial highways
TAX	full-value property-tax rate per $ 10,000
PTRATIO	pupil-teacher ratio by town
BLACK	proportion of blacks by town
LOWSTAT	per cent lower status of the population
MEDV	median value of owner-occupied homes in $ 1000's

Table 16.1. Explanatory variables (above) and target (below) of the Boston housing data set [21].

16.2 Prediction with a single model

16.2.1 Methodology

In a preliminary experiment, the performance of a single-network predictor was studied. A GM-RVFL network was employed, which contained $H = 70$ units in the S-layer, $K = 5$ kernel nodes in the G-layer, and direct weights between the input and the G-layer. This was motivated by the results of an earlier study, summarised in Fig.16.1, which suggested that architectures with direct weights achieve better results on this problem. All the random weights, i.e. the weights between the input and the S-layer, were drawn from the same distribution $N(0, \sigma_{rand})$ (i.e. a zero-mean Gaussian with standard deviation σ_{rand}). The networks were trained with the EM-algorithm and regularised with the simple Bayesian scheme of Chapter 9. The update of the network parameters thus followed equations (9.42), (9.43), and (9.46). The data were normalised, as in [44], and split into two disjoint sets of equal size ($N = 253$): one for training, and the other for estimating the generalisation capability. The simulations were repeated several times, using different values for σ_{rand}, and different hyperparameters ρ, ξ, ν, and α_k (which determine the prior distribution on the parameters, see Section 9.3).

---

[1] See http://www.cs.utoronto.ca/delve for an overview. The original data set is available by anonymous ftp from StatLib (lib.stat.cmu.edu), directory datasets.

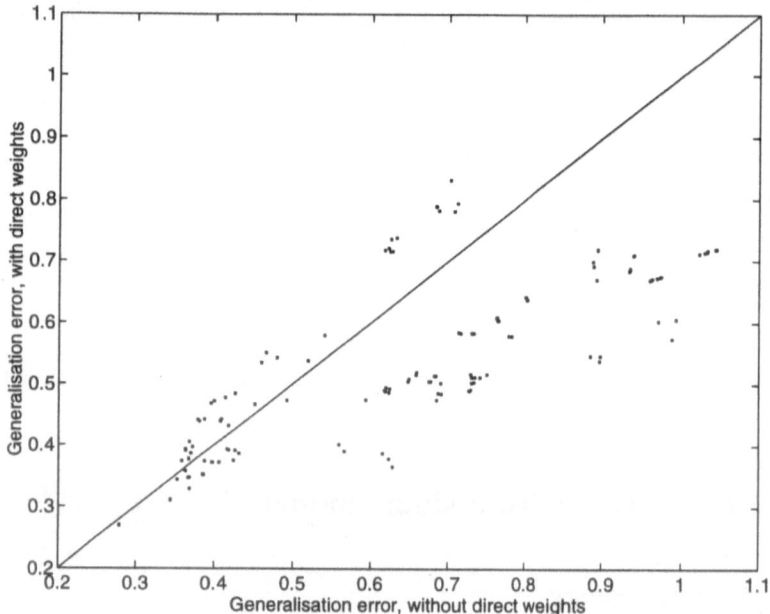

Fig. 16.1. Influence of direct weights. In order to study the influence of direct weights between the input layer and the $\mathcal{G}$-layer, two architectures were compared. Network 1 contained $H = 70$ nodes in the $\mathcal{S}$-layer, $K = 5$ nodes in the $\mathcal{G}$-layer, and direct weights between the input and the $\mathcal{G}$-layer. In network 2, the direct weights were removed, and the size of the $\mathcal{S}$-layer was increased by the same number of parameters (leading to $H = 83$). 200 combinations of different hyperparameters for the Bayesian prior (α_k, ξ, ρ, and ν), different values for σ_{rand}, and different random number generator seeds were selected. For each of these combinations, the simulations (using a training set of size $N = 253$) were carried out twice - once for each architecture. The figure shows a scatter plot of the achieved generalisation 'error', estimated on a test set of the same size as the training set. This suggests that the prediction performance, on the average, can be significantly increased by employing direct weights.

16.2.2 Results

The results are summarised in Fig.16.2. The graphs show the dependence of the training and generalisation 'errors'[2], E_{train} and E_{gen}, on the number of EM-steps for three simulations with three different selections of the random weights. In the first row, training was quasi-unregularised[3]. Since the number

---

[2] Recall (8.2), (8.3), and (8.4) for a definition of the terms 'training error', 'cross-validation error', and 'generalisation error'

[3] This means that the hyperparameters α_k were set to small positive values, in this case $\alpha_k = 10^{-4} \; \forall k$ (necessary since otherwise the Hessian may be singular),

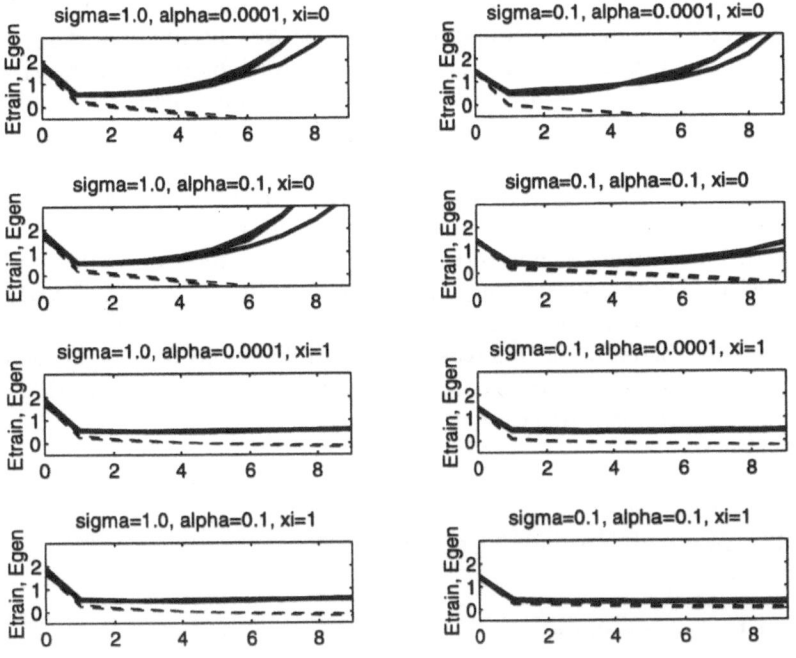

Fig. 16.2. Bayesian regularisation of a single-network predictor. A GM-RVFL network was trained with the Bayesian-regularised EM algorithm, according to the update equations given in Section 9.5. The graphs show the dependence of the training 'error' (*narrow dashed lines*) and the generalisation 'error' (*bold solid lines*) on the number of EM-steps. Three simulations with different values for the random weights were carried out. *Left column:* $\sigma_{rand} = 1.0$; *right column:* $\sigma_{rand} = 0.1$. *First row:* quasi-unregularised ($\alpha_k = 10^{-4} \, \forall k$, $\xi = 0$); *second row:* regularisation of only the weights $\mathbf{w}$ ($\alpha_k = 0.1 \, \forall k$, $\xi = 0$); *third row:* regularisation of only the kernel widths ($\alpha_k = 10^{-4} \, \forall k$, $\xi = 1$); *fourth row:* regularisation of both weights $\mathbf{w}$ and kernel widths ($\alpha_k = 0.1 \, \forall k$, $\xi = 1$). The hyperparameters $\bar{p}$ and $\bar{\nu}$ were set to neutral values $\bar{p} = \bar{\nu} = 0$.

of network parameters is larger than the number of training samples, it is not surprising that serious overfitting occurs. Regularisation of only the weights $\mathbf{w}$ (second row) hardly alleviated this problem, whereas regularisation of the kernel widths led to a substantial improvement (third row). The best results were obtained when both the weights $\mathbf{w}$ and the kernel widths were regularised (fourth row). The exact values of the hyperparameters are given in the caption of Fig.16.2. Regularisation of the output weights (or priors) was found to hardly have any influence on the results, as long as ν was chosen sufficiently small (for $\bar{\nu} > 10$, the performance deteriorated). For studying the

---

and the remaining hyperparameters were set to neutral values $\xi = \bar{p} = \bar{\nu} = 0$. See Section 12.1.2 for details.

influence of the random weight distribution, the left column of Fig.16.2 shows the results of simulations obtained with a 'broad' distribution, $\sigma_{rand} = 1.0$, which is compared with the results for a 'narrow' distribution, $\sigma_{rand} = 0.1$, shown on the right. Apparently overfitting is always slightly worse for the broader distribution. Note that in nearly all the simulations, even with regularisation, the generalisation error increases as training continues. It is only in the bottom graph on the right, that is for a narrow random weight distribution and the regularisation of both the weights $\mathbf{w}$ and the kernel precisions β_k, that overfitting is prevented completely. The generalisation 'error' obtained in this case (mean and standard deviation of three simulations) was $E_{gen} = 0.31 \pm 0.03$. This is still considerably worse than the results obtained by Neal (Tab.16.9). For this reason, two possible improvements of the training/prediction scheme were studied: application of network committees and automatic relevance determination (ARD).

16.3 Test of the ARD scheme

16.3.1 Methodology

In order to test the ARD scheme, the input vector was augmented by two irrelevant variables, drawn from a standard normal distribution $N(0, 1)$. $N_{Com} = 200$ networks were trained, following the ARD paradigm that the ρ_gs of the $(n + 1)$th network are obtained from those of the nth network by application of (15.3). The gradient of E with respect to ρ was calculated according to scheme ARD2 (15.15). The training simulation was split up into two parts, and after 100 ARD-updates the ρ_gs were reset to their initial value of $\rho_g = 0$. The network employed in the simulation contained $H = 60$ nodes in the S-layer, $K = 5$ kernels in the $\mathcal{G}$-layer, and direct weights between the input-layer and the $\mathcal{G}$-layer. The upper limit for the ρ_gs was set to $\rho_{max} = 0$, and a Manhattan stepsize of $\triangle\rho = 0.25$ was employed.

16.3.2 Results

The results of the simulation are shown in Fig.16.3. It is seen that the ARD scheme leads to a drastic decay in the ρ_gs for the irrelevant inputs, so that in fact these inputs are effectively switched off after about 50 ARD updates. (Note that for $\rho_g = -4$ the standard deviation $\sigma_g = \exp(-\rho_g)$ is less than 0.02). Concerning the other inputs, the following tendency can be observed. The variables ROOMS (average number of rooms per dwelling), DIS (distance to employment centres), RAD (accessibility to radial highways), TAX

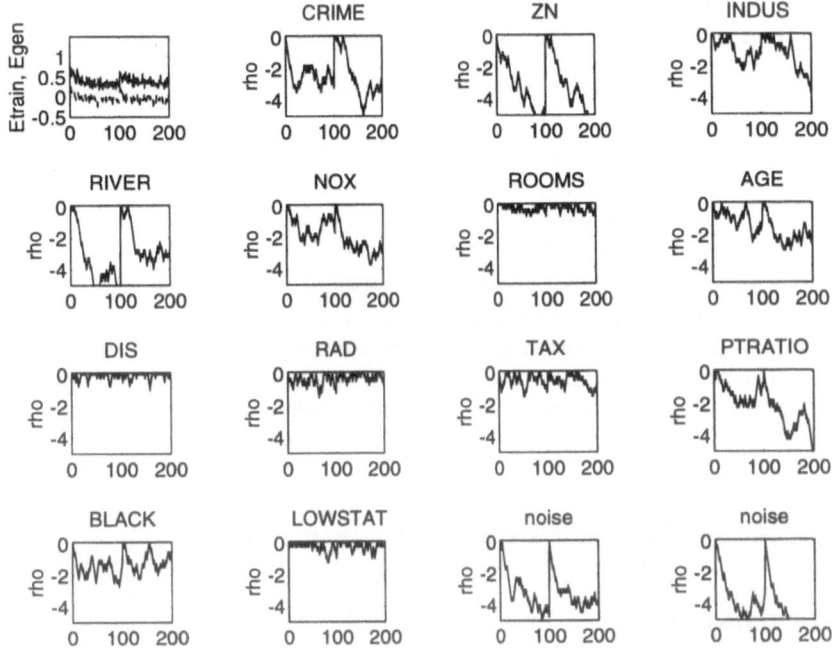

Fig. 16.3. Test of the ARD scheme. $N_{Com} = 200$ GM-RVFL networks were trained on the Boston housing data with two added irrelevant inputs drawn from a standard normal distribution. The ρ_gs of the $(n+1)$th network were obtained from those of the nth network by applying scheme ARD2 with $\rho_{max} = 0$ and $\Delta\rho = 0.25$. After 100 ARD-steps, the ρ_gs were reset to their initial values of $\rho_g = 0$. The graphs in the top row on the left show the evolution of the training 'error' (narrow line) and the test 'error' (bold line). The remaining graphs show the evolution of the ρ_gs for the different weight groups. The names over the graphs indicate the input variable, as listed in Tab.16.1. The last two curves correspond to the irrelevant inputs. Note that the ARD scheme succeeds in detecting their irrelevance, and switches them effectively off in both parts of the simulation. Reprinted from [31] with permission from Elsevier Science.

(property tax rate) and LOWSTAT (percentage of lower status in the population) are relevant, with their ρ_gs most of the time close to the allowed maximum value of $\rho_{max} = 0$. In contrast, the variables ZN (proportion of residential land), RIVER (closeness to river) and, slightly less substantial, CRIME (crime rate - surprisingly!) seem to be irrelevant, with a decay of the ρ_gs in both parts of the simulation. The variables INDUS (proportion of non-retail business), NOX (nitric oxides concentration) and PTRATIO (pupil-teacher ratio) show an ambiguous behaviour, with a decay of their respective ρ_g only in one out of the two parts of the simulations. For this reason it seems to be advisable to break up the ARD scheme into fragments and reset the ρ_gs in certain intervals, rather than follow a single trajectory[4].

A further subtlety is the introduction of an upper limit ρ_{max} (see equation (15.2)). Without such a limit, the ARD-scheme tends to increase the ρ_gs of the relevant inputs to unreasonably large values. As a consequence of this excessive nonlinearity, the prediction performance of the network deteriorates considerably. This can easily be detected if a cross-validation set is available, since the performance on this set will degrade as ρ is adapted. Even without such a set, it was found that basing the decision on the training set gives a reliable indication for too large an upper limit ρ_{max}, since excessively large network weights were found to have also an adverse effect on the training-set performance. A proper choice of ρ_{max} should lead to a decrease in both E_{train} and E_{cross}, as shown in the top left sub-figure of Fig.16.3.

16.4 Prediction with network committees

16.4.1 Objective

The previous section has demonstrated that ARD succeeds in indentifying and effectively switching off irrelevant inputs. This suggests that its incorporation into the training process is likely to enhance the generalisation performance of a network committee. However, from Section 15.2 it is known that there exist two alternative procedures, according to whether direct weights between the input layer and the $\mathcal{G}$-layer are included in the scheme or not. Since both methods have certain merits and drawbacks, as discussed in Section 15.2, it is desirable to carry out comparative simulations on the benchmark problem considered here. Moreover, several other issues need to be clarified:

Temperature: How does the prediction performance of the network committee depend on the temperature of the weighting scheme (see Section 13.3)?

Complexity: How complex a network architecture should be chosen? Is the prediction performance improved or degraded by limiting the network complexity?

Bayes: Can the generalisation performance be improved by applying Bayesian regularisation?

---

[4] Once a ρ_g has decayed to a small value, the corresponding input has effectively been switched off. This makes it hardly possible for the ARD scheme to resurrect it at a later stage.

Bagging: Does bagging (bootstrapping and aggregating) lead to any improvement in the generalisation performance?

Diversity: How does the performance of the committee depend on the diversity of its members?

16.4.2 Methodology

In order to find a significant answer to these questions, which does not depend on the arbitrary partition of the data, a four-fold cross-validation scheme was applied. That is, the available data were divided into four disjoint subsets of equal size. One subset was used for cross-validation (25% of the data), one subset for testing the generalisation performance (25% of the data), and two subsets were used for training (altogether 50% of the data). The simulations were repeated four times for all cyclic permutations of this configuration. The tables in this chapter show, for the respective entities of interest, the mean and the standard deviation obtained in this way. In each simulation, $N_{Com} = 200$ networks were trained, with different sets of the fixed weights between the input layer and the S-layer. The individual networks were trained with the EM algorithm and regularised with the simple Bayesian scheme of Section 9.5. That is, the output weights a_k were updated according to (9.46), the inverse kernel widths β_k according to (9.43), and the adaptation of the remaining weights followed (9.42). In most simulations, the hyperparameters α_k were set to a small value of $\alpha_k = 10^{-4}$ (identical for all weight groups), and the remaining hyperparameters were set to neutral values $\overline{\nu} = \overline{\rho} = \xi = 0$ (for which the priors on a_k and β_k are constant). This type of simulation will be referred to as *quasi-unregularised*[5] and was applied unless explicitly stated otherwise. A training process was terminated when the performance on the cross-validation set started to deteriorate (method of *early stopping*). Various architectures were employed, which differed with respect to the number of nodes in the hidden layers, H and K, but otherwise were of the same topology in that they all contained direct weights between the input layer and the G-layer.

Following the ARD-scheme, random weights exiting the same input node constituted a weight group, i.e. all the weights in this group were drawn from the same distribution. The standard deviations σ^g_{rand} of these distributions were updated according to (15.3). For initialisation, all standard deviations were set to $\sigma^g_{rand} := 1$, and were reset to this value after 100 ARD updates (i.e. for the 101st network). Both ARD schemes (with and without inclusion

---

[5] For the *quasi-unregularised* scheme, the update equations are identical to the *unregularised* EM algorithm, (7.23), (7.24) (7.36), except that diagonal incrementation ensures the non-singularity of the matrix on the left of Equation (7.36). See also Section 12.1.2.

of the direct weights) were applied, with different values for the Manhattan stepsize $\Delta\rho$ and the upper limit ρ_{max}.

After all the networks had been trained, the individual networks were weighted on the basis of their cross-validation 'error'. Recall from Section 13.3 that the weighting factors are given by expression (13.33), which depends on a temperature-like parameter T. As a small modification of this scheme, a theshold term E_{cross}^{thresh} was introduced, that is, the weighting scheme defined by Equation (13.33) was only applied to networks with $E_{cross} \leq E_{cross}^{thresh}$, whereas networks with $E_{cross} > E_{cross}^{thresh}$ were discarded. Of special interest is the case $E_{cross}^{thresh} = \infty$, $T = \infty$, which is identical to assigning uniform weights to all ensemble members and indicates the performance that is achieved if no cross-validation set is available. (Note that the study presented in this chapter was carried out *before* the derivations in Chapter 11 had been completed, otherwise a weighting scheme based on the Bayesian evidence could also have been tested.)

16.4.3 Weighting scheme and temperature

Table 16.2 shows the dependence of the committee prediction on (i) the temperature and (ii) the cutoff value for the cross-validation 'error', E_{cross}^{thresh}. These results are compared with the average single-network performance (averaged over the committee with equal weights). A GM-RVFL network with $H = 70$ nodes in the $\mathcal{S}$-layer and $K = 5$ kernels in the $\mathcal{G}$-layer was employed. Scheme ARD2 was applied for adapting the ρ_gs, and the simulations were carried out twice with (A) $\rho_{max} = 0$, $\Delta\rho = 1.0$, and (B) $\rho_{max} = 0$, $\Delta\rho = 0.25$. Recall from (13.19) that the generalisation error of the committee is given by

$$E_{gen}^{Com} = \overline{E_{gen}} - A \qquad (16.1)$$

in which $\overline{E_{gen}}$ denotes the mean generalisation error of the individual network, averaged over the ensemble, and A the *ambiguity* or *diversity* term, defined in (13.18). Moreover, recall that A is non-negative, and therefore (16.1) implies that the generalisation error of the ensemble is never greater than the average generalisation error of the individual network. This is confirmed in Tab.16.2.

When selecting a small temperature T or a small threshold E_{cross}^{thresh}, as in the first two rows of Tab.16.2, the average generalisation error of the individual network, $\overline{E_{gen}}$, is small because only the best networks contribute with significant weights. However, the diversity term A is also small, because the effective committee size is small, and the selected networks give similar outputs. Consequently, $\overline{E_{gen}}$ will only be about as good as the generalisation error of the best individual networks in the ensemble, and nothing is gained from using an *ensemble*.

When the temperature and the threshold E_{cross}^{thresh} are increased, $\overline{E_{gen}}$ also increases since larger weights are assigned to poorer models. However, this deterioration in the performance is more than compensated for by an increase of the diversity term A, which leads to an overall net decrease of E_{gen}^{Com}. This is seen from the third, fourth and fifth row of Tab.16.2.

When T is increased beyond some optimal value, the deterioration in $\overline{E_{gen}}$ will have a stronger influence than the diversity term A, and E_{gen}^{Com} will increase again. This is shown in the last two rows of Tab.16.2. However, even for $T \to \infty$, which corresponds to uniform weights, the deterioration of the generalisation performance E_{gen}^{Com} is rather moderate, and the results obtained in this case are still significantly better than those obtained with the best individual models. This suggests that even without a cross-validation set for weighting the individual members, a considerable improvement of the generalisation performance can be achieved by using network committees.

T	E_{cross}^{thresh}	E_{gen}^{Com} (A)	E_{gen}^{single} (A)	E_{gen}^{Com} (B)	E_{gen}^{single} (B)
0.01	–	0.385 ±0.059	0.644 ±0.051	0.356 ±0.084	0.603 ±0.059
0.1	0.4	0.251 ±0.081	0.494 ±0.048	0.243 ±0.056	0.523 ±0.059
0.1	0.7	0.155 ±0.016	0.572 ±0.054	0.157 ±0.026	0.543 ±0.068
0.1	–	0.152 ±0.015	0.644 ±0.051	0.157 ±0.026	0.603 ±0.055
1.0	–	0.154 ±0.020	0.644 ±0.051	0.168 ±0.025	0.603 ±0.059
10.0-∞	–	0.156 ±0.020	0.644 ±0.051	0.170 ±0.026	0.603 ±0.059

Table 16.2. Dependence of the committee error on T. $N_{Com} = 200$ GM-RVFL networks of the architecture $H = 70, K = 5$ were trained by applying scheme ARD2 with (A) $\rho_{max} = 0, \Delta\rho = 1.0$, and (B) $\rho_{max} = 0, \Delta\rho = 0.25$. The table shows the dependence of the estimated generalisation 'error' of the committee, E_{gen}^{Com}, on (i) the temperature T and (ii) the cutoff value for E_{cross}. This is compared with the mean 'error' of a single-network prediction, $\overline{E_{gen}^{single}}$ (averaged over the committee with equal weights). Listed are the mean and the standard deviation obtained from a 4-fold cross-validation scheme.

16.4.4 ARD parameters

The results listed in Tab.16.3 suggest that the influence of the Manhattan stepsize $\Delta\rho$ and the upper limit ρ_{max} is not dramatic as long as these parameters are varied in a 'reasonable' range. The best results were obtained with $\rho_{max} = -1$ and $\Delta\rho = 0.25$. As mentioned before, care must be taken with respect to ρ_{max}, since for too large a value of this parameter the prediction performance deteriorates considerably. In the problem studied here, this happened for values larger than about $\rho_{max} \geq 1$. Recall that this is indicated by an increase of both E_{train} and E_{cross} during training.

scheme	ρ_{max}	$\Delta\rho$	$E_{gen}^{Com}(T=0.1)$	$E_{gen}^{Com}(T=1.0)$	E_{gen}^{single}
ARD2	0	1.0	0.152 ± 0.015	0.154 ± 0.020	0.644 ± 0.051
ARD2	0	0.25	0.157 ± 0.026	0.168 ± 0.025	0.603 ± 0.059
ARD2	-1	1.0	0.156 ± 0.015	0.159 ± 0.019	0.513 ± 0.047
ARD2	-1	0.25	0.146 ± 0.031	0.139 ± 0.030	0.528 ± 0.063
ARD1	0	1.0	0.138 ± 0.016	0.138 ± 0.016	0.589 ± 0.058
ARD1	-1	0.25	0.132 ± 0.025	0.137 ± 0.025	0.582 ± 0.060

Table 16.3. Comparison between the two ARD schemes. The table shows the dependence of the estimated generalisation 'error' of the committee, E_{gen}^{Com}, on the ARD scheme, the Manhattan stepsize $\Delta\rho$, and the upper limit ρ_{max}. For comparison, the right column shows the mean single-network generalisation 'error'. In each case, 200 GM-RVFL networks (K=5, H=70) were trained as described in the text. The values shown are the mean and the standard deviation obtained from 4-fold cross-validation.

Number of kernels	$E_{gen}^{Com}(T=0.1)$	$E_{gen}^{Com}(T=1.0)$	E_{gen}^{single}
5	0.142 ± 0.029	0.153 ± 0.025	0.465 ± 0.053
3	0.168 ± 0.015	0.175 ± 0.014	0.517 ± 0.041
1	0.335 ± 0.027	0.342 ± 0.015	0.681 ± 0.054

Table 16.4. Number of kernels. The table shows the dependence of the generalisation error (for the committee, E_{gen}^{Com}, and the average single-network prediction, $\overline{E_{gen}^{single}}$) on the number of kernel functions K. All models employed $H = 45$ nodes in the $\mathcal{S}$-layer, and the training process followed scheme ARD1 over 200 networks ($\rho_{max} = 0$, $\Delta\rho = 0.25$).

16.4.5 Comparison between the two ARD schemes

A comparison between the two ARD schemes can be found in Tab.16.3 and Tab.16.6. It can be seen that scheme ARD1, which does not include direct weights, shows a slightly but consistently better prediction performance than scheme ARD2. This might be due to the fact that in the given problem none of the inputs are completely redundant, so that the violation of scaling properties by ARD2 has a larger influence than its capability of switching the *linear* contributions of irrelevant inputs off.

16.4.6 Number of kernels

The dependence of the prediction performance on the number of kernels is shown in Tab.16.4. If the noise on the target were homoscedastic and Gaussian distributed, the addition of more than one kernel node in the $\mathcal{G}$-layer, $k > 1$, would not lead to any improvement in the generalisation performance. However, Tab.16.4 shows that the generalisation 'error' decreases with increasing number of kernels k. This strongly suggests that the distribution

$P(y|\mathbf{x})$ is in fact non-Gaussian, and provides another example for the superiority of the Gaussian mixture network over a conventional MLP (which corresponds to a prediction with $k = 1$).

16.4.7 Bayesian regularisation

So far, all training simulations have been carried out with the quasi-unregularised model. This is to be compared with the Bayesian regularisation scheme. The hyperparameters were selected in preliminary simulations so as to optimise the single-network performance on the cross-validation set. The values finally adopted were $\alpha_k = 0.1\,\forall k$, $\xi = 1.0$, $\bar{p} = 0$, $\bar{\nu} = 0$. Table 16.5 shows a comparison between simulations carried out with and without Bayesian regularisation. It can be observed that for the single-network prediction a significant improvement in the generalisation performance can be achieved as a result of applying *Bayesian regularisation* (right column). This is not surprising since, due to the large number of network parameters (relative to the data set size), the unregularised individual network can be expected to overfit drastically even when early stopping is applied. However, such an improvement cannot be found for the *committee* prediction, which, in contrast, shows a slight degradation when the regularisation scheme is applied.

16.4.8 Network complexity

Table 16.6 and Fig.16.4 show the dependence of the generalisation error on H, the number of nodes in the $\mathcal{S}$-layer. For the single-network prediction (right column), there exists an optimal degree of network complexity. When the number of hidden nodes is too small, the network is not sufficiently flexible to model the data-generating function; when the number of hidden nodes becomes too large, the network becomes too flexible and overfits. Such a behaviour is well-known in the neural network community, and has been found in many earlier studies before (see, for example, [42]). However, the interesting result found in the present study is that for a network *committee* this *principle of parsimony* does not hold, and that the generalisation error steadily decreases with increasing network complexity. This is similar to the results of the previous section (Bayesian regularisation) and suggests that for a network committee, the best generalisation performance is achieved with a complex under-regularised model[6].

---

[6] The number $H = 70$ as an upper limit for the network complexity was only chosen for practical reasons, since for larger values of H numerical roundoff errors led to convergence difficulties in the singular-value decomposition of the matrix on the left of (9.42) (performed with the subroutine listed in [51], pp.67-70).

16.4.9 Cross-validation

A standard regularisation scheme is to limit the network complexity dynamically with the method of *cross-validation*. The generalisation performance is monitored on the cross-validation set, and the training process is terminated as soon as overfitting occurs. However, if the data are sparse, this approach is not feasible since all the data are needed for training. In this case the best that can be done is to determine, in a preliminary study, the average time $n^\star$ after which overfitting usually sets on, and then limit the training process to this fixed number of $n^\star$ adaptation steps. The results obtained with this technique are indicated in Tab.16.7. The table shows the generalisation 'error' of the committee (for two temperatures, middle columns) and the single-network predictor (right column) for different numbers of EM steps. CV stands for 'cross-validation', and shows the performance achieved if the training process is terminated at the point where the cross-validation 'error' starts to increase (which usually happened after between 3 and 5 EM steps). From the right column it can be seen that, as far as the single-network prediction is concerned, the choice of a fixed number of training steps leads to an increase of the generalisation 'error' over that obtained with cross-validation. This confirms the trivial fact that with a fixed number of adaptation steps networks either overfit or underfit the training data, both of which deteriorates the prediction results. Perhaps surprisingly, but in support of the findings in the previous section, this does not hold for the *network committee*. Here, the optimum generalisation performance is *not* obtained by terminating the training process with cross-validation, but by choosing a fixed number of EM-steps that is slightly *larger* than the average number of steps resulting from cross-validation. This suggests, again, that some degree of overfitting is 'useful' and results in an improved generalisation performance of the committee.

16.5 Discussion: How overfitting can be useful

The results of sections 16.4.7-16.4.9 suggest that regularisation methods that aim at optimising the generalisation performance of a single-model predictor do not necessarily optimise the generalisation performance of an ensemble of networks. In contrast, it was observed that under-regularisation and overfitting of the individual networks can lead to a considerable decrease in the generalisation 'error' of the committee. This can be understood from Equation 16.1 in a similar way as already discussed in Section 16.4.3. While under-regularisation leads to a deterioration of the average single-model generalisation 'error' $\overline{E_{gen}}$, it also gives rise to a larger diversity A, from which an overall enhancement of the committee performance E_{gen}^{Com} can result. Similar results supporting this conjecture were discussed in Chapter 14. A recent

Bagging	Bayes	ρ_{max}	$\Delta\rho$	$E_{gen}^{Com}(T=0.1)$	$E_{gen}^{Com}(T=1.0)$	E_{gen}^{single}
no	no	0	1	0.152 ±0.015	0.154 ±0.020	0.644 ±0.051
no	no	-1	0.25	0.146 ±0.031	0.139 ±0.030	0.528 ±0.063
no	yes	0	1	0.178 ±0.019	0.193 ±0.017	0.405 ±0.033
no	yes	-1	0.25	0.155 ±0.021	0.160 ±0.020	0.315 ±0.028
yes	no	0	1	0.251 ±0.039	0.246 ±0.028	1.483 ±0.120
yes	yes	0	1	0.250 ±0.018	0.281 ±0.018	0.738 ±0.043

Table 16.5. Bayesian regularisation and bagging. The table compares quasi-unregularised training ($\alpha_k = 10^{-4}\,\forall k, \xi = 0$) with Bayesian regularisation ($\alpha_k = 0.1\,\forall k, \xi = 1.0$) and bagging. In each case, 200 networks ($K = 5$, $H = 70$) were trained, applying scheme ARD2 for adaptation of ρ. E_{gen}^{Com}: generalisation error of the committee. $\overline{E_{gen}^{single}}$: average generalisation error of a single network. While Bayesian regularisation improves the generalisation performance of the single-network predictor, it leads to a degradation of the results obtained with the network committee. Bagging achieves better results than those obtained with a single predictor, but is not as good as a committee trained on the unmanipulated data.

H	ARD scheme	$E_{gen}^{Com}(T=0.1)$	$E_{gen}^{Com}(T=1.0)$	$\overline{E_{gen}^{single}}$
70	ARD2	0.152 ±0.015	0.154 ±0.020	0.644 ±0.051
45	ARD2	0.165 ±0.019	0.178 ±0.021	0.474 ±0.045
20	ARD2	0.223 ±0.012	0.234 ±0.014	0.435 ±0.042
5	ARD2	0.312 ±0.015	0.332 ±0.015	0.466 ±0.026
70	ARD1	0.138 ±0.016	0.138 ±0.016	0.589 ±0.058
45	ARD1	0.142 ±0.029	0.153 ±0.025	0.465 ±0.053
20	ARD1	0.200 ±0.023	0.209 ±0.023	0.406 ±0.046
5	ARD1	0.299 ±0.016	0.316 ±0.015	0.458 ±0.029

Table 16.6. Network complexity. The table shows the dependence of the generalisation performance on H, the number of nodes in the $\mathcal{S}$-layer. The average single-network generalisation 'error', $\overline{E_{gen}^{single}}$, has a minimum for a certain value of H, and increases as the network complexity becomes larger. In contrast, the generalisation 'error' of the committee, E_{gen}^{Com}, steadily improves with increasing H. All simulations were carried out with $K=5$ nodes in the $\mathcal{G}$-layer, and the same parameters $\rho_{max} = 0$, $\Delta\rho = 1$. The committee size was $N_{Com} = 200$. A comparison between the ARD schemes suggests that ARD1 is superior to ARD2, as also seen from Tab.16.3.

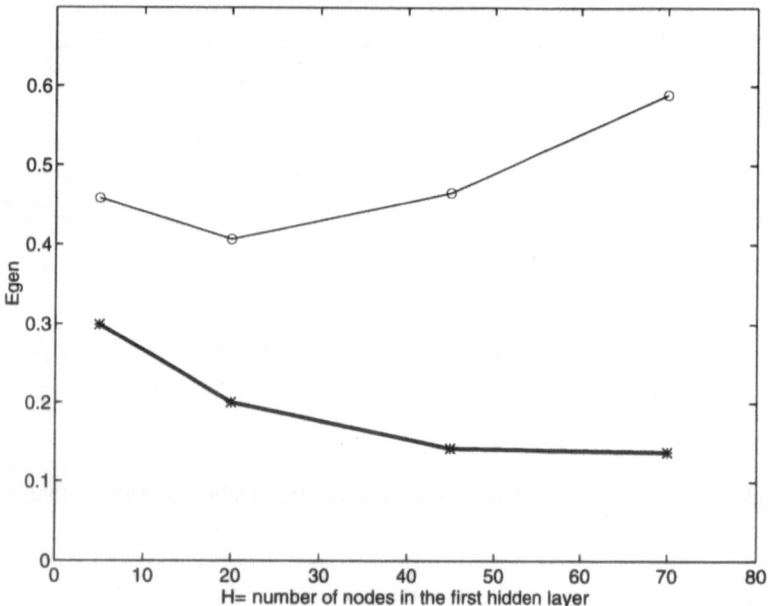

Fig. 16.4. Dependence of the generalisation error on H, the number of nodes in the S-layer. The figure displays the results of Tab.16.6 for the committee trained with ARD1. The abscissa represents H, the number of nodes in the S-layer. The ordinate shows the estimated generalisation 'error', E_{gen}. For details, see caption of Tab.16.6. *Narrow line:* single-network prediction. *Bold line:* committee prediction. Whereas there exists an optimum number of parameters for the single-network prediction, beyond which the performance deteriorates due to overfitting, the committee performance improves steadily with increasing model complexity. Reprinted with permission from [31].

theoretical study of this phenomenon, although restricted to linear interpolation models under the assumption of uniform Gaussian noise, was performed by Krogh and Sollich [35], [57].

16.6 Increasing diversity

The importance of the diversity term for improving the generalisation performance raises the question if there are systematic ways of increasing it. Two approaches were studied in the simulations.

Number of EM steps	$E_{gen}^{Com}(T = 0.1)$	$E_{gen}^{Com}(T = \infty)$	E_{gen}^{single}
3	0.158 ±0.028	0.161 ±0.024	0.611 ±0.057
CV (3-5)	0.132 ±0.025	0.139 ±0.025	**0.582** ±0.060
5	**0.102** ±0.033	**0.112** ±0.025	0.695 ±0.060
7	0.134 ±0.026	0.149 ±0.034	1.340 ±0.086

Table 16.7. Early stopping. The table shows the generalisation 'error' of the committee, E_{gen}^{Com}, and the average generalisation 'error' of a single network, E_{gen}^{single}, as a function of the number of EM-steps. 'CV' stands for cross-validation, that is, the training process was terminated as soon as the performance on the cross-validation set started to decrease (which happened usually after between 3 and 5 EM-steps). It can be seen that the optimal performance of the *single* network is achieved when cross-validation is applied for determining the stopping criterion, whereas for a network *committee* a slightly longer training time (leading to over-fitting of the individual members) is superior. Further details of the simulation: Scheme ARD1 with $\rho_{max} = -1, \Delta\rho = 0.25$ was applied for adapting ρ. Network architecture: $H = 70, K = 5$. The size of the committee was $N_{Com} = 200$.

16.6.1 Bagging

By training a model, or a committee of models, on a limited data set $\mathbf{D}$, the prediction is conditional on this particular data set, $P(y|\mathbf{x}, \mathbf{D})$. The correct distribution is formally obtained by integrating over all possible data sets,

$$P(y|\mathbf{x}) = \int P(y|\mathbf{x}, \mathbf{D})P(\mathbf{D})d\mathbf{D}. \tag{16.2}$$

In practice, of course, this cannot be accomplished since the distribution $P(\mathbf{D})$ is generally not known. A widely accepted approach, known as the *bootstrap* method, is to approximate the unknown distribution $P(\mathbf{D})$ by an empirical estimate $P^{\star}(\mathbf{D})$, which is obtained by drawing a sample of size $N = |\mathbf{D}|$ (number of elements in $\mathbf{D}$) with replacement from $\mathbf{D}$ (see, for example, [16], [50]). While in conventional applications bootstrapping is basically restricted to estimating only the *prediction error* of a model (see, for instance, [62]), Breiman suggested the use of this scheme for obtaining a committee of predictors [10]. His method of *bagging*, an acronym for *bootstrapping and aggregating*, is the following approximation to Equation (16.2):

$$P(y|\mathbf{x}) \approx \sum_i P(y|\mathbf{x}, \mathbf{D}_i^{\star}), \tag{16.3}$$

where the $\mathbf{D}_i^{\star}$ are drawn from the bootstrap distribution, $P^{\star}(\mathbf{D}^{\star})$. In this way a further increase of the diversity term A can be achieved. Ormoneit and Tresp [47] applied this scheme to mixture models for predicting *unconditional* probability densities and found a significant improvement in the generalisation performance over that of a single-network predictor. In the present study, bagging was applied to predicting *conditional* probability densities. That is,

200 GM-RVFL networks were trained on 200 bootstrap samples. The parameters of the distributions for the random weights, $\sigma^g_{rand} = \exp(\rho_g)$, were adjusted by ARD2 every time a network was trained, and they were reset to their initial values after 100 networks had been trained. The results of the simulations can be found in Tab.16.5. The average single-network prediction $\overline{E_{gen}}$, shown in the right column, has deteriorated considerably as a consequence of training the networks on 'wrong' data sets. The generalisation performance of the committee, E^{Com}_{gen}, has become significantly better than the single-network predictions in the non-bootstrapping case (first four rows, right column). However, compared to the committee prediction obtained *without* bootstrapping, a degradation in the performance is found. This result is different from that obtained by Ormoneit and Tresp [47], who found a noticeable increase in the generalisation performance as a result of bagging. However, the simulations reported here differ substantially from those in [47] in that, by adopting the RVFL approach, the training scheme contains, per se, a large degree of inherent stochasticity. While Ormoneit and Tresp need a diversity-boosting scheme like bagging to ensure that the committee members show a sufficient degree of ambiguity in the first place, the present study suggests that when adopting the RVFL concept, a further diversification of networks beyond that already achieved by the inherent stochasticity of the training scheme is wasteful. Interestingly, this is in accordance with theoretical results obtained by Krogh and Sollich [35], [57], who found that the combination of two different diversity-increasing techniques only led to a deterioration in the overall generalisation performance (Figure 2 in both [35] and [57]). The authors write:

> *Nothing can be gained by making the students more diverse by training them on smaller, less overlapping training sets. One would also expect this kind of 'diversification' to be unnecessary or even counterproductive when the training noise is high enough to provide sufficient 'inherent' diversity of the students.*

Again, this can be explained in terms of Equation (13.19), which suggests that the increase in the diversity A is not sufficient to compensate for the deterioration in $\overline{E_{gen}}$, and overall, nothing is gained by applying bagging.

16.6.2 Nonlinear Preprocessing

Sharkey et al. [55], [54] recommend to subject the training data to different nonlinear transformations in order to increase the diversity of the networks in the committee. When applying this technique to the problem of fault detection in a ship's engine with neural networks, they found a significant improvement in the classification results. Following their suggestion, three different nonlinear transformations were applied. In the first case, all input variables, except for the binary indicator variable RIVER, were subjected to the logarithmic transformation

$$Tr1(x_t^i) \quad = \quad \ln\left(x_t^i - \min_t\{x_t^i\} + \epsilon\right), \tag{16.4}$$

where $\epsilon := 0.1$. The motivation is that since $\ln(\prod_i x^i) = \sum_i \ln x^i$, the application of a logarithmic transformation might make it easier for the network to deal with correlations of the form $(x^i x^j x^k \ldots)$. However, as the function (16.4) compresses the output range with increasing argument, it is advisable to introduce a 'conjugate' function that inflates the output range with increasing argument:

$$Tr2(x_t^i) \quad = \quad \ln\left(\max_t\{x_t^i\} - x_t^i + \epsilon\right), \tag{16.5}$$

again with $\epsilon := 0.1$. The third transformation was taken from [21]. The variables DIS, RAD and LOWSTAT are transformed logarithmically, the input NOX is squared, whereas the remaining variables are left unchanged:

$$Tr3(x_t^i) = (\delta_{i,8} + \delta_{i,9} + \delta_{i,13}) \ln(x_t^i) + \delta_{i,5}(x_t^i)^2 + (1 - \delta_{i,5} - \delta_{i,8} - \delta_{i,9} - \delta_{i,13}) x_t^i. \tag{16.6}$$

After the application of each of these transformations, the data were normalised again. In the simulation, 600 networks were trained, each for a fixed number of 5 EM steps. The ρ_gs were adapted with scheme ARD1, and reset in intervals of 100 steps. The first 200 networks were trained on the 'un-processed' data (that is, the data were only normalised, which is a *linear* transformation). The next 200 networks were trained on data subjected to transformation $Tr1$. The last 200 networks were trained on data that had been preprocessed by application of either $Tr2$ or $Tr3$. Table 16.8 summarises the results and further details of the training process. The first two rows show the prediction performance obtained *without* nonlinear pre-processing for a committee of (i) $N_{Com} = 200$ and (ii) $N_{Com} = 600$ networks. The comparison suggests that beyond a certain committee size the diversity and, consequently, the generalisation performance cannot be further improved by adding more networks to the committee. This is different when (several) nonlinear preprocessing transformations are applied, as shown in rows 2 and 3 of Tab.16.8. It can be seen that in both cases a slight enhancement of the generalisation performance has been achieved. The fact that the decrease of the average single-network generalisation error $\overline{E_{gen}}$, shown in the right column of Tab.16.8, is noticeably smaller than that of E_{gen}^{Com} suggests that this can in fact be attributed to an increase in the ambiguity A. However, the error bars (that is, the standard deviations resulting from the four-fold cross-validation scheme) are rather large, and the results are therefore not necessarily significant.

Preprocessing	N_{Com}	$E_{gen}^{Com}(T=0.1)$	$E_{gen}^{Com}(T=\infty)$	E_{gen}^{single}
linear	200	0.102 ±0.033	0.112 ±0.025	0.695 ±0.060
linear	600	0.103 ±0.030	0.111 ±0.029	0.715 ±0.068
linear, $Tr1$, $Tr2$	600	0.087 ±0.020	0.098 ±0.019	0.711 ±0.054
linear, $Tr1$, $Tr3$	600	0.080 ±0.020	0.089 ±0.022	0.703 ±0.053

Table 16.8. Nonlinear preprocessing. The table shows the effect of applying the nonlinear preprocessing transformations defined in equation (16.4-16.6). N_{Com} denotes the number of networks in the committee. Scheme ARD1 was applied for adapting the ρ_gs, with $\rho_{max} = -1$ and $\Delta\rho = 0.25$. All ρ_gs were reset to the initial value of $\rho_g = -1$ in intervals of 100 ARD updates. Network architecture: $H = 70, K = 5$.

16.7 Comparison with Neal's results

	Details	E_{gen}
Neal	No hidden layer	0.670
Neal	1 hidden layer (8 hidden units)	0.173
Neal	1 hidden layer (16 hidden unit)	0.189
Neal	2 hidden layers	0.085
GM-RVFL	mean single-network prediction	0.715 ±0.068
GM-RVFL	committee ($N_{Com} = 200$)	0.112 ±0.025
GM-RVFL	committee ($N_{Com} = 600$), nonlinear preprocessing	0.089 ±0.022

Table 16.9. Comparison with Neal. The first four rows show the generalisation 'error' E_{gen} obtained by R.Neal on the same problem, taken from Tab.4.10 in [44] and transformed to normalised data. This is compared with (i) the average performance of a single GM-RVFL network (averaged over a committee of size $N_{Com} = 600$), (ii) the performance of a committee of 200 networks, and (iii) the results obtained with a committee of 600 networks trained on data subjected to different nonlinear transformations (compare with Tab. 16.7 and Tab. 16.8). No cross-validation was used, i.e. the individual networks were trained over a fixed number of 5 EM-steps, and the reported committee 'errors' correspond to $T = \infty$. (Further details: $H = 70$, $K = 5$, $\rho_{max} = -1$, $\Delta\rho = 0.25$.)

The best results so far on this data set have been obtained by Neal [44], who adopted a Bayesian approach and applied a hybrid Monte-Carlo scheme to sample the network parameters from the posterior distribution (conditional on the training data). Table 16.9 shows the results of his thesis, transformed to normalised data[7]. Since these results were obtained with a training set of $N = 256$ samples, a fair comparison requires that no cross-validation be

---

[7] Recall that if $z = z(y)$ is invertible, then $P(y) = |\frac{dz}{dy}|P(z)$. Normalisation is a linear operation of the form $z = \frac{y-a}{b}$, therefore $P(z) = bP(y)$. The generalisation error is given by $E = -\frac{1}{N}\sum_t \ln P(y_t|x_t, q)$ for the original data, and by $\tilde{E} = -\frac{1}{N}\sum_t \ln P(z_t|x_t, q)$ after normalisation. Applying the transformation rule for

used. This implies that the GM-RVFL networks need to be trained over a fixed number of EM steps, and that all members in the committee contribute with equal weights (which is equivalent to an infinite temperature, $T = \infty$). Table 16.9 compares Neal's results with those obtained for a complex GM-RVFL network, $H = 70$, trained over a fixed number of 5 EM steps (Tab.16.7 and Tab.16.8). It can be seen that whereas a single GM-RVFL leads to a very poor performance, the committee of GM-RVFLs achieves results that are significantly better than Neal's one-hidden-layer networks. When different nonlinear transformations are applied to the data, the generalisation performance even reaches that of Neal's two-hidden-layer network. This result is most encouraging, since Neal's approach gives the best performance on this data set to date. The reason for the absence of an *improvement* in the performance suggests that the conditional distribution is *not* multi-modal and *without* a strong skewness. However, employing a GM-RVFL network rather than a standard MLP allows long tails in the conditional distribution to be taken into account, which in fact achieved an improvement over a model that assumes Gaussian distributed noise (see Section 16.4.6). Note that by modelling the noise with a t-distribution, Neal also takes long tails in the conditional distribution into account, and this fact is certainly another reason for the superior prediction performance achieved in his study [44].

16.8 Conclusions

- **ARD:** The ARD scheme has proven to be capable of detecting irrelevant inputs and adjusting the respective ρ_gs accordingly. If scheme ARD2 is applied, irrelevant inputs can be completely switched off. With scheme ARD1, it is only the nonlinear contributions that are affected. However, since ARD2 is inconsistent with certain scaling properties, ARD1 should be the superior choice. This was confirmed empirically in the simulations.

- **Network committees:** The simulations confirmed that the generalisation performance of a network *committee* is always better than the *average performance* of an individual network [see (13.19)].

- **Weighting scheme:** The networks in the committee were weighted by the temperature-dependent weighting scheme of Section 13.3.4, which is based on the cross-validation 'error' E_{cross} (13.33). For low temperatures, only the best models give significant contributions. As a consequence, the average generalisation error of the individual network, $\overline{E_{gen}^{single}}$, is small,

---

probabilities stated above, this gives $\tilde{E} = E - \ln b$, where b is the standard deviation of the target, found to be $b = 9.188$.

but the network diversity is also small and does not lead to any significant enhancement. With an increase of T, $\overline{E_{gen}^{single}}$ deteriorates, but this is more than counteracted by an increase of the diversity term A, which leads to an overall improvement of the generalisation performance. Optimal values for the temperature were found about $T \approx 0.1$. However, an increase of $T \to \infty$ did *not* result in any serious deterioration, and left the generalisation performance still markedly better than obtained with the best individual models. This result implies that network committees can achieve a considerable decrease of the generalisation 'error' even in the absence of a cross-validation set for model weighting.

– **Network complexity and overfitting:** While Bayesian regularisation and a limitation of the model complexity typically improve the generalisation performance of the *single-model predictor*, they lead to a deterioration of the *committee* performance. The lowest generalisation 'error' of the committee was obtained with under-regularised, complex models. This can be understood from (13.19), (16.1): the average single-model performance deteriorates as a consequence of overfitting, but this is counteracted by an increase of the diversity, which leads to an overall reduction of E_{gen}^{Com}. Note that these observations are in accordance with recent theoretical studies by Krogh and Sollich on linear models with Gaussian noise [35], [57].

– **Diversity:** Two strategies for increasing diversity were studied. A committee trained with *bagging* gave better results than the best single-network predictor obtained on the original data, but showed a significantly larger generalisation error than a committee of networks trained on the unmanipulated data. This suggests that a further diversification beyond that already achieved by the inherent stochasticity of the RVFL scheme has only an adverse effect on the predictions. Subjecting the training data to different *nonlinear pre-processing transformations* was found to be more successful, and led to a noticeable, though possibly not significant improvement of the generalisation performance.

– **Comparison:** The prediction results obtained with a committee of GM-RVFL networks is comparable to the results obtained in [44], which gives the best performance to date. Both approaches allow the modelling of long tails in the conditional distribution of the target. As shown in Section 16.4.6, this leads to a significant improvement over a model that is restricted to Gaussian distributed noise. The fact that no improvement over the results in [44] has been obtained suggests that the true conditional distribution, though being long-tailed, is *not* multimodal or strongly skewed. If it were, the committee of GM-RVFLs would most likely have outperformed the approach in [44][8].

---

[8] Note that the model presented in [44] inherently assumes symmetric unimodal noise.

17. Summary

The motivation for the work presented in this book results from the problem of time series prediction. A standard method is to train a neural network to predict a single future value as a function of a so-called lag vector of m past observations or measurements. The crucial requirement for the successful application of such a scheme is that the probability distribution of the targets conditional on the inputs is unimodal and symmetric. However, even when a series of past measurements is only subjected to *Gaussian* observational noise, this assumption does not necessarily hold. On the contrary, for reasons discussed in Chapter 1, the distribution is likely to be distorted and may be multimodal. This suggests that, in general, it is not sufficient to train a network to predict only a single value, but that the complete probability distribution of the target conditional on the input vector should be modelled.

The structure of a universal approximator network for accomplishing this task was derived in Chapter 2. The network contains at least *two* hidden layers in order to preserve its universal approximation capability in the singular low-noise limit. An error function for network training was derived from a maximum likelihood approach, and allowed backpropagation-like learning rules to be derived (Chapter 3). The application to a set of benchmark problems, introduced in Chapter 4, has demonstrated the effectiveness of this approach in modelling multimodal stochastic time series. However, due to the two-hidden-layer structure, the required training times were rather long. This motivated the search for a more efficient training scheme.

A straightforward idea was the application of the EM algorithm, which is known from statistics to be efficient in adapting the parameters of mixture models and was found to decrease convergence times considerably. However, the direct application of this scheme to the proposed network architecture was impossible since the M-step of the algorithm was intractable with respect to one subset of parameters. This slowed down the convergence speed considerably and did not lead to any improvement over the backpropagation-like training algorithm of Chapter 3.

So that the EM-algorithm could be applied, the network architecture was modified such that certain parameters were chosen at random and then held at these values during the training process (Chapter 7). A similar method, termed *random vector functional link* (RVFL) net approach and applied to avoid excessive training times in standard multilayer perceptrons, was studied earlier by Pao et al. A recent theoretical study, summarised in Chapter 6, showed that the imposed constraints have no impact on the universal approximation capability, and that the *curse of dimensionality*, that is, an exponential increase of the model complexity with increasing input dimension, is avoided. The incorporation of the RVFL concept into the proposed network model allowed the M-step of the EM-algorithm to be carried out for all the remaining adaptable parameters (Chapter 7). Chapter 8 has demonstrated that this speeds up training by about two orders of magnitude.

Whereas the first 8 chapters concerned *learning*, the subject of the subsequent chapters was the problem of *generalisation*. When the training data are sparse, finding one particular 'optimal' parameter configuration, which corresponds to a maximum-likelihood estimate in statistics, usually leads to serious overfitting. This problem is caused by the fact that the distribution in parameter space is likely to be strongly skewed, with a mode that is not representative of the distribution as a whole. The proper approach is, therefore, to integrate over the entire parameter space. This was pursued in Chapter 10, where the integration was carried out by Gaussian approximation, and the skewness problem was alleviated by optimising certain parameters, termed *hyperparameters*, only after others had been integrated out. The resulting regularisation scheme led to an extension of the Bayesian evidence approach of MacKay, and an empirical test on a stochastic time series, presented in Chapter 12, demonstrated that in fact this new algorithm could achieve a considerable reduction of network overfitting (Chapter 12) and stabilised the training process with respect to variations in the number of adaptation steps.

A further improvement in the generalisation performance could be achieved by applying an *ensemble* of predictors rather than only *one* optimised single network. From a Bayesian viewpoint, this procedure arises naturally from the requirement to integrate over the model space, giving rise to the *model evidence* as a weighting factor for the individual predictors. In practice, however, the integration cannot be carried out analytically, and the model evidence, although analytically derivable under the assumption of a Gaussian distribution in parameter space, as shown in Chapter 11, is rather susceptible to numerical round-off errors. Chapter 13 therefore introduced a modified weighting scheme, which can be based on both the *evidence* and the *cross-validation* 'error', and whose flexibility is increased by a further temperature-like parameter T that allows smooth transitions between the regimes of uniform weights and best-model selection. Simulations on an artificial (Chapter 14) as well as a real-world problem (Chapter 16) confirmed that a considerable improvement in the generalisation ability can be achieved with weighted committees, where in the latter case the performance was further improved by the so-called ARD method for automatically determining the relevance of various inputs during training (introduced in Chapter 15).

An interesting observation made in the simulations concerns the question of regularisation. While an adequate regularisation scheme is crucial for achieving satisfactory prediction results with an *individual network*, this does not necessarily carry over to network *ensembles*. The best results were obtained with under-regularised models, which individually overfitted the training data but, when combined in a committee, achieved a lower generalisation 'error' than their regularised counterparts. An explanation for this curiosity was given, in Chapter 13, in terms of a modified bias-variance decomposition, which suggests that model *diversity* is a crucial prerequisite for increasing the generalisation ability of a committee, and that this diversity is boosted by increased model flexibility resulting from under-regularisation.

It is here that a further advantage of the RVFL approach becomes apparent. While the original motivation for the incorporation of this approach was the acceleration of the training process – which allowed the generation of an ensemble of predictors in the first place – the intrinsic stochasticity of this scheme leads to a further diversification of the models in the committee. This results in a significantly improved generalisation performance.

18. Appendix:
Derivation of the Hessian for the Bayesian Evidence Scheme

This chapter provides a derivation of the Hessian of the error function E, which is required for the Bayesian evidence scheme of chapters 10 and 11. The derivation is based on an extended version of the EM algorithm, which allows the full Hessian to be decomposed into three additive components. The derivation of the first term, the Hessian of the EM error function U, is straightforward. The second term, the outer product of the gradient of the EM error function, is found to be cancelled out. An approximation is made for the third term, the expectation value for the outer product of the gradient of Ψ, which is approximated by a diagonal block matrix. The justification for this simplification is given in the text.

18.1 Introduction and notation

The Bayesian evidence scheme of chapters 10 and 11 requires the calculation of the Hessian of the error function $E = -\ln P(\mathbf{D}|\mathbf{q})$. This can be accomplished with an extended version of the EM algorithm. Recall, from Chapter 7, the following definitions:

$$E(\mathbf{q}) = -\ln P(\mathbf{D}|\mathbf{q}) \tag{18.1}$$

$$\Psi(\mathbf{q}, \mathbf{\Lambda}) = -\ln P(\mathbf{D}, \mathbf{\Lambda}|\mathbf{q}) \tag{18.2}$$

$$U(\mathbf{q}|\mathbf{q}') = \left\langle \Psi(\mathbf{q}, \mathbf{\Lambda}) \right\rangle_{\mathbf{\Lambda}|\mathbf{D},\mathbf{q}'} \tag{18.3}$$

$$S(\mathbf{q}|\mathbf{q}') = \left\langle \ln(P(\mathbf{\Lambda}|\mathbf{D}, \mathbf{q})) \right\rangle_{\mathbf{\Lambda}|\mathbf{D},\mathbf{q}'} \tag{18.4}$$

where $\left\langle X \right\rangle_{\mathbf{\Lambda}|\mathbf{D},\mathbf{q}'} := \int X(\mathbf{\Lambda}) P(\mathbf{\Lambda}|\mathbf{D}, \mathbf{q}') d\mathbf{\Lambda}$, and $\mathbf{\Lambda}$ is a set of hidden indicator variables, defined in (7.10).

The first part of this chapter, Section 18.2, will prove a theorem that states a relation between the actual Hessian $\mathbf{H} = \nabla_{\mathbf{q}} \nabla_{\mathbf{q}}^{\dagger} E$ and the Hessian of the EM error function U. This theorem will then be applied in the second part, Section 18.3, to derive an explicit expression for $\mathbf{H}$.

In what follows, the vector of all network parameters will be denoted by $\mathbf{q} = (\mathbf{w}, \boldsymbol{\beta}, \mathbf{a})$, and the Hessian with respect to $\mathbf{q}$ by $\mathbf{H}$. Note that this deviates slightly from the notation in chapters 10 and 11, where $\mathbf{q}$ contained only the network weights feeding into the $\mathcal{G}$-nodes, $\mathbf{q} = \mathbf{w}$, and $\mathbf{H}$ the Hessian with respect to $\mathbf{w}$. The remaining parameters were included in the hyperparameter vector $\mathbf{r}$, whose Hessian was denoted by $\mathbf{A}$. Hence the matrix denoted by $\mathbf{H}$ in this chapter includes the matrices $\mathbf{H}$ and $\mathbf{A}$ of chapters 10 and 11 as sub-matrices.

18.2 A decomposition of the Hessian using EM

The derivation of the Hessian is based on the following Lemma:

LEMMA

$$\text{If } \mathbf{q} = \mathbf{q}', \quad \text{then} \quad \frac{\partial^2 S(\mathbf{q}|\mathbf{q}')}{\partial q_i \partial q_k} = -\left\langle \frac{\partial \Psi}{\partial q_i} \frac{\partial \Psi}{\partial q_k} \right\rangle_{\Lambda|D,\mathbf{q}'} + \frac{\partial E}{\partial q_i} \frac{\partial E}{\partial q_k} \quad (18.5)$$

PROOF
From the definition of S, (18.4), we obtain

$$\frac{\partial^2 S(\mathbf{q}|\mathbf{q}')}{\partial q_i \partial q_k} = \frac{\partial^2}{\partial q_i \partial q_k} \int \left(\ln P(\Lambda|D, \mathbf{q}) \right) P(\Lambda|D, \mathbf{q}') d\Lambda$$

$$= \frac{\partial}{\partial q_i} \int \frac{1}{P(\Lambda|D, \mathbf{q})} \frac{\partial P(\Lambda|D, \mathbf{q})}{\partial q_k} P(\Lambda|D, \mathbf{q}') d\Lambda$$

$$= \int \frac{1}{P(\Lambda|D, \mathbf{q})} \frac{\partial^2 P(\Lambda|D, \mathbf{q})}{\partial q_i \partial q_k} P(\Lambda|D, \mathbf{q}') d\Lambda$$

$$- \int \frac{1}{\left(P(\Lambda|D, \mathbf{q}) \right)^2} \frac{\partial P(\Lambda|D, \mathbf{q})}{\partial q_i} \frac{\partial P(\Lambda|D, \mathbf{q})}{\partial q_k} P(\Lambda|D, \mathbf{q}') d\Lambda$$

For $\mathbf{q}=\mathbf{q}'$, the first integral in the last equation is zero:

$$\int \frac{1}{P(\Lambda|D, \mathbf{q})} \frac{\partial^2 P(\Lambda|D, \mathbf{q})}{\partial q_i \partial q_k} P(\Lambda|D, \mathbf{q}') d\Lambda$$

$$= \int \frac{\partial^2 P(\Lambda|D, \mathbf{q})}{\partial q_i \partial q_k} d\Lambda = \frac{\partial^2}{\partial q_i \partial q_k} \int P(\Lambda|D, \mathbf{q}) d\Lambda = \frac{\partial^2}{\partial q_i \partial q_k} 1 = 0$$

With $\frac{1}{P(\Lambda|D,\mathbf{q})} \frac{\partial P(\Lambda|D,\mathbf{q})}{\partial q_i} = \frac{\partial \ln P(\Lambda|D,\mathbf{q})}{\partial q_i}$ one therefore obtains

$$\frac{\partial^2 S(\mathbf{q}|\mathbf{q}')}{\partial q_i \partial q_k} = -\int \frac{\partial \ln P(\Lambda|\mathbf{D}, \mathbf{q})}{\partial q_i} \frac{\partial \ln P(\Lambda|\mathbf{D}, \mathbf{q})}{\partial q_k} P(\Lambda|\mathbf{D}, \mathbf{q}')d\Lambda$$

Now use (i) the identity $P(\Lambda|\mathbf{D}, \mathbf{q}) = \frac{P(\Lambda, \mathbf{D}|\mathbf{q})}{P(\mathbf{D}|\mathbf{q})} \Rightarrow \ln P(\Lambda|\mathbf{D}, \mathbf{q}) = \ln P(\Lambda, \mathbf{D}|\mathbf{q}) -$ $\ln P(\mathbf{D}|\mathbf{q})$ and (ii) the normalisation condition $\int P(\Lambda|\mathbf{D}, \mathbf{q}')d\Lambda = 1$ to arrive at

$$\begin{aligned}
\frac{\partial^2 S(\mathbf{q}|\mathbf{q}')}{\partial q_i \partial q_k} =\ & -\int \frac{\partial \ln P(\Lambda, \mathbf{D}|\mathbf{q})}{\partial q_i} \frac{\partial \ln P(\Lambda, \mathbf{D}|\mathbf{q})}{\partial q_k} P(\Lambda|\mathbf{D}, \mathbf{q}')d\Lambda \\
& -\frac{\partial \ln P(\mathbf{D}|\mathbf{q})}{\partial q_i} \frac{\partial \ln P(\mathbf{D}|\mathbf{q})}{\partial q_k} \\
& +\frac{\partial \ln P(\mathbf{D}|\mathbf{q})}{\partial q_i} \int \frac{\partial \ln P(\Lambda, \mathbf{D}|\mathbf{q})}{\partial q_k} P(\Lambda|\mathbf{D}, \mathbf{q}')d\Lambda \\
& +\frac{\partial \ln P(\mathbf{D}|\mathbf{q})}{\partial q_k} \int \frac{\partial \ln P(\Lambda, \mathbf{D}|\mathbf{q})}{\partial q_i} P(\Lambda|\mathbf{D}, \mathbf{q}')d\Lambda
\end{aligned}$$

The integral in the last two terms gives (recall that $\mathbf{q} = \mathbf{q}'$)

$$\begin{aligned}
\int \frac{\partial \ln P(\Lambda, \mathbf{D}|\mathbf{q})}{\partial q_i} P(\Lambda|\mathbf{D}, \mathbf{q}')d\Lambda &= \int \frac{1}{P(\Lambda, \mathbf{D}|\mathbf{q})} \frac{\partial P(\Lambda, \mathbf{D}|\mathbf{q})}{\partial q_i} P(\Lambda|\mathbf{D}, \mathbf{q}')d\Lambda \\
&= \int \frac{1}{P(\Lambda|\mathbf{D}, \mathbf{q})P(\mathbf{D}|\mathbf{q})} \frac{\partial P(\Lambda, \mathbf{D}|\mathbf{q})}{\partial q_i} P(\Lambda|\mathbf{D}, \mathbf{q}')d\Lambda \\
&= \frac{1}{P(\mathbf{D}|\mathbf{q})} \int \frac{\partial P(\Lambda, \mathbf{D}|\mathbf{q})}{\partial q_i} d\Lambda \\
&= \frac{1}{P(\mathbf{D}|\mathbf{q})} \frac{\partial}{\partial q_i} \int P(\Lambda, \mathbf{D}|\mathbf{q})d\Lambda \\
&= \frac{1}{P(\mathbf{D}|\mathbf{q})} \frac{\partial}{\partial q_i} P(\mathbf{D}|\mathbf{q}) = \frac{\partial \ln P(\mathbf{D}|\mathbf{q})}{\partial q_i}
\end{aligned}$$

Therefore

$$\begin{aligned}
\frac{\partial^2 S(\mathbf{q}|\mathbf{q}')}{\partial q_i \partial q_k} =\ & -\int \frac{\partial \ln P(\Lambda, \mathbf{D}|\mathbf{q})}{\partial q_i} \frac{\partial \ln P(\Lambda, \mathbf{D}|\mathbf{q})}{\partial q_k} P(\Lambda|\mathbf{D}, \mathbf{q}')d\Lambda \\
& +\frac{\partial \ln P(\mathbf{D}|\mathbf{q})}{\partial q_i} \frac{\partial \ln P(\mathbf{D}|\mathbf{q})}{\partial q_k}
\end{aligned}$$

With the definitions (18.1) and (18.2), $\ln P(\Lambda, \mathbf{D}|\mathbf{q}) = -\Psi(\mathbf{q}, \Lambda)$ and $\ln P(\mathbf{D}|\mathbf{q}) = -E(\mathbf{q})$, and the notation $\langle f \rangle_{\Lambda|\mathbf{D},\mathbf{q}'} = \int f(\Lambda)p(\Lambda|\mathbf{D}, \mathbf{q}')d\Lambda$, this leads to the expression (18.5).

◇

It is now easy to prove the following relation, which allows the calculation of the Hessian of the error function E from the EM error function U:

THEOREM

If $\mathbf{q} = \mathbf{q}'$ then $\dfrac{\partial^2 E}{\partial q_i \partial q_k} = \dfrac{\partial^2 U}{\partial q_i \partial q_k} + \dfrac{\partial U}{\partial q_i}\dfrac{\partial U}{\partial q_k} - \left\langle \dfrac{\partial \Psi}{\partial q_i}\dfrac{\partial \Psi}{\partial q_k} \right\rangle_{\Lambda|D,\mathbf{q}'}$ (18.6)

PROOF
From (7.6) we get

$$E(\mathbf{q}) = U(\mathbf{q}|\mathbf{q}') + S(\mathbf{q}|\mathbf{q}') \quad \Rightarrow \quad \dfrac{\partial^2 E(\mathbf{q})}{\partial q_i \partial q_k} = \dfrac{\partial^2 U(\mathbf{q}|\mathbf{q}')}{\partial q_i \partial q_k} + \dfrac{\partial^2 S(\mathbf{q}|\mathbf{q}')}{\partial q_i \partial q_k}.$$
(18.7)

Making now use of the lemma, (18.5), this leads to

$$\dfrac{\partial^2 E(\mathbf{q})}{\partial q_i \partial q_k} = \dfrac{\partial^2 U(\mathbf{q}|\mathbf{q}')}{\partial q_i \partial q_k} - \left\langle \dfrac{\partial \Psi}{\partial q_i}\dfrac{\partial \Psi}{\partial q_k} \right\rangle_{\Lambda|D,\mathbf{q}'} + \dfrac{\partial E}{\partial q_i}\dfrac{\partial E}{\partial q_k}.$$
(18.8)

Since $\mathbf{q} = \mathbf{q}'$, $\frac{\partial E}{\partial q_i}$ can be replaced by $\frac{\partial U(\mathbf{q}|\mathbf{q}')}{\partial q_i}$ due to (7.9). This completes the proof.

◇

18.3 Explicit calculation of the Hessian

Let us now apply Equation (18.6) to derive an expression for the Hessian $\mathbf{H} = \nabla_{\mathbf{q}}\nabla_{\mathbf{q}}^\dagger E$, which splits up into a sum of three sub-matrices: the Hessian of the EM error function U, and the two outer gradient products $\nabla_{\mathbf{q}}U(\nabla_{\mathbf{q}}U)^\dagger$ and $\left\langle \nabla_{\mathbf{q}}\Psi(\nabla_{\mathbf{q}}\Psi)^\dagger \right\rangle_{\Lambda|D,\mathbf{q}'}$. To simplify the notation, the latter expression will be written without subscript henceforth: $\left\langle \nabla_{\mathbf{q}}\Psi(\nabla_{\mathbf{q}}\Psi)^\dagger \right\rangle :=$ $\left\langle \nabla_{\mathbf{q}}\Psi(\nabla_{\mathbf{q}}\Psi)^\dagger \right\rangle_{\Lambda|D,\mathbf{q}'}$.

First term: The Hessian of the EM error function U
From (7.18) and (7.34) we obtain

$$\nabla_{\mathbf{w}_i}\nabla_{\mathbf{w}_k}^\dagger U = \delta_{ik}\beta_k \mathbf{G}\boldsymbol{\Pi}_k \mathbf{G}^\dagger$$
(18.9)

$$\dfrac{\partial^2 U}{\partial \beta_i \partial \beta_k} = \dfrac{\delta_{ik}}{2(\beta_k)^2}\sum_{t=1}^{N}\pi_k(t)$$
(18.10)

$$\dfrac{\partial}{\partial \beta_i}\nabla_{\mathbf{w}_k}U = \dfrac{\delta_{ik}}{\beta_k}\nabla_{\mathbf{w}_k}U$$
(18.11)

When taking the second derivatives of U with respect to the output weights a_k, recall that $\frac{\partial a_K}{\partial a_k} = -1$ due to the constraint $a_K = 1 - \sum_{k=1}^{K-1} a_k$. Therefore (7.19) gives:

$$\frac{\partial^2 U}{\partial a_i \partial a_k} = \frac{\delta_{ik}}{(a_k)^2} \sum_{t=1}^{N} \pi_k(t) + \frac{1}{(a_K)^2} \sum_{t=1}^{N} \pi_K(t) \tag{18.12}$$

$$\frac{\partial^2 U}{\partial a_i \partial \beta_k} = \frac{\partial}{\partial a_i} \nabla_{\mathbf{w}_k} U = 0 \tag{18.13}$$

Second term: The outer product $\left\langle \nabla_{\mathbf{q}} \Psi (\nabla_{\mathbf{q}} \Psi)^\dagger \right\rangle$

Let us now turn to the calculation of the outer product $\left\langle \nabla_{\mathbf{q}} \Psi (\nabla_{\mathbf{q}} \Psi)^\dagger \right\rangle$. Recall from (7.15) that

$$\Psi(\mathbf{q}, \mathbf{\Lambda}) = \sum_{t=1}^{N} \sum_{k=1}^{K} \lambda_k(t) \left[\frac{\beta_k}{2} \Big(y_t - \mu(\mathbf{x}_t; \mathbf{w}) \Big)^2 - \ln a_k - \frac{1}{2} \ln \left(\frac{\beta_k}{2\pi} \right) \right]. \tag{18.14}$$

Moreover, recall from (7.28) that for the GM-RVFL network

$$\mu_k(\mathbf{x}_t; \mathbf{w}) = \mu(\mathbf{x}_t; \mathbf{w}_k) := \mathbf{w}_k^\dagger \mathbf{g}(\mathbf{x}_t). \tag{18.15}$$

This leads to the gradient terms

$$\nabla_{\mathbf{w}_k} \Psi = -\sum_{t=1}^{N} \lambda_k(t) \beta_k \Big(y_t - \mu(\mathbf{x}_t; \mathbf{w}_k) \Big) \mathbf{g}(\mathbf{x}_t) \tag{18.16}$$

$$\frac{\partial \Psi}{\partial \beta_k} = \sum_{t=1}^{N} \frac{\lambda_k(t)}{2} \left[\Big(y_t - \mu(\mathbf{x}_t; \mathbf{w}_k) \Big)^2 - \frac{1}{\beta_k} \right] \tag{18.17}$$

$$\frac{\partial \Psi}{\partial a_k} = -\frac{1}{a_k} \sum_{t=1}^{N} \lambda_k(t) + \frac{1}{a_K} \sum_{t=1}^{N} \lambda_K(t) \tag{18.18}$$

In what follows, we will find expressions of the form $\left\langle \left[\sum_t \lambda_i(t) \Phi_i(t) \right] \left[\sum_{t'} \lambda_k(t') \tilde{\Phi}_k(t') \right] \right\rangle$. For independent training data, events at different times $t \neq t'$ are independent. Therefore $\left\langle \lambda_i(t) \lambda_k(t') \right\rangle = \left\langle \lambda_i(t) \right\rangle \left\langle \lambda_k(t') \right\rangle = \pi_i(t) \pi_k(t')$. For $t = t'$, however, the events are *not* independent since at any particular time only one of the indicator variables $\lambda_k(t)$ is 'switched on'. Moreover, since the $\lambda_k(t)$ are binary, $\left[\lambda_k(t) \right]^2 = \lambda_k(t)$. This can be summarised as $\lambda_i(t) \lambda_k(t) = \delta_{ik} \lambda_k(t)$, yielding the overall relation

$$\left\langle \lambda_i(t) \lambda_k(t') \right\rangle = (1 - \delta_{tt'}) \pi_i(t) \pi_k(t') + \delta_{tt'} \delta_{ik} \pi_k(t). \tag{18.19}$$

Using this result, we obtain

$$\left\langle \sum_t \lambda_i(t) \Phi_i(t) \sum_{t'} \lambda_k(t') \tilde{\Phi}_k(t') \right\rangle =$$

$$= \left\langle \sum_t \sum_{t'} \lambda_i(t)\lambda_k(t')\Phi_i(t)\tilde{\Phi}_k(t') \right\rangle = \sum_t \sum_{t'} \left\langle \lambda_i(t)\lambda_k(t') \right\rangle \Phi_i(t)\tilde{\Phi}_k(t')$$

$$= \sum_t \sum_{t'} \left((1-\delta_{tt'})\pi_i(t)\pi_k(t') + \delta_{tt'}\delta_{ik}\pi_k(t) \right)\Phi_i(t)\tilde{\Phi}_k(t')$$

$$= \sum_t \sum_{t'} \pi_i(t)\pi_k(t')\Phi_i(t)\tilde{\Phi}_k(t') + \delta_{ik}\sum_t \pi_k(t)\Phi_k(t)\tilde{\Phi}_k(t)$$

$$= \sum_t \sum_{t'} \pi_i(t)\pi_k(t')\Phi_i(t)\tilde{\Phi}_k(t') - \sum_t \pi_i(t)\pi_k(t)\Phi_i(t)\tilde{\Phi}_k(t)$$

$$+ \delta_{ik}\sum_t \pi_k(t)\Phi_k(t)\tilde{\Phi}_k(t)$$

$$= \left\langle \sum_t \lambda_i(t)\Phi_i(t) \right\rangle \left\langle \sum_t \lambda_k(t)\tilde{\Phi}_k(t) \right\rangle - \sum_t \pi_i(t)\pi_k(t)\Phi_i(t)\tilde{\Phi}_k(t)$$

$$+ \delta_{ik}\sum_t \pi_k(t)\Phi_k(t)\tilde{\Phi}_k(t)$$

and finally, after adding and subtracting a term $\delta_{ik}\sum \left(\pi_k(t)\right)^2 \Phi_k\tilde{\Phi}_k$,

$$\left\langle \sum_t \lambda_i(t)\Phi_i(t) \sum_{t'} \lambda_k(t')\tilde{\Phi}_k(t') \right\rangle =$$

$$\left\langle \sum_t \lambda_i(t)\Phi_i(t) \right\rangle \left\langle \sum_t \lambda_k(t)\tilde{\Phi}_k(t) \right\rangle - (1-\delta_{ik})\sum_t \pi_i(t)\pi_k(t)\Phi_i(t)\tilde{\Phi}_k(t)$$

$$+ \delta_{ik}\sum_t \pi_k(t)\left[1 - \pi_k(t)\right]\Phi_k(t)\tilde{\Phi}_k(t). \tag{18.20}$$

Applying (18.17) and (18.20), we obtain with

$$\Phi_k(t) = \tilde{\Phi}_k(t) := \frac{1}{2}\left[(y_t - \mu(\mathbf{x}_t;\mathbf{w}_k))^2 - \frac{1}{\beta_k} \right] \tag{18.21}$$

and the identity $\left\langle \Psi \right\rangle = U$ (from (18.3))

$$\left\langle \frac{\partial \Psi}{\partial \beta_i}\frac{\partial \Psi}{\partial \beta_k} \right\rangle =$$

$$\frac{\partial U}{\partial \beta_i}\frac{\partial U}{\partial \beta_k}$$

$$-\frac{1-\delta_{ik}}{4}\sum_{t=1}^{N} \pi_i(t)\pi_k(t)\left[\left(y_t - \mu(\mathbf{x}_t;\mathbf{w}_i)\right)^2 - \frac{1}{\beta_i} \right]\left[\left(y_t - \mu(\mathbf{x}_t;\mathbf{w}_k)\right)^2 - \frac{1}{\beta_k} \right]$$

$$+\frac{\delta_{ik}}{4}\sum_{t=1}^{N} \pi_k(t)\left[1 - \pi_k(t)\right]\left[\left(y_t - \mu(\mathbf{x}_t;\mathbf{w}_k)\right)^2 - \frac{1}{\beta_k} \right]^2. \tag{18.22}$$

The argument of the second sum can be both positive and negative. Its elements therefore tend to cancel each other out so that this sum can be expected to be much smaller than the other two terms. It will therefore be neglected:

$$\left\langle \frac{\partial \Psi}{\partial \beta_i} \frac{\partial \Psi}{\partial \beta_k} \right\rangle \approx \frac{\partial U}{\partial \beta_i} \frac{\partial U}{\partial \beta_k} + \frac{\delta_{ik}}{4} \sum_{t=1}^{N} \pi_k(t) \left[1 - \pi_k(t)\right] \left[\left(y_t - \mu(\mathbf{x}_t; \mathbf{w}_k)\right)^2 - \frac{1}{\beta_k}\right]^2$$

(18.23)

In the same way, applying (18.3), (18.16), (18.20), and defining $\Phi_k(t) = \tilde{\Phi}_k(t) := \beta_k(y_t - \mu(\mathbf{x}_t; \mathbf{w}_k))\mathbf{g}(\mathbf{x}_t)$, leads to

$$\left\langle (\nabla_{\mathbf{w}_i} \Psi)(\nabla_{\mathbf{w}_k} \Psi)^\dagger \right\rangle = (\nabla_{\mathbf{w}_i} U)(\nabla_{\mathbf{w}_k} U)^\dagger$$

$$- (1 - \delta_{ik})\beta_i\beta_k \sum_{t=1}^{N} \pi_i(t)\pi_k(t)\left(y_t - \mu(\mathbf{x}_t; \mathbf{w}_i)\right)\left(y_t - \mu(\mathbf{x}_t; \mathbf{w}_k)\right)\mathbf{g}(\mathbf{x}_t)\mathbf{g}^\dagger(\mathbf{x}_t)$$

$$+ \delta_{ik}(\beta_k)^2 \sum_{t=1}^{N} \pi_k(t)\left[1 - \pi_k(t)\right]\left(y_t - \mu(\mathbf{x}_t; \mathbf{w}_k)\right)^2 \mathbf{g}(\mathbf{x}_t)\mathbf{g}^\dagger(\mathbf{x}_t).$$

(18.24)

Since the argument of the second sum has a varying sign again, this term will be neglected:

$$\left\langle (\nabla_{\mathbf{w}_i} \Psi)(\nabla_{\mathbf{w}_k} \Psi)^\dagger \right\rangle \approx (\nabla_{\mathbf{w}_i} U)(\nabla_{\mathbf{w}_k} U)^\dagger$$

(18.25)

$$+ \delta_{ik}(\beta_k)^2 \sum_{t=1}^{N} \pi_k(t)\left[1 - \pi_k(t)\right]\left(y_t - \mu(\mathbf{x}_t; \mathbf{w}_k)\right)^2 \mathbf{g}(\mathbf{x}_t)\mathbf{g}^\dagger(\mathbf{x}_t)$$

For obtaining an expression for $\left\langle \frac{\partial}{\partial \beta_i} \nabla_{\mathbf{w}_k} \Psi \right\rangle$, define

$$\Phi_i(t) := \frac{1}{2}\left[\left(y_t - \mu(\mathbf{x}_t; \mathbf{w}_i)\right)^2 - \frac{1}{\beta_i}\right]$$

(18.26)

$$\tilde{\Phi}_k(t) := \beta_k(y_t - \mu(\mathbf{x}_t; \mathbf{w}_k))\mathbf{g}(\mathbf{x}_t),$$

(18.27)

and use (18.3), (18.16), (18.17), and (18.20):

$$\left\langle \frac{\partial \Psi}{\partial \beta_i} \nabla_{\mathbf{w}_k} \Psi \right\rangle = \frac{\partial U}{\partial \beta_i} \nabla_{\mathbf{w}_k} U$$

(18.28)

$$+ \frac{(1 - \delta_{ik})\beta_k}{2} \sum_{t=1}^{N} \pi_i(t)\pi_k(t)\left[\left(y_t - \mu(\mathbf{x}_t; \mathbf{w}_i)\right)^2 - \frac{1}{\beta_i}\right]\left(y_t - \mu(\mathbf{x}_t; \mathbf{w}_k)\right)\mathbf{g}(\mathbf{x}_t)$$

$$- \frac{\delta_{ik}\beta_k}{2} \sum_{t=1}^{N} \pi_k(t)\left[1 - \pi_k(t)\right]\left[\left(y_t - \mu(\mathbf{x}_t; \mathbf{w}_k)\right)^2 - \frac{1}{\beta_k}\right]\left(y_t - \mu(\mathbf{x}_t; \mathbf{w}_k)\right)\mathbf{g}(\mathbf{x}_t)$$

Now the signs of the arguments in both sums vary randomly (all powers in $[y_t - \mu(\mathbf{x}_t; \mathbf{w}_k)]$ are odd), suggesting to neglect all but the first term:

$$\left\langle \frac{\partial \Psi}{\partial \beta_i} \nabla_{\mathbf{w}_k} \Psi \right\rangle \approx \frac{\partial U}{\partial \beta_i} \nabla_{\mathbf{w}_k} U. \tag{18.29}$$

In a similar way it can be shown that

$$\left\langle \frac{\partial \Psi}{\partial a_i} \nabla_{\mathbf{w}_k} \Psi \right\rangle \approx \frac{\partial U}{\partial a_i} \nabla_{\mathbf{w}_k} U, \tag{18.30}$$

$$\left\langle \frac{\partial \Psi}{\partial a_i} \frac{\partial \Psi}{\partial \beta_k} \right\rangle \approx \frac{\partial U}{\partial a_i} \frac{\partial U}{\partial \beta_k}. \tag{18.31}$$

Finally, the expression for $\left\langle \frac{\partial \Psi}{\partial a_i} \frac{\partial \Psi}{\partial a_k} \right\rangle$ has to be derived:

$$\left\langle \frac{\partial \Psi}{\partial a_i} \frac{\partial \Psi}{\partial a_k} \right\rangle = \tag{18.32}$$

$$\left\langle \sum_t \left(\frac{\lambda_K(t)}{a_K} - \frac{\lambda_i(t)}{a_i} \right) \sum_{t'} \left(\frac{\lambda_K(t')}{a_K} - \frac{\lambda_k(t')}{a_k} \right) \right\rangle =$$

$$\sum_t \sum_{t'} \left(\frac{\langle \lambda_i(t)\lambda_k(t') \rangle}{a_i a_k} - \frac{\langle \lambda_i(t)\lambda_K(t') \rangle}{a_i a_K} - \frac{\langle \lambda_k(t)\lambda_K(t') \rangle}{a_k a_K} + \frac{\langle \lambda_K(t)\lambda_K(t') \rangle}{(a_K)^2} \right)$$

where (18.18) has been used. Applying (18.19) leads to

$$\left\langle \frac{\partial \Psi}{\partial a_i} \frac{\partial \Psi}{\partial a_k} \right\rangle =$$

$$= \sum_t \sum_{t' \neq t} \left(\frac{\pi_i(t)\pi_k(t')}{a_i a_k} - \frac{\pi_i(t)\pi_K(t')}{a_i a_K} - \frac{\pi_k(t)\pi_K(t')}{a_k a_K} + \frac{\pi_K(t)\pi_K(t')}{(a_K)^2} \right)$$

$$+ \sum_t \left(\frac{\delta_{ik}\pi_k(t)}{(a_k)^2} - \frac{\delta_{iK}\pi_K(t)}{(a_K)^2} - \frac{\delta_{kK}\pi_K(t)}{(a_K)^2} + \frac{\pi_K(t)}{(a_K)^2} \right) \tag{18.33}$$

Recall that due to the constraint $\sum_k a_k = 1$ only $(K - 1)$ of the a_k are free variables, which without loss of generality were chosen to be $a_1, \ldots, a_{K-1}$. Therefore $\delta_{iK} = \delta_{kK} = 0$, and the two middle terms in the second sum disappear. Applying (18.12), this sum therefore becomes

$$\sum_t \left(\frac{\delta_{ik}\pi_k(t)}{(a_k)^2} - \frac{\delta_{iK}\pi_K(t)}{(a_K)^2} - \frac{\delta_{kK}\pi_K(t)}{(a_K)^2} + \frac{\pi_K(t)}{(a_K)^2} \right) = \sum_t \left(\frac{\delta_{ik}\pi_k(t)}{(a_k)^2} + \frac{\pi_K(t)}{(a_K)^2} \right)$$

$$= \frac{\partial^2 U}{\partial a_i \partial a_k} \tag{18.34}$$

For the first sum we obtain

$$\sum_t \sum_{t' \neq t} \left(\frac{\pi_i(t)\pi_k(t')}{a_i a_k} - \frac{\pi_i(t)\pi_K(t')}{a_i a_K} - \frac{\pi_k(t)\pi_K(t')}{a_k a_K} + \frac{\pi_K(t)\pi_K(t')}{(a_K)^2} \right)$$

$$= \sum_t \sum_{t' \neq t} \left(\frac{\pi_K(t)}{a_K} - \frac{\pi_i(t)}{a_i} \right) \left(\frac{\pi_K(t')}{a_K} - \frac{\pi_k(t')}{a_k} \right)$$

$$= \sum_t \sum_{t'} \left(\frac{\pi_K(t)}{a_K} - \frac{\pi_i(t)}{a_i} \right) \left(\frac{\pi_K(t')}{a_K} - \frac{\pi_k(t')}{a_k} \right)$$

$$- \sum_t \left(\frac{\pi_K(t)}{a_K} - \frac{\pi_i(t)}{a_i} \right) \left(\frac{\pi_K(t)}{a_K} - \frac{\pi_k(t)}{a_k} \right)$$

$$= \frac{\partial U}{\partial a_i} \frac{\partial U}{\partial a_k} - \sum_t \left(\frac{\pi_K(t)}{a_K} - \frac{\pi_i(t)}{a_i} \right) \left(\frac{\pi_K(t)}{a_K} - \frac{\pi_k(t)}{a_k} \right) \qquad (18.35)$$

where in the last step (7.19) has been used. Inserting (18.34) and (18.35) into (18.33) yields

$$\left\langle \frac{\partial \Psi}{\partial a_i} \frac{\partial \Psi}{\partial a_k} \right\rangle = \frac{\partial U}{\partial a_i} \frac{\partial U}{\partial a_k} - \sum_t \left(\frac{\pi_K(t)}{a_K} - \frac{\pi_i(t)}{a_i} \right) \left(\frac{\pi_K(t)}{a_K} - \frac{\pi_k(t)}{a_k} \right) + \frac{\partial^2 U}{\partial a_i \partial a_k}$$
$$(18.36)$$

For $i \neq k$, the argument of the sum changes signs randomly so that

$$(1 - \delta_{ik}) \sum_t \left(\frac{\pi_K(t)}{a_K} - \frac{\pi_i(t)}{a_i} \right) \left(\frac{\pi_K(t)}{a_K} - \frac{\pi_k(t)}{a_k} \right) \ll \sum_t \left(\frac{\pi_K(t)}{a_K} - \frac{\pi_k(t)}{a_k} \right)^2$$

This justifies the following approximation:

$$\left\langle \frac{\partial \Psi}{\partial a_i} \frac{\partial \Psi}{\partial a_k} \right\rangle \approx \frac{\partial U}{\partial a_i} \frac{\partial U}{\partial a_k} + \frac{\partial^2 U}{\partial a_i \partial a_k} - \delta_{ik} \sum_t \left(\frac{\pi_K(t)}{a_K} - \frac{\pi_k(t)}{a_k} \right)^2 (18.37)$$

Third term: Outer product of the gradient ∇U

Note that all the expressions for $\left\langle \frac{\partial \Psi}{\partial q_i} \frac{\partial \Psi}{\partial q_k} \right\rangle$, that is, equations (18.23), (18.25), (18.29), (18.30), (18.31), and (18.37), contain a term of the form $\frac{\partial U}{\partial q_i} \frac{\partial U}{\partial q_k}$. When combining all the contributions according to (18.6),

$$\frac{\partial^2 E}{\partial q_i \partial q_k} = \frac{\partial^2 U}{\partial q_i \partial q_k} + \frac{\partial U}{\partial q_i} \frac{\partial U}{\partial q_k} - \left\langle \frac{\partial \Psi}{\partial q_i} \frac{\partial \Psi}{\partial q_k} \right\rangle, \qquad (18.38)$$

these gradient terms thus cancel each other exactly out.

Putting it all together

We therefore obtain from (18.6),(18.9), and (18.25):

$$\nabla_{\mathbf{w}_i} \nabla_{\mathbf{w}_k}^\dagger E = \qquad (18.39)$$

$$\delta_{ik} \left[\beta_k \mathbf{G} \Pi_k \mathbf{G}^\dagger - (\beta_k)^2 \sum_{t=1}^N \pi_k(t) \left[1 - \pi_k(t) \right] \left(y_t - \mu(\mathbf{x}_t; \mathbf{w}_k) \right)^2 \mathbf{g}(\mathbf{x}_t) \mathbf{g}^\dagger(\mathbf{x}_t) \right]$$

From (18.6), (18.10), and (18.23) we obtain

$$\frac{\partial^2 E}{\partial \beta_i \partial \beta_k} = \tag{18.40}$$

$$\delta_{ik}\left(\frac{1}{2(\beta_k)^2}\sum_{t=1}^{N}\pi_k(t) - \frac{1}{4}\sum_{t=1}^{N}\pi_k(t)\left[1 - \pi_k(t)\right]\left[\left(y_t - \mu(\mathbf{x}_t;\mathbf{w}_k)\right)^2 - \frac{1}{\beta_k}\right]^2\right)$$

Equations (18.6) and (18.37) lead to

$$\frac{\partial^2 E}{\partial a_i \partial a_k} = \delta_{ik}\sum_{t=1}^{N}\left(\frac{\pi_K(t)}{a_K} - \frac{\pi_k(t)}{a_k}\right)^2. \tag{18.41}$$

For the off-diagonal terms, we see from (18.29), (18.30), and (18.31) that $\left\langle\frac{\partial\Psi}{\partial q_i}\frac{\partial\Psi}{\partial q_k}\right\rangle = \frac{\partial U}{\partial q_i}\frac{\partial U}{\partial q_k}$. Due to (18.6) the outer products thus cancel each other out completely, and the only non-vanishing terms are the second derivatives of U, given by (18.11) and (18.13):

$$\frac{\partial^2 E}{\partial a_i \partial \beta_k} = \frac{\partial^2 U}{\partial a_i \partial \beta_k} = 0 \tag{18.42}$$

$$\frac{\partial}{\partial a_i}\nabla_{\mathbf{w}_k}E = \frac{\partial}{\partial a_i}\nabla_{\mathbf{w}_k}U = 0 \tag{18.43}$$

$$\frac{\partial}{\partial \beta_i}\nabla_{\mathbf{w}_k}E = \frac{\partial}{\partial \beta_i}\nabla_{\mathbf{w}_k}U = \frac{\delta_{ik}}{\beta_k}\nabla_{\mathbf{w}_k}U \tag{18.44}$$

It is seen that were it not for the last term, (18.44), the Hessian would have a diagonal block structure. However, recall from Section 10.3, equation (10.33), that we are only interested in the Hessian at the *mode*, for which $\nabla_{\mathbf{w}_k}E^* = 0$. For the unregularised case, that is, when $E^* = E$, we see from (7.9) that $\nabla_{\mathbf{w}_k}U = 0$, hence the off-diagonal terms (18.44) disappear. For the regularised case, equation (9.33) leads to $\nabla_{\mathbf{w}_k}U = -\nabla_{\mathbf{w}_k}R$. With (9.39) this gives

$$\mathbf{H}_{\beta_i,\mathbf{w}_k} = \frac{\delta_{ik}}{\beta_k}\nabla_{\mathbf{w}_k}U = -\frac{\delta_{ik}}{\beta_k}\nabla_{\mathbf{w}_k}R = -\delta_{ik}\frac{\alpha_k}{\beta_k}\mathbf{w}_k \approx 0, \tag{18.45}$$

where the approximation on the right follows from the fact that the fraction $\frac{\alpha_k}{\beta_k}$ corresponds to a weight decay constant, which in general is much smaller than unity[1]. Let us introduce the definitions

$$\mathbf{H}_k := \beta_k\mathbf{G}\mathbf{\Pi}_k\mathbf{G}^\dagger - (\beta_k)^2\sum_{t=1}^{N}\pi_k(t)\left[1 - \pi_k(t)\right]\left(y_t - \mu(\mathbf{x}_t;\mathbf{w}_k)\right)^2\mathbf{g}(\mathbf{x}_t)\mathbf{g}^\dagger(\mathbf{x}_t) \tag{18.46}$$

$$B_k := \frac{(\beta_k)^2}{2}\sum_{t=1}^{N}\pi_k(t)\left[1 - \pi_k(t)\right]\left[\left(y_t - \mu(\mathbf{x}_t;\mathbf{w}_k)\right)^2 - \frac{1}{\beta_k}\right]^2 \tag{18.47}$$

---

[1] This approximation is not always good, but recall from chapters 10 and 11 that the off-diagonal terms $\mathbf{H}_{a_i,\mathbf{w}_k}$ and $\mathbf{H}_{\beta_i,\mathbf{w}_k}$ are not needed for the evidence scheme.

and make use of Equation (10.68), $a_k = \frac{1}{N} \sum_{t=1}^{N} \pi_k(t)$, which holds at the mode. This allows (18.39)-(18.44) to be written in the more concise form (writing $\mathbf{H}_{q_i,q_k} := \frac{\partial^2 E}{\partial q_i \partial q_k}$)

$$\mathbf{H}_{\mathbf{w}_i,\mathbf{w}_k} = \delta_{ik}\mathbf{H}_k \tag{18.48}$$

$$\mathbf{H}_{\beta_i,\beta_k} = \frac{\delta_{ik}}{2(\beta_k)^2}(Na_k - B_k) \tag{18.49}$$

$$\mathbf{H}_{a_i,a_k} = \delta_{ik}\sum_{t=1}^{N}\left(\frac{\pi_K(t)}{a_K} - \frac{\pi_k(t)}{a_k}\right)^2 \tag{18.50}$$

$$\mathbf{H}_{a_i,\beta_k} = 0 \tag{18.51}$$

$$\mathbf{H}_{a_i,\mathbf{w}_k} = 0 \tag{18.52}$$

$$\mathbf{H}_{\beta_i,\mathbf{w}_k} \approx 0 \tag{18.53}$$

Note that this matrix is of a diagonal block structure. Its determinant is thus given by the product of the determinants of the individual block matrices,

$$\det \mathbf{H} = \prod_k \det \mathbf{H}_{\mathbf{w}_k,\mathbf{w}_k} \prod_k \mathbf{H}_{a_k,a_k} \prod_k \mathbf{H}_{\beta_k,\beta_k} \tag{18.54}$$

which reduces the computational costs considerably.

18.4 Discussion

Recall that, according to Equation (18.6), the expression for the Hessian splits up into three components. The derivation of the first term, the Hessian of the EM error function U, is straightforward. The matrix is nearly of a diagonal block structure, where the remaining off-diagonal terms (18.11) are not needed for the evidence scheme presented in this book[2]. The second term, the outer product of the gradient of the EM error function U, is found to be cancelled out. The last term, the expectation value for the outer product of the gradient of Ψ, is approximated by a diagonal block matrix. This simplification was justified by the fact that off-diagonal terms contain sums with arguments of varying sign, whereas the sums for the diagonal blocks contain arguments with uniform signs. This suggests that off-diagonal contributions are considerably smaller than diagonal blocks and can therefore be neglected.

The derivation of the Hessian presented in this section made use of Equation (18.15), which is exact for the GM-RVFL network. For the GM network, Equation (18.15) gives only the first-order term in a Taylor series expansion of $\mu_k(\mathbf{x}; \mathbf{w})$, and the expression for the Hessian is compromised to the extent that the validity of this approximation is violated.

---

[2] For alternatives schemes, that make use of the whole Hessian (e.g. [26]), approximation (18.45) can be applied.

References

[1] ALBANO, A. M., PASSAMENT, A., HEDIGER, T., AND FARRELL, M. E. Using neural nets to look for chaos. *Physica D 58* (1992), 1–9.

[2] ALLEN, D., AND TAYLOR, J. G. Learning time series by neural networks. In *Proceedings of the 4th International conference on Artificial Neural Networks (ICANN)* (1994), M. Marinaro and P. Morasso, Eds., Springer, pp. 529–532.

[3] BARON, A. R. Universal approximation bounds for superpositions of a sigmoidal function. *IEEE Transactions on Information Theory 39* (1993), 930–945.

[4] BERENDSEN, H. J. C. Biophysical applications of molecular dynamics. *Computer Physics Communications 44*, 3 (1987), 233–242.

[5] BERENDSEN, H. J. C., AND VAN GUNSTEREN, W. F. Practical algorithms for dynamic simulations. In *Molecular-Dynamics Simulation of Statistical-Mechanical Systems*, G. Ciccotti and W. Hoover, Eds. North-Holland, Amsterdam, 1986.

[6] BISHOP, C. M. Mixture density networks. Tech. rep., Department of Computer Science and Applied Mathematics, Aston University, Birmingham, UK, 1994.

[7] BISHOP, C. M. *Neural Networks for Pattern Recognition*. Oxford University Press, New York, 1995.

[8] BISHOP, C. M., AND QAZAZ, C. S. Bayesian inference of noise levels in regression. *Proceedings ICANN 95* (1995), 59–64.

[9] BOX, G. E. P., AND TIAO, G. C. *Bayesian Inference in Statistical Analysis*. Addison Wesley, 1977.

[10] BREIMAN, L. Bagging predictors. Tech. rep., Department of Statistics, University of Berkley,California, 1994.

[11] BRONSTEIN, I. N., AND SEMENDJAJEW, K. A. *Taschenbuch der Mathematik*. Verlag Harri Deutsch, 1984.

[12] CASDAGLI, M., EUBANK, S., FARMER, J. D., AND GIBSON, J. State space reconstruction in the presence of noise. *Physica D 51* (1991), 52–98.

[13] DEGROOT, M. H. *Optimal statistical decisions*. McGraw-Hill, 1970.

[14] DEMPSTER, A. P., LAIRD, N. M., AND RUBIN, D. B. Maximum likelihood from incomplete data via the EM algorithm. *Journal of the Royal Statistical Society B39*, 1 (1977), 1–38.

[15] DEVANEY, R. L. *An Introduction to Chaotic Dynamical Systems.* Addison-Wesley, 1989.

[16] EFRON, B., AND GONG, G. A leisurely look at the bootstrap, the jacknife, and cross-validation. *The American Statistician 37*, 1 (1983), 36–47.

[17] ERDMANN, G. Limits of predictability in evolutionary systems. Tech. rep., ETH Zürich, 1992.

[18] GEMAN, S., BIENENSTOCK, E., AND DOURSAT, R. Neural networks and the bias/variance dilemma. *Neural Computation 4* (1992), 1–58.

[19] GORSE, D., SHEPHERD, A., AND TAYLOR, J. G. A classical algorithm for avoidung local minima. *Proceedings of WCNN'94* (1994), 364–369.

[20] GRASSBERGER, P., AND PROCACCIA, I. Characterization of strange attractors. *Physical Review Letters 50*, 5 (1983), 346–349.

[21] HARRISON, D., AND RUBINFELD, D. L. Hedonic prices and the demand for clean air. *Journal of Environmental Economics and Management 5* (1978), 81–102.

[22] HOEL, P. G. *Introduction to Mathematical Statistics.* John Wiley and Sons, Singapore, 1984.

[23] HORNICK, K. Approximation capabilities of multilayer feedforward networks. *Neural Networks 4* (1991), 251–257.

[24] HORNICK, K., MAXWELL, S., AND HALBERT, W. Multilayer feedforward networks are universal approximators. *Neural Networks 2* (1989), 359–366.

[25] HUSMEIER, D. Numerische Bestimmung des Lösungsmitteleinflusses auf die thermodynamischen Größen einer intramolekularen Proteinreaktion. Master's thesis, Department of Biophysics, University of Bochum, 1991.

[26] HUSMEIER, D. *Modelling Conditional Probability Densities with Neural Networks.* PhD thesis, Department of Mathematics, King's College London, 1998. http://www.ee.ic.ac.uk/research/neural/Husmeier.html.

[27] HUSMEIER, D., ALLEN, D., AND TAYLOR, J. G. A universal approximator for learning conditional probability densities. In *Mathematics of Neural Networks: Models, Algorithms, and Applications* (Boston, 1997),

S. W. Ellacott, J. C. Mason, and I. J. Anderson, Eds., Kluwer Academic Press, pp. 198–203. ISBN 0-7923-9933-1.

[28] HUSMEIER, D., AND ALTHOEFER, K. Modelling conditional probabilities with network committees: How overfitting can be useful. *Neural Network World 4/98* (1998), 417–439.

[29] HUSMEIER, D., AND TAYLOR, J. G. A neural network approach to predicting noisy time series. In *Annals of the Third Brazilian Symposium on Neural Networks* (Recife, 1995), T. B. Ludermir, Ed., pp. 221–226.

[30] HUSMEIER, D., AND TAYLOR, J. G. Predicting conditional probability densities of stationary stochastic time series. *Neural Networks 10*, 3 (1997), 479–497.

[31] HUSMEIER, D., AND TAYLOR, J. G. Neural networks for predicting conditional probability densities: Improved training scheme combining EM and RVFL. *Neural Networks 11*, 1 (1998), 89–116.

[32] IGELNIK, B., AND PAO, Y. H. Stochastic choice of basis functions in adaptive functional approximation and the functional-link net. *IEEE Transactions on Neural Networks 6* (1995), 1320–1329.

[33] JETSCHKE, G. *Mathematik der Selbstorganisation*. Friedr. Vierweg und Sohn, Braunschweig, 1987.

[34] JORDAN, M. I., AND JACOBS, R. A. Hierarchical mixtures of experts and the EM algorithm. *Neural Computation 6* (1994), 181–214.

[35] KROGH, A., AND SOLLICH, P. Statistical mechanics of ensemble learning. *Physical Review E 55* (1997), 811–825.

[36] KROGH, A., AND VEDELSBY, J. Neural network ensembles, cross validation, and active learning. In *Advances in Neural Information Processing Systems* (Cambridge, MA, 1995), D. S. Touretzky, G. Tesauro, and T. K. Leen, Eds., vol. 7, MIT Press, pp. 231–238.

[37] MACKAY, D. J. C. Bayesian interpolation. *Neural Computation 4* (1992), 415–447.

[38] MACKAY, D. J. C. A practical Bayesian framework for backpropagation networks. *Neural Computation 4* (1992), 448–472.

[39] MACKAY, D. J. C. Bayesian non-linear modeling for the prediction competition. In *Maximum Entropy and Bayesian method*, H. G.R., Ed. Kluwer Academic Publisher, Santa Barbara, 1993, pp. 221–234.

[40] MacKay, D. J. C. Hyperparameters: optimize, or integrate out. In *Maximum Entropy and Bayesian Methods*, G. Heidbreder, Ed. Kluwer Academic Publisher, Santa Barbara, 1993, pp. 43–59.

[41] MacKay, D. J. C. Developments in probabilistic modelling with neural networks - ensemble learning. Tech. rep., Cavendish Laboratory, University of Cambridge, 1995.

[42] Moody, J. E., and Utans, J. Principled architecture selection for neural networks: Application to corporate bond rating prediction. In *Advances in Neural Information Processing Systems* (San Mateo, CA, 1992), J. E. Moody, S. J. Hanson, and R. P. Lippman, Eds., vol. 4, Morgan Kaufmann Publishers, pp. 683–690.

[43] Murata, N. An integral representation of functions using three-layered networks and their approximation bounds. *Neural Networks* (1996).

[44] Neal, R. M. *Bayesian Learning for Neural Networks*, vol. 118 of *Lecture Notes in Statistics*. Springer, New York, 1996. ISBN 0-387-94724-8.

[45] Neuneier, R., Hergert, F., Finnoff, W., and Ormoneit, D. Estimation of conditional densities: A comparison of neural network approaches. In *Proceedings of ICANN 94* (1994), M. Marinaro and P. Morasso, Eds., Springer, pp. 689–692.

[46] Ormoneit, D. Estimation of probability densities using neural networks. Master's thesis, Technische Universität München, Fachbereich Informatik,, 1993.

[47] Ormoneit, D., and Tresp, V. Improved gaussian mixture density estimates using bayesian penalty terms and network averaging. In *Advances in Neural Information Processing Systems* (San Mateo, CA, 1996), D. S. Touretzky, M. C. Mozer, and M. E. Hasselmo, Eds., vol. 8, Morgan Kaufmann Publishers, pp. 542–548.

[48] Pao, Y. H., Park, G. H., and Sobajic, D. J. Learning and generalization characteristics of the random vector functional-link net. *Neurocomputing 6* (1994), 163–180.

[49] Papoulis, A. *Probability, Random Variables, and Stochastic Processes*, 3 ed. McGraw-Hill, Singapore, 1991.

[50] Pass, G. Assessing and improving neural network predictions by the bootstrap algorithm. In *Advances in Neural Information Processing Systems* (San Mateo, CA, 1993), S. J. Hanson, J. D. Cowan, and L. C. Giles, Eds., vol. 5, Morgan Kaufmann Publishers, pp. 196–203.

[51] PRESS, W. H., TEUKOLSKY, S. A., VETTERLING, W. T., AND FLAN-
NERY, B. P. *Numerical Recipes in C.* Cambridge University Press, New
York, 1992.

[52] RÖGNVALDSSON, T. On langevin updating in multilayer perceptrons.
Neural Computation 6 (1994)), 916–926.

[53] SCHLITTER, J. Zur Rolle der intramolekularen Entropie von Proteinen.
Experimente und Simulationsrechnungen zu einer internen Bindung an
Alpha-Methämoglobin. Habilitationsschrift, Department of Biophysics,
University of Bochum, 1992.

[54] SHARKEY, A. J. C., AND SHARKEY, N. E. How to improve the relia-
bility of artificial neural networks. Tech. rep., Department of Computer
Science, University of Sheffield, U.K., 1996.

[55] SHARKEY, A. J. C., SHARKEY, N. E., AND GOPINATH, O. C. Diversity, neural nets and safety critical applications. In *Current Trends in
Connectionism* (Hillsdale, New Jersey, 1995), L. Niklasson and M. Bo-
den, Eds., Lawrence Erlbaum Associates, pp. 165–178.

[56] SINGH, S. *Fermat's Last Theorem.* Fourth Estate, London, 1998.

[57] SOLLICH, P., AND KROGH, A. Learning with ensembles: How over-
fitting can be useful. In *Advances in Neural Information Processing
Systems* (1996), D. S. Tourezky, M. C. Mozer, and M. E. Hasselmo,
Eds., vol. 8, pp. 190–196.

[58] SPECHT, D. F. Probabilistic neural networks. *Neural Networks 3*, 1
(1990), 109–118.

[59] SRIVASTAVA, A. N., AND WEIGEND, A. S. Computing the probability
density in connectionist regression. In *Proceedings of ICANN '94* (1994),
M. Marinaro and P. Morasso, Eds., Springer.

[60] TAKENS, F. Detecting strange attractors in fluid turbulence. In *Dynamical Systems and Turbulence* (Berlin, 1981), D. Rand and L. S. Young,
Eds., Springer.

[61] THODBERG, H. H. A review of Bayesian neural networks with an application to near infrared spectroscopy. Tech. rep., The Danish Meat
Research Insitute, Maglegaardsvej 2, DK-4000 Roskilde, Denmark, 1995.

[62] TIBSHIRANI, R. A comparison of some error estimates for neural networks. *Neural Computation 8*, 1 (1996), 152–163.

[63] VANGUNSTEREN, W. F., AND BERENDSEN, H. J. C. Computer simulation of molecular dynamics: Methodology, applications, and perspectives

in chemistry. *Angewandte Chemie - International Edition in English 9* (1990), 992–1023.

[64] WEIGEND, A., AND NIX, A. N. Predictions with confidence intervals (local error bars). *Proceedings of the International Conference on Neural Information Processing* (1994), 847–852.

[65] WEIGEND, A. S., AND GERSHENFIELD, N. A. *Time Series Prediction: Forecasting the Future and Understanding the Past.* Addison-Wesley, Reading, MA, 1993.

[66] WEIGEND, A. S., AND SRIVASTAVA, A. N. Predicting conditional probability distributions: A connectionist approach. *International Journal of Neural Systems 6*, 2 (1995), 109–118.

Index